Hybrid Bulk Metal Components

Hybrid Bulk Metal Components

Editor

Bernd-Arno Behrens

MDPI • Basel • Beijing • Wuhan • Barcelona • Belgrade • Manchester • Tokyo • Cluj • Tianjin

Editor
Bernd-Arno Behrens
Institute of Forming Technology
and Machines,
Leibniz Universität Hannover
Germany

Editorial Office
MDPI
St. Alban-Anlage 66
4052 Basel, Switzerland

This is a reprint of articles from the Special Issue published online in the open access journal *Metals* (ISSN 2075-4701) (available at: https://www.mdpi.com/journal/metals/special_issues/ hybrid_bulk_metal_components).

For citation purposes, cite each article independently as indicated on the article page online and as indicated below:

LastName, A.A.; LastName, B.B.; LastName, C.C. Article Title. *Journal Name* **Year**, *Volume Number*, Page Range.

ISBN 978-3-0365-0884-9 (Hbk)
ISBN 978-3-0365-0885-6 (PDF)

Cover image courtesy of Bernd-Arno Behrens.

Contents

About the Editor

Bernd-Arno Behrens studied mechanical engineering at the University of Hannover and received his doctorate in 1997 in the field of bulk metal forming. Subsequently, he took over a leading position in the industry. Since October 2003, he has been director of the Institute of Forming Technology and Machines (IFUM). In addition, in 2004 he was appointed spokesman of the board of the Material Testing Institute for Materials and Production Technology (MPA) in Hannover. In 2005 he also assumed the function of managing partner of IPH—Institut für Integrierte Produktion Hannover gemeinnützige GmbH. He is, among others, speaker of the Collaborative Research Center 1153 "Tailored Forming". He is a member of the Scientific Council of the AIF and the Arbeitsgemeinschaft Umformtechnik. His research interests include sheet and bulk metal forming technology and machines, machine development (drives, actuators, control), thermo-mechanical material characterization and FE simulation.

Preface to "Hybrid Bulk Metal Components"

In recent years, the requirements for technical components have steadily been increasing. This development is intensified by the desire for products with a lower weight, smaller size, and extended functionality, but also with a higher resistance against specific stresses. Mono-material components, which are produced by established processes, feature limited properties according to their respective material characteristics. Thus, a significant increase in production quality and efficiency can only be reached by combining different materials in a hybrid metal component. In this way, components with tailored properties can be manufactured that meet the locally varying requirements. Through the local use of different materials within a component, for example, the weight or the use of expensive alloying elements can be reduced.

This Special Issue includes research articles on investigations considering the production of hybrid components, most of which were prepared within the framework of Collaborative Research Center 1153 ("Process Chain for Manufacturing of Hybrid High Performance Components by Tailored Forming", www.sfb1153.de) or by international scientists.

The aim of this Special Issue is to cover the recent progress and new developments regarding all aspects of hybrid bulk metal components. This includes fundamental questions regarding the joining, forming, finishing, simulation, and testing of hybrid metal parts.

Bernd-Arno Behrens
Editor

Article

Contact Geometry Modification of Friction-Welded Semi-Finished Products to Improve the Bonding of Hybrid Components

Bernd-Arno Behrens [1], Johanna Uhe [1], Tom Petersen [1], Florian Nürnberger [2], Christoph Kahra [2], Ingo Ross [1] and René Laeger [1,*]

[1]   Institute of Forming Technology and Machines, Leibniz Universität Hannover, 30823 Garbsen, Germany; behrens@ifum.uni-hannover.de (B.-A.B.); uhe@ifum.uni-hannover.de (J.U.); petersen@ifum.uni-hannover.de (T.P.); ross@ifum.uni-hannover.de (I.R.)
[2]   Institute of Materials Science, Leibniz Universität Hannover, 30823 Garbsen, Germany; nuernberger@iw.uni-hannover.de (F.N.); kahra@iw.uni-hannover.de (C.K.)
*   Correspondence: laeger@ifum.uni-hannover.de; Tel.: +49-5117624958

Abstract: To improve the bond strength of hybrid components when joined by friction welding, specimens with various front end surface geometries were evaluated. Rods made of aluminum AA6082 (AlSi1MgMn/EN AW-6082) and the case-hardening steel 20MnCr5 (AISI 5120) with adapted joining surface geometries were investigated to create both a form-locked and material-bonded joint. Eight different geometries were selected and tested. Subsequently, the joined components were metallographically examined to analyze the bonding and the resulting microstructures. The mechanical properties were tested by means of tensile tests and hardness measurements. Three geometrical variants with different locking types were identified as the most promising for further processing in a forming process chain due to the observed material bond and tensile strengths above 220 MPa. The hardness tests revealed an increase in the steel's hardness and a softening of the aluminum near the transition area. Apparent intermetallic phases in the joining zone were analyzed by scanning electron microscopy (SEM) and an accumulation of silicon in the joining zone was detected by energy-dispersive X-ray spectroscopy (EDS).

Keywords: friction welding; hybrid components; tailored forming; surface geometry modification

Citation: Behrens, B.-A.; Uhe, J.; Petersen, T.; Nürnberger, F.; Kahra, C.; Ross, I.; Laeger, R. Contact Geometry Modification of Friction-Welded Semi-Finished Products to Improve the Bonding of Hybrid Components. *Metals* **2021**, *11*, 115. https://1doi.org/0.3390/met11010115

Received: 17 November 2020
Accepted: 4 January 2021
Published: 8 January 2021

Publisher's Note: MDPI stays neutral with regard to jurisdictional claims in published maps and institutional affiliations.

1. Introduction

If a component has to withstand diverse local loads or a lightweight design is demanded [1], the combination of different materials offers the use of a load-adapted component. Components consisting of at least two materials are called hybrid components. Due to different material-specific properties such as melting points or flow stresses, these components require adapted joining methods. Depending on the specific material combination, this can be, for example, a fusion welding process or a friction welding process.

The most important technical advantages of friction welding compared to fusion welding are the high reproducibility and the wide variety of possible material combinations, such as aluminum and steel, since the joining process is based on plastic deformation instead of melting. Compared to friction welding, fusion-welded products have much larger heat-affected zones which can result in undesired microstructures and reduces the resilience of parts [2]. The molten phase may cause defects such as gas porosity, which leads to brittle fracture.

Common multimaterial components are produced by joining several individual parts that are already in a near-net shape. Therefore, the joining process takes place at the end of the process chain—for example, splicing or riveting of sheet metal components in the production of automobile chassis [3]. Another approach is joining by forming, such as the

1

consolidation of powder with simultaneous bonding with steel during forming to produce hybrid gears [4] or the application of ultrasound enhanced friction stir welding to join different materials [5].

As part of the collaborative research center 1153 (CRC 1153) "Tailored Forming", a novel process chain was developed, in which various materials are joined at an initial stage before being subjected to further processing [6]. The aim of this concept is to further improve the joining zone by the subsequent processing steps resulting in a load-adapted component. The CRC 1153 maps several process chains in their entirety, to improve components such as shafts, bevel gears or bearing disks [6]. The process chain for manufacturing hybrid shafts by applying friction welding is depicted in Figure 1. Within this process chain, joining is followed by impact extrusion, which requires a homogenous formability in both material sections. Hence, an inhomogeneous temperature distribution in the joined parts prior to the forming has to be ensured to align the flow curves of the investigated 20MnCr5 steel and of the AA6082 aluminum alloy (EN AW-6082). Therefore, a customized inductive heating strategy was developed to achieve the material-specific forming temperatures of 900 °C in steel and 20 °C in aluminum simultaneously [7].

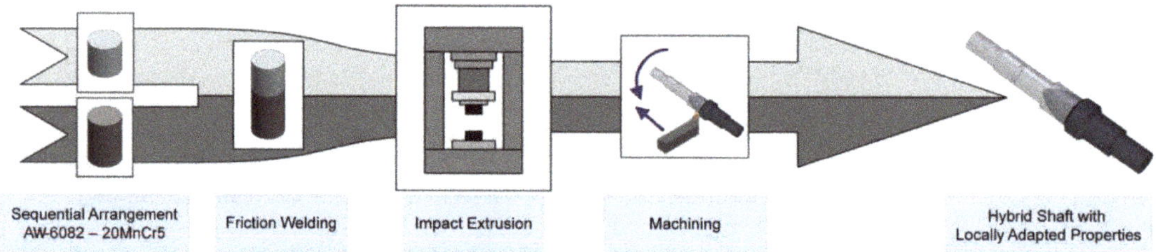

Figure 1. Schematic tailored forming-process chain of the collaborative research center 1153 (CRC 1153) [8], material combination EN AW-6082 (AA6082) and 20MnCr5.

Friction welding was selected based on various reports which concluded that the successful joining of aluminum alloys and steels and the free designability of joining zone geometry—e.g., Ashfaq et al. detected an increased bond strength when using a conical geometry instead of flat surface. They found that this modification benefits material flow and results in an improved bond quality [9]. Fukumoto et al. investigated the influence of different parameters on the completeness of the bond. The most significant result was that the highest bond strength is achieved by certain friction times of 1 s with a pressure of 50 MPa and 6 s with a pressure of 150 MPa. Higher or lower friction times resulted in lower bond strength [10]. Lee et al. focused on the resulting microstructures and their correlations with the friction parameters. Besides the base metals, they identified different regions—that is, a region of dynamic recrystallization—a heat-affected zone (HAZ) and a deformation zone, and how these are formed due to different forming pressures (70 to 150 MPa) and friction times (0.1 to 3.0 s) [11]. Fukumoto et al. studied the properties of the bonds created by a friction welding process of the aluminum alloy EN AW 1050 and the stainless steel 1.4301 (AISI 304). They were able to show that the extension of the frictional time from 0.1 to 0.2 s increased the bond strength from 85 to 96 MPa [12]. Sahin characterized the bond by different test methods such as tensile tests and hardness measurements and found a significant influence of contaminants at the interface on the joint quality. He recommended a statistical analysis as an economical and reliable method for selecting optimized welding parameters [13]. Behrens et al. investigated the influence of surface geometry by using a conical shape. They found out that at room temperature a sharper shape with an increased friction path results in a higher bond strength. Compared to specimens with flat surfaces, bond strength could be improved from 252 to 294 MPa using a conical surface of 30° [14]. So far, only a few studies such as [9], [14] or [15] took an adaption of the surface geometry into account. In [15], the effects of frictional contact surfaces on the formation of an intermetallic phase were studied. Since most investigations are focused on flat surfaces, which often show compound defects in the zone around the central axis after joining [10], on other material combinations or without focusing the bonding strength [15], further research is required regarding alternatives such as a combination of material bond and form locking by varying the friction contact geometries. Comparison these results with additional references is only possible to a limited extent, since parameters of the friction welding process differ as well as the material combinations.

Friction welding processes are divided into three sequences: contact phase, friction phase and deformation phase [16]. In the contact phase, the geometries are aligned and brought into contact with a specific pressure. The heat is generated in the friction phase, in which one component begins to rotate—in this case, the steel side. This phase can be adjusted by controlling the friction time or the relative friction path of the welding components covering in the axial direction. In the deformation phase, the rotation stops and the welding components are joined by generating high axial pressure.

To improve the bonding strength of the steel–aluminum specimen and thus manufacture semifinished parts suited to subsequent impact forging, this work is mainly concerned with varying the contact geometries. In addition to increasing the contact areas between both materials or increasing the contact times and contact pressures in the sample center, possibilities for generating a form closure are also investigated in addition to the pure material bond. Different combinations of friction surface geometries are tested experimentally in the following and their impact on the bond strength is determined. For example, the applicability of undercuts is examined to implement the additional bonding mechanisms such as form locking.

2. Materials and Methods

The following subsections describe the applied materials and the performed methods of the investigation. For this purpose, the basic conditions are explained and clarified with the help of illustrations.

2.1. Materials

For the friction welding, the aluminum alloy AA6082 (EN AW-6082) and the case hardening steel 20MnCr5 were chosen. 20MnCr5 is a chromium–manganese alloyed steel. During friction welding, the steel was employed in its delivery condition (soft annealed) with a tensile strength of 554 MPa. The aluminum alloy used featured the T6 condition with a tensile strength above 360 MPa. The mechanical properties, tested in prior investigations, are listed in Table 1 and the chemical compositions are given in the content lists in Table 2, measured by optical emission spectrometry.

Table 1. Mechanical properties of the 20MnCr5 steel and AA6082 aluminum alloy.

Material	Tensile Strength R_m in MPa	Uniform Elongation A_g in %	Elongation at Fracture A in %	Hardness in HV0.1
20MnCr5	554 ± 5	111 ± 1	276 ± 6	170 ± 13
AA6082	364 ± 0	45 ± 1	104 ± 6	113 ± 1

Table 2. Chemical composition of the 20MnCr5 steel and AA6082 aluminum alloy in wt.%., measured by optical emission spectrometry.

Element	C	Si	Mn	P	Cr	S	Al	Fe	Cu	Mg	Zn	Ti
20MnCr5	0.195	0.275	1.190	0.013	1.050	0.010	0.030	96.85	0.164	-	0.023	-
AA6082	-	1.040	0.451	0.001	0.035	-	97.60	0.152	0.031	0.620	0.011	0.018

The microstructures of both base materials prior to friction welding are shown in Figure 2. On the left side (a) the ferritic–pearlitic microstructure of the steel 20MnCr5 is depicted. To visualize the grain boundaries and the different microstructures, the sample was etched with Nital, a solution of nitric acid (3%) and alcohol. A micrograph of the aluminum alloy in its T6 condition is shown on the right side (b).

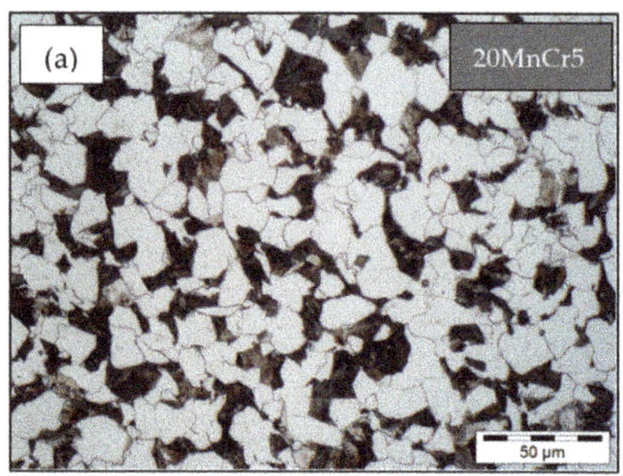
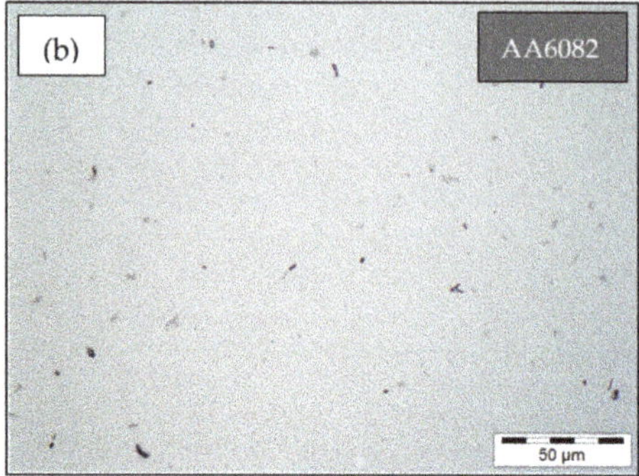

Figure 2. (**a**) Ferritic–pearlitic microstructure of the 20MnCr5 steel in the soft annealed condition, etched with Nital (3%), and (**b**) unetched micrograph of the AA6082 aluminum alloy (T6 condition).

2.2. Friction Surface Geometries

The different surface geometries of the semifinished products investigated were chosen to improve the joining zone properties by surface enlargements, undercuts and shrinkage. An increase in the friction surface leads to a higher temperature generation, from which a lower demand for the yield forces results. As described in the literature [10], air pockets can occur in the center of the specimen for flat surfaces. With higher temperatures, these can be avoided. An undercut results in a form fit or force fit, depending on whether the aluminum fills a hole by flowing or encloses a shape by shrinking. The geometries, manufactured by machining, are depicted in Figure 3.

Geometry A (Figure 3) was selected for a form-locking connection to enhance the bonding strength. During friction welding, the undercut of the cavity located in the steel part with an angle of 75° was filled with aluminum. On the basis of preliminary tests, an angle of 75° was determined to be optimum, since at this angle complete mold-filling can be ensured, despite a relatively concise form fitting. Geometry B offers an enlarged friction surface due to the hemispherical geometry, which results in a higher heat generation due to friction. The shoulder at the transition from the hemisphere was designed with a beveled edge to improve material flow. Geometry C features four drilled holes intended to increase the torsional stiffness by means of flowing aluminum entering the holes, thus achieving a form lock. Geometry D forms a hemispherical surface, resulting in an enlarged friction surface analogous to Geometry B. The difference to Geometry B is the absence of a shoulder to examine its necessity for the material flow.

Conical geometries were welded with varying angles of 30° (Geometry E) and 45° (Geometry G) using an increased friction contact surface and reduced manufacturing effort compared to the hemispherical Geometry B. The conical Geometry G is additionally truncated to simplify production and to combine an axial force with directed material flow during the forming process. Compared to Geometry A, Geometry F has no cavity in the steel component. The undercut was formed by a protruding elevation with an angle of 80°, while the aluminum is of a flat geometry. The aluminum was intended to flow around the shoulder and shrink to the steel due to the greater thermal expansion coefficient. In addition to the form lock and material bond, this geometry provides a force-locked connection to enhance the bonding strength. Preliminary tests have shown that too large a pin or an angle smaller than 80° will result in air pockets. Geometry H has a pin on the aluminum side to investigate the influence of the expected deformation ratio and the high friction path on the bonding strength (Table 3).

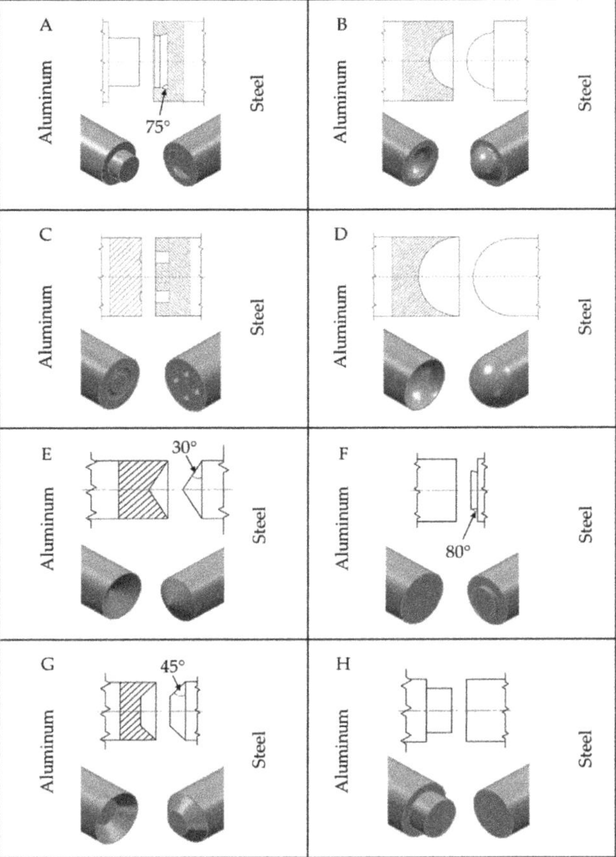

Figure 3. Geometries of friction surfaces (**A–H**), outer diameters of 40 mm.

Table 3. Main parameters of the friction welding process.

Geometry	A	B	C	D	E	F	G	H
Friction speed in 1/min	1500	1500	1500	1500	2000	2000	2000	2000
Frictional force in kN	150	100	80	80	70	75	70	75
Friction time in s	-	2	2	0.05	0.1	-	0.1	-
Relative friction path in mm	4	-	-	-	-	4	-	10
Press force in kN	240	120	150	120	120	150	120	120
Press time in s	2	1	2	1	2	2	2	2

2.3. Friction Welding

At first, the geometries presented above were cleaned in an ultrasonic bath filled with ethanol. After drying, these were friction-welded on a KUKA Genius Plus (Kuka AG, Augsburg, Germany). The most important process parameters are listed in Table 3. The parameters were selected according to prior investigations. For comparability, most parameters were chosen to be identical or limited to a few varying values according to the different geometries. Parameters with varying values were selected since these resulted in similar shapes and qualities of the bonding, according to first visual examinations.

The major differences between the performed welding processes of the first four geometries and the second four are the following: The friction speed was increased from

1500 to 2000 rpm for Geometries E, F, G and H to ensure a convenient heat generation at lower friction forces. Lower frictional forces were chosen for the second four geometries to prevent undesired deformations during the friction phase. In order to investigate the influence of high true strain, the friction path of Geometry H was increased in comparison to the path-controlled processes of Geometries A and F.

2.4. Metallographic Analysis

Following the friction welding process, the samples were cut along the axis of rotation and the cross section of the joining zones were prepared for metallographic examinations by grinding, polishing and etching. The quality of the joining zone was analyzed on micrographs by detecting phases and inclusions. In addition, the Vickers hardness was measured according to DIN EN ISO 6507-1 [17] (HV0.1) to compare the mechanical properties of the joining zone and of the heat-affected zones with those of the base materials. Furthermore, the joining zone was analyzed by scanning electron microscopy (SEM) (AURIGA from Zeiss, Oberkochen, Germany) and energy dispersive X-ray spectroscopy (EDS) (Oxford Instruments, Abingdon, UK). A slope cut was prepared by applying a focused ion beam (FIB) to excavate a cross section not influenced by prior conventional steps of metallographic preparations.

2.5. Mechanical Testing

Tensile tests were carried out for all geometries to determine the tensile strength and to evaluate the bonding strength. For each geometry two samples were tested on a Zwick Z250 kN (ZwickRoell GmbH & Co. KG, Ulm, Germany) with the preload force of 300 N, the clamping pressure of 30 MPa and the strain rate of 0.002 s^{-1}. The geometry of the tested tensile specimen is depicted in Figure 4, which was manufactured out of the friction welding products with a reduced diameter for the gauge length. The joining zone is not located at the center of the tensile specimens due to the geometry of the welding products. Besides the decentralized joining zone, the results of the tensile tests reveal no irregularities, since necking did not occur or was located within the gauge length.

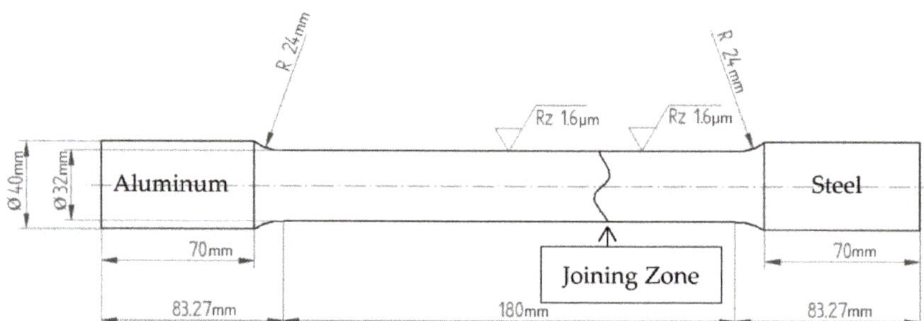

Figure 4. Geometry of tensile samples with highlighted position of the joining zone, according to DIN EN ISO 6892-1 [18], in millimeters.

3. Results

The following sections present the results of the different testing methods. These include the determination of mechanical parameters as well as metallographic investigations.

3.1. Tensile Test

In Figure 5 the stress–strain curve of a sample of Geometry A is exemplarily depicted on the left side and a comparison of the samples with the highest tensile strengths R_m is presented on the right side.

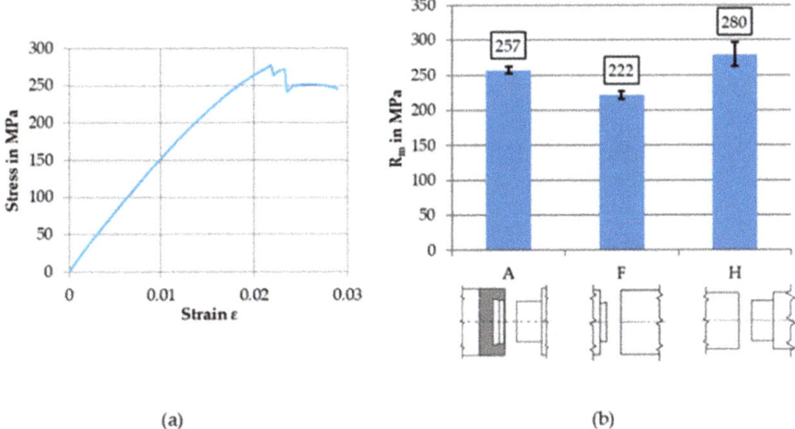

(a)   (b)

Figure 5. (**a**) stress–strain curve of Geometry A; (**b**) schematic drafts of the specimen geometries featuring the highest tensile strengths values with the standard deviation.

The tensile tests show that though the tensile strength R_m achievable with the various geometries differs, the qualitative shapes of the curves are almost identical. Fracture in the joining zone occurs due to brittleness of all geometries except Geometry A (confer Figure 6). An increased elongation at fracture is only visible in the stress–strain curve of Geometry A; here, reaching the stress maximum (Figure 5, at a strain of 0.022) a lateral contraction of the aluminum alloy can be observed.

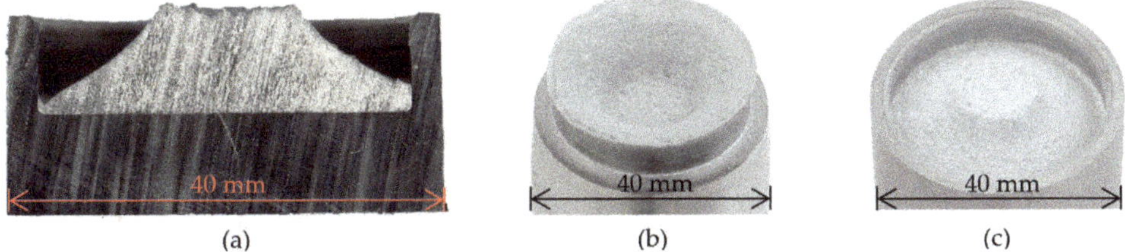

(a)   (b)   (c)

Figure 6. Specimen of Geometry A after tensile testing: (**a**) cross section; (**b**) aluminum side; (**c**) steel side.

All specimens have brittle fractures. Just specimens of Geometry A have a necking (Figure 5) and more remains of the aluminum (Figure 6) than other specimens. Figure 6 shows one specimen of Geometry A after tensile testing.

3.2. Metallography

In the following, cross sections of the specimens of Geometries B, C, D, E and F are depicted to show exemplary bonding defects. The Geometries A, G and H feature the desired bonding quality and visible defects such as gas pores, inclusions or cracks are not present in the joining zone.

Figure 7 gives an overview of the Geometries B (a) and D (b). The plotted angles mark the direction of the material flow when the aluminum alloy is detached from the steel. The bond of sample Geometry B is almost complete. At an angle of approximately 15°, the bond starts detaching and closes again in the shoulder area. This results in air inclusions and is a weak point over the complete circumference of the joint. The reason for this is the material flow which is indicated schematically by the violet arrows in Figure 7a.

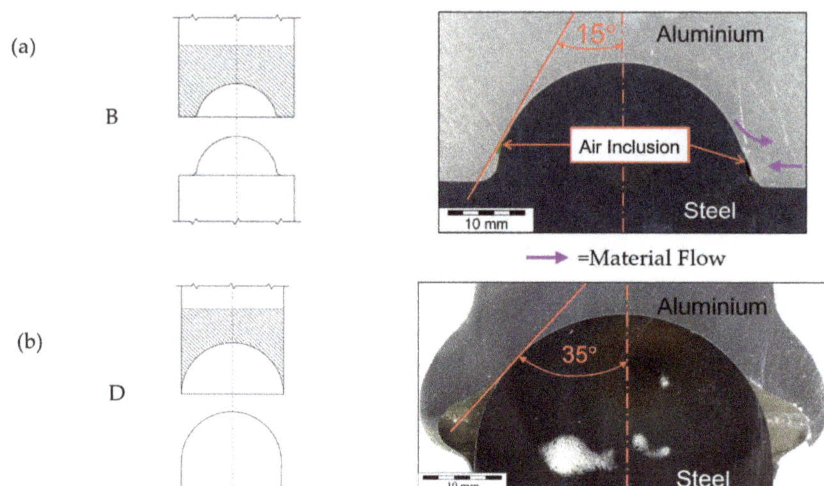

Figure 7. (**a**) Cross section of Geometry B and (**b**) cross section of Geometry D; the angles mark the detachment of the aluminum.

At Geometry D (Figure 7b), the aluminum alloy peels off at an angle of 35° and does not get into further contact. The material flow and the applied forces possibly cause the detachment in both geometries.

Geometry C is depicted in Figure 8. A complete filling of the holes was not achieved and gaps on the circumference occur with increasing depth; additionally, fragments of the aluminum alloy are visible.

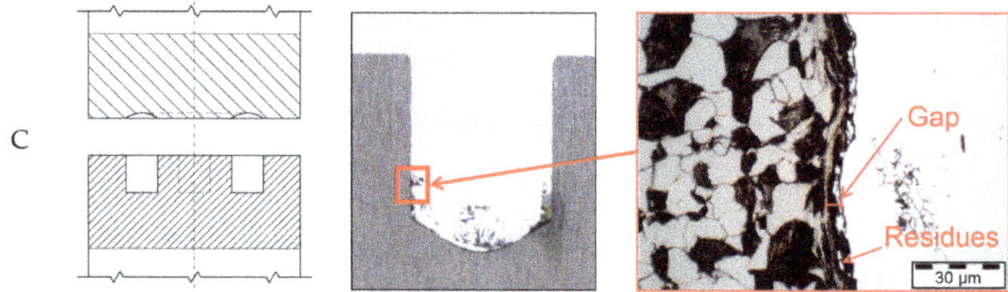

Figure 8. Geometry C, (**left**) schematic draft, (**middle**) overview of a filled borehole, and (**right**) exemplary gap at the borehole flank.

A section of Geometry E is depicted in Figure 9. The bond is complete except for higher radii, where air inclusions at diameters of 37 to 40 mm can be seen. For Geometry F, a small air inclusion appears near the undercut. This area is displayed in Figure 10.

The hardness of the samples was measured at different distances across the joining zone to characterize the influence of the generated heat and the forming during the friction welding process. It can be assumed that the size of the grains and the concentration of elements are influenced by the heat resulting in varying hardnesses compared to the basic materials. The space between two recording points in the aluminum alloy was chosen according to DIN EN ISO 6507-1 [17]. For simplification, the same distance of 0.5 mm was used in the steel. Figure 11 gives an example (Geometry F) of the measurements. The transition area could not be narrowed down due to the limiting conditions.

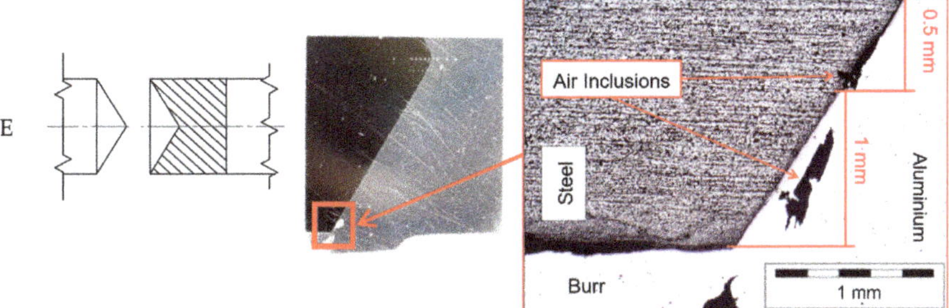

Figure 9. Geometry E, (**left**) schematic draft, (**middle**) overview of a half-cut sample, and (**right**) example of air inclusions in regions of the increased diameter.

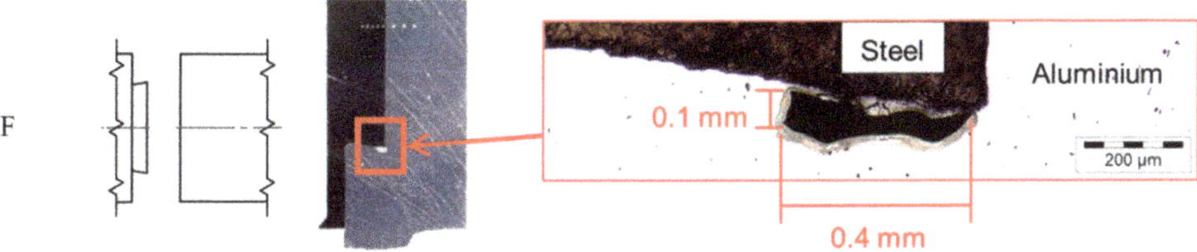

Figure 10. Geometry F, (**left**) schematic draft, (**middle**) overview of a nearly half-cut sample, and (**right**) exemplary air inclusion at the undercut.

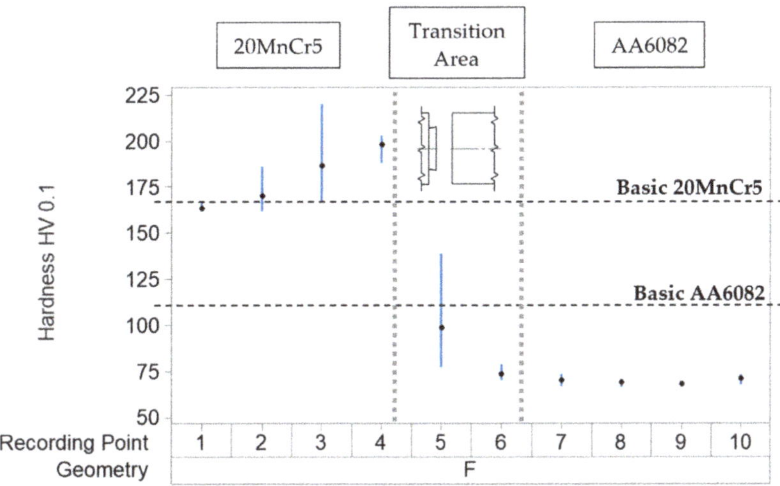

Figure 11. Diagram of the hardness of a sample with Geometry F including the variance, range between recording points is approximately 0.5 mm.

On the steel side, almost all samples show a small increase in hardness for the measuring point closest to the transition area—for example, recording point 4 in Figure 11. The soft annealed base material has an average hardness of 170 HV0.1 and is marked in Figure 11 as a horizontal dotted line. It can be concluded that some samples have experienced a slight softening and others an increase in steel hardness further from the interface in the axial direction.

The aluminum alloy has an average hardness of 113 HV0.1 in the T6 condition. Close to the joining zone the aluminum becomes softened and has an average hardness below 75 HV0.1, as can be seen for Geometry F in Figure 11. Geometry E is the only exception where a hardness of 103 HV0.1 was determined, possibly caused by a lower heat generation.

To investigate the joining zone, micrographs were examined. Two different types of interlayers between steel and aluminum alloys were found in the metallographic analyses. The first layer is located on the aluminum side near the friction welding surface and has a darker color. Figure 12 depicts an analyzed example of such a layer. Its thickness varies up to 1.5 μm. It is mainly found on flat areas of the friction surfaces—for example, in the undercut in Geometry A around the central axis. Since it was not possible to characterize the layer in detail by light microscopy, scanning electron microscopy was applied (confer Section 3.3).

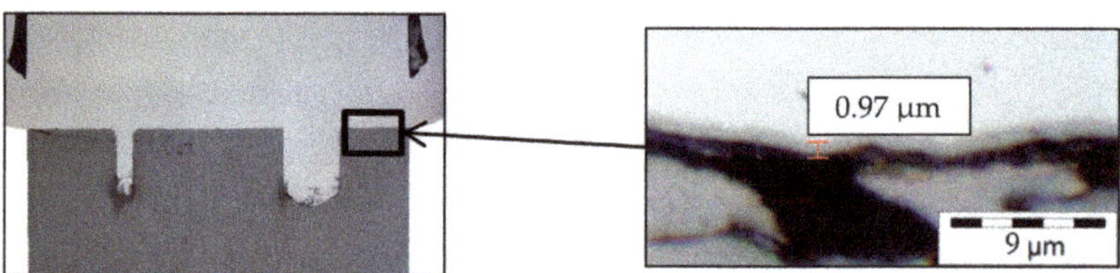

Figure 12. (**Left**) overview of Geometry C, and (**right**) dark layer on the aluminum side close to the joining zone.

The second layer found close to the joining zone, is a layer of fine-grained steel microstructures with increasing degree of fineness from the basic steel to the interface. Its thickness increases with the diameter from about 0.5 up to 3 μm (Figure 13).

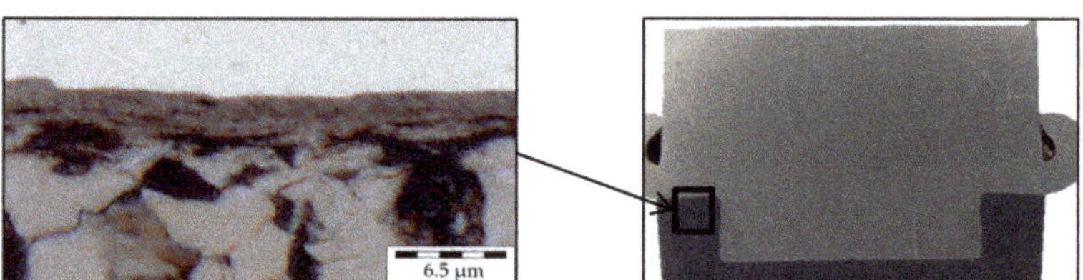

Figure 13. (**Left**) fine-grained layer on the steel side close to the joining zone and (**right**) overview of Geometry A.

3.3. Scanning Electron Microscopy

To identify the darkened layers described in Section 3.2, EDS analyses were carried out via a scanning electron microscope using a sample of Geometry C. This sample was chosen due to the clear formation of the darkened layer (Figure 12). Figure 14 depicts the cross section prepared by a FIB with the highlighted recording line of the EDS measurement. The results of the EDS analysis are given in Figure 15 and illustrate the chemical composition of the elements along the marked line.

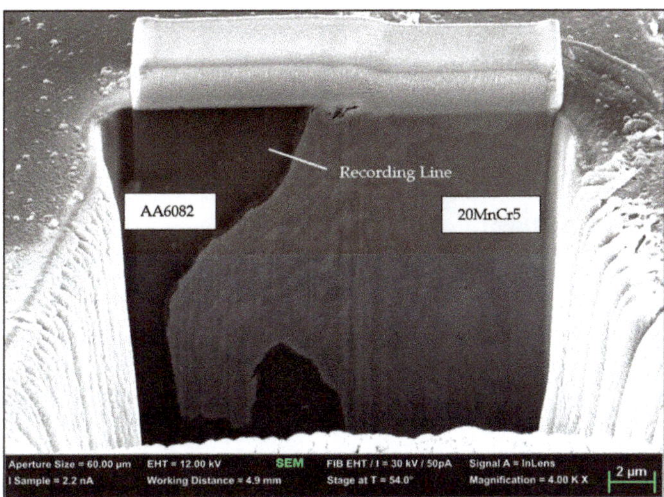

Figure 14. Scanning electron microscopy (SEM) image of a cross section prepared by a focused ion beam (FIB) in-lens detector (sample of Geometry C).

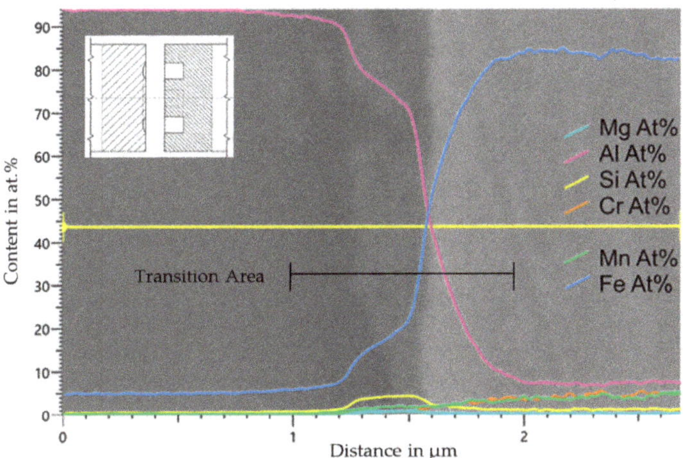

Figure 15. Energy-dispersive X-ray spectroscopy (EDS) analysis of the transition area in a specimen of Geometry C, SEM image of the recording line (yellow), Electron High Tension (EHT) = 12 kV, probe current = 1.7 nA, and working distance = 4.9 mm.

The left side of the graph in Figure 15 depicts the base material composition of the aluminum alloy. A content of almost 5 at.% of diffused iron is noticeable. On the right side of the graph, the composition of the steel base material is displayed, which contains a certain amount of alloying elements. Akin to the diffused iron on the aluminum side, aluminum diffused into the steel side with a content of about 5 at.%. Furthermore, an increased occurrence of manganese, magnesium and silicon can be observed in the transition zone. A mapping of the silicon content reveals its enrichment within a zone of about 0.5 µm as can be seen in Figure 16.

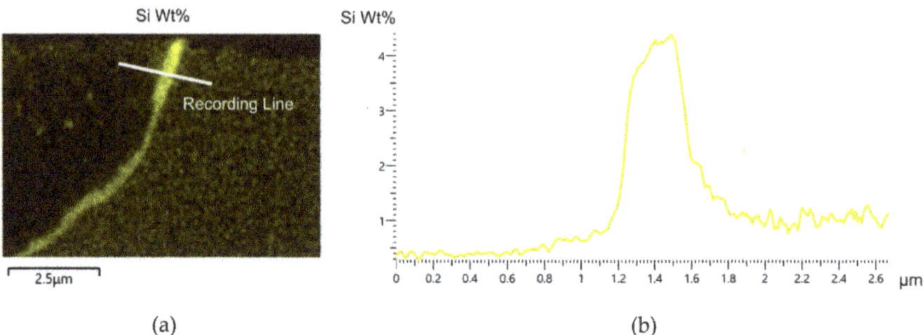

(a) (b)

Figure 16. EDS analysis of a sample with Geometry C, (**a**) distribution of silicon, and (**b**) silicon content in wt.% near the joining zone.

High silicon contents on sample surfaces can result from conventional sample preparation with silicon carbide grinding discs when a slope is formed in the joining zone during preparation due to the large differences in strength between aluminum and steel [19]. Silicon carbide particles can thus accumulate at the slope. Since in this case the EDS measurement was carried out on a cross section prepared using a FIB, such an influence of the preparation can be excluded.

Here, the increased silicon content measured in the joining zone by an EDS analysis is in accordance with observations of Liu et al. and Wang et al., who also reported increased silicon concentrations in the intermetallic compound (IMC) layer in the joining zone in investigations on friction welding of aluminum and steel [20,21]. The silicon is incorporated into the IMC layer and slows down the growth of the IMC layer [22]. With increasing silicon content in the aluminum alloy, the thickness of the IMC layer is reduced, and the phase constitution of the aluminide layers is altered [23].

4. Discussion

The presented results reveal that the surface geometry of friction-welded semifinished products and the parameters of the friction welding process have a decisive impact on the resulting bond and its strength. For example, Figure 14 shows microinterlocking, which may result from the different contact areas of the various geometries generating different temperatures during the friction process. This interlocking can result in an improvement of the bonding strength. Additionally, the different generated temperatures can lead to different microstructures, such as the grain size or the thickness of a possible intermetallic phase, and further in different bonding strengths.

Micrographs of the cross sections reveal that the material flow has an impact on the completeness of the bond. For example, for Geometry B (Figure 7a), air inclusions occur due to the resulting material flow. The aluminum alloy flowed outwards over the dome and detached from the 20MnCr5 steel. In addition, it can be assumed that some of the aluminum alloy was pushed back inwards onto the shoulder due to the colder outer zone, which has a higher deformation resistance. Another example of the importance of the material flow are the boreholes of Geometry C (Figure 8), which were not fully filled by flowing aluminum. The pressure of the enclosed air inside the boreholes inhibited a complete filling. Higher temperatures generated by rotational speed or pressure would increase the degree of deformation. This could lead to a better material flow.

The hardness tests show a decisive influence of the processing as a small increase in the hardness on the steel side near transition area (confer Figure 11). This hardness increase is probably caused by a combination of strain hardening due to the deformation process and grain refinement in the joining zone or by the occurrence of harder phases such as an intermetallic phases. An indication for the latter might be the multiple changes of the slope visible in the EDS line scan at the aluminum side in Figure 15 and the similarity of the darker layer in Figure 12 compared to the literature, such as [15].

Contrary to the hardness increase in the steel, the aluminum became softened close to the transition area which can probably be attributed to recrystallization or overageing of the T6 state due to heat generation during the friction welding process. In the transition area, the hardness decreased gradually between steel and aluminum, caused by mutual diffusion of aluminum and iron. An example of the concentration profile across the transition area is depicted in Section 3.3, Figure 15.

Hardness itself does not account for the quality of a bond, but it correlates with the tensile strength. Summarized, the hardness measurements reveal a heat-affected zone in both materials and in between them but could not be narrowed down due to the limiting conditions.

For the intended following impact extrusion process, Geometries A, F and H show the most promising results in mechanical tests and metallographic analyses. They feature a nearly complete bonding and high tensile strengths above 220 MPa. Only Geometry F contains a small volume of air inclusions at the undercut, which can possibly be avoided using modified parameters of the friction welding process. All specimens exhibited brittle fractures except for Geometry A. Brittle fractures underline the possible presence of (brittle) intermetallic phases [15]. Geometry H has the highest tensile strength of almost 280 MPa which is 100 MPa lower than the tensile strength of the base material (over 360 MPa), thus reaching about 77% of the base strength. Regarding the future processing by subsequent heating and impact extrusion, a recrystallization of the microstructure in the transition area can be expected and might further increase the bonding strength. With the evaluated bonding strength, Geometries A, F and H are hence suited for the subsequent inductive heating and forming process.

5. Summary and Conclusions

Based on the presented results, the following conclusions can be drawn:

- The highest tensile strengths values have been achieved using Geometries A (257 MPa), F (222 MPa) and H (280 MPa) (flat surface: 252 MPa in [14]);
- The completeness of the joint differs depending on the geometry and the correlation to the parameters of the friction welding process;
- The hardness close to the transition area was influenced by thermal effects of the friction welding process, resulting in a softening of the aluminum and an increased hardness in the steel;
- The EDS analysis showed what is most likely an intermetallic phase at the joining zone with a high content of silicon.

Table 4 presents the geometries investigated, their bond strengths and the main comments with regard to the further process chain of the CRC 1153.

Table 4. Results of the different geometries with regard to the further process.

	Geometry	Bond Strength (Average)	Comments
A		257 ± 5	• Complete form-filling and bonding resulted in form locking • Necking in aluminum at tensile tests
B		180 ± 14	• Air inclusions at radius to shoulder • No further investigations recommended
C		120 ± 8	• Boreholes were not filled • No further investigations recommended

Table 4. *Cont.*

	Geometry	Bond Strength (Average)	Comments
D		190 ± 15	• Incomplete bonding • No further investigations recommended
E		180 ± 21	• Air inclusions at outer diameter could be eliminated by using different friction parameters
F		222 ± 5	• Small air inclusion at the undercut • Suitable for inductive heating • Shrinkage resulted in force- and form-locking
G		215 ± 4	• Complete bonding • Bond strength could be enhanced by using different friction parameters
H		280 ± 17	• Complete bonding

The influence of various factors, such as friction welding parameters and material choice, leads to a large spectrum of possible improvements for enhancing the bonding strength. For example, the integrity of the joining zone might be improved by increasing heat generation during processing and thus diffusion of the alloying elements, though grain growth is to be expected. Hence, further investigations will focus on specimens with fixed surface geometries but varied friction welding parameters—e.g., a modification of Geometry F with an undercut angle of 80° on the steel side is to be expected promising regarding a further increase in the bonding strength. With this geometry, not only material bonds but force and form locks as well can be accomplished without a significant penetration of the aluminum alloy on the steel side, which otherwise could lead to a premature melting of the aluminum during induction heating in the further processes within the process chain of the CRC 1153 [8].

Author Contributions: Conceptualization, J.U. and B.-A.B.; methodology, R.L. and I.R.; validation, I.R., T.P., F.N. and J.U.; investigation, R.L. and I.R.; writing—original draft preparation, R.L.; writing—review and editing, J.U., T.P., F.N., C.K. and I.R.; visualization, R.L. and I.R.; supervision, B.-A.B., T.P. and J.U.; project administration, B.-A.B. and J.U.; funding acquisition, B.-A.B. All authors have read and agreed to the published version of the manuscript.

Funding: Funded by the Deutsche Forschungsgemeinschaft (DFG, German Research Foundation)—CRC 1153, subproject B3—252662854. The authors thank the German Research Foundation (DFG) for their financial support of this project.

Institutional Review Board Statement: Not applicable.

Informed Consent Statement: Not applicable.

Data Availability Statement: The data presented in this study are available in the article.

Acknowledgments: The authors appreciate the support of the subproject A2 regarding tensile testing and of Torsten Heidenblut regarding the SEM investigations.

Conflicts of Interest: The authors declare that they have no known competing financial interests or personal relationships that could have appeared to influence the work reported in this paper.

References

1. Erman Tekkaya, A. Energy saving by manufacturing technology. *Procedia Manuf.* **2018**, *21*, 392–396. [CrossRef]
2. Lancaster, J.F. *Metallurgy of Welding*; Springer: Dordrecht, The Netherlands, 1980; ISBN 978-94-010-9508-2.
3. Friedrich, H.E. *Leichtbau in der Fahrzeugtechnik*; Springer Fachmedien Wiesbaden: Wiesbaden, Germany, 2013; ISBN 978-3-8348-1467-8.
4. Hootak, M.; Kuwert, P.; Behrens, B.-A. (Eds.) Numerical Process Design for Compound Forging of Powder—Metallurgical and Solid Dissimilar Workpieces. In Proceedings of the 8th Congress of the German Academic Association for Production Technology (WGP), Aachen, Germany, 19–20 November 2018.
5. Thomä, M.; Gester, A.; Wagner, G.; Straß, B.; Wolter, B.; Benfer, S.; Gowda, D.K.; Fürbeth, W. Application of the hybrid process ultrasound enhanced friction stir welding on dissimilar aluminum/dual-phase steel and aluminum/magnesium joints. *Mater. Werkst.* **2019**, *50*, 893–912. [CrossRef]
6. Behrens, B.-A.; Huskic, A.; Bouguecha, A.; Frischkorn, C.; Chugreeva, A.; Duran, D. Tailored Forming Technology for Three Dimensional Components: Approaches to Heating and Forming. In Proceedings of the 5th International Conference on Thermomechanical Processing, Milan, Italy, 26–28 October 2016.
7. Behrens, B.-A.; Bonhage, M.; Bohr, D.; Duran, D. Simulation Assisted Process Development for Tailored Forming. *Mater. Sci. Forum* **2019**, *949*, 101–111. [CrossRef]
8. Behrens, B.-A.; Goldstein, R.; Duran, D. Role of Thermal Processing in Tailored Forming Technology for Manufacturing Multi-Material Components. In Proceedings of the Heat Treat 2017: Proceedings of the 29th ASM Treating Society Conference, Columbus, OH, USA, 24–26 October 2017.
9. Ashfaq, M.; Sajja, N.; Khalid Rafi, H.; Prasad Rao, K. Improving Strength of Stainless Steel/Aluminum Alloy Friction Welds by Modifying Faying Surface Design. *J. Mater. Eng. Perform.* **2013**, *22*, 376–383. [CrossRef]
10. Fukumoto, S.; Tsubakino, H.; Okita, K.; Aritoshi, M.; Tomita, T. Friction welding process of 5052 aluminium alloy to 304 stainless steel. *Mater. Sci. Technol.* **1999**, *15*, 1080–1086. [CrossRef]
11. Lee, W.B.; Yeon, Y.M.; Kim, D.U.; Jung, S.B. Effect of friction welding parameters on mechanical and metallurgical properties of aluminium alloy 5052–A36 steel joint. *Mater. Sci. Technol.* **2003**, *19*, 773–778. [CrossRef]
12. Fukumoto, S.; Inuki, T.; Tsubakino, H.; Okita, K.; Aritoshi, M.; Tomita, T. Evaluation of friction weld interface of aluminium to austenitic stainless steel joint. *Mater. Sci. Technol.* **1997**, *13*, 679–686. [CrossRef]
13. Sahin, M. Joining of stainless-steel and aluminium materials by friction welding. *Int. J. Adv. Manuf. Technol.* **2009**, *41*, 487–497. [CrossRef]
14. Behrens, B.-A.; Chugreev, A.; Selinski, M.; Matthias, T. Joining zone shape optimisation for hybrid components made of aluminium-steel by geometrically adapted joining surfaces in the friction welding process. In Proceedings of the 22nd International ESAFORM Conference on Material Forming: ESAFORM 2019, Vitoria-Gasteiz, Spain, 8–10 May 2019; AIP Publishing: College Park, MD, USA, 2019; p. 40027.
15. Ambroziak, A.; Korzeniowski, M.; Kustroń, P.; Winnicki, M.; Sokołowski, P.; Harapińska, E. Friction Welding of Aluminium and Aluminium Alloys with Steel. *Adv. Mater. Sci. Eng.* **2014**, *2014*, 981653. [CrossRef]
16. Schuler, V.; Twrdek, J. *Praxiswissen Schweißtechnik*; Springer: Wiesbaden, Germany, 2019; ISBN 978-3-658-24265-7.
17. DIN Deutsches Institut für Normung E.V. *DIN EN ISO 6507-1: Metallische Werkstoffe—Härteprüfung nach Vickers—Teil 1: Prüfverfahren (ISO 6507-1:2018)*; Deutsche Fassung EN ISO 6507-1:2018; Beuth Verlag GmbH: Berlin, Germany, 2018.
18. DIN Deutsches Institut für Normung E.V. *DIN EN ISO 6892-1: Metallische Werkstoffe—Zugversuch—Teil 1: Prüfverfahren bei Raumtemperatur (ISO 6892-1:2016)*; Deutsche Fassung EN ISO 6892-1:2016; Beuth Verlag GmbH: Berlin, Germany, 2017.
19. Barienti, K.; Kahra, C.; Herbst, S.; Nürnberger, F.; Maier, H.J. Ion Beam Processing for Sample Preparation of Hybrid Materials with Strongly Differing Mechanical Properties. *Metallogr. Microstruct. Anal.* **2020**, *9*, 54–60. [CrossRef]
20. Liu, Y.; Zhao, H.; Peng, Y.; Ma, X. Microstructure and tensile strength of aluminum/stainless steel joint welded by inertia friction and continuous drive friction. *Weld World* **2020**, *64*, 1799–1809. [CrossRef]
21. Wang, T.; Sidhar, H.; Mishra, R.S.; Hovanski, Y.; Upadhyay, P.; Carlson, B. Evaluation of intermetallic compound layer at aluminum/steel interface joined by friction stir scribe technology. *Mater. Des.* **2019**, *174*, 107795. [CrossRef]
22. Eggeler, G.; Auer, W.; Kaesche, H. On the influence of silicon on the growth of the alloy layer during hot dip aluminizing. *J. Mater. Sci.* **1986**, *21*, 3348–3350. [CrossRef]
23. Cheng, W.-J.; Wang, C.-J. Effect of silicon on the formation of intermetallic phases in aluminide coating on mild steel. *Intermetallics* **2011**, *19*, 1455–1460. [CrossRef]

Article

Lateral Angular Co-Extrusion: Geometrical and Mechanical Properties of Compound Profiles

Susanne Elisabeth Thürer [1,*], **Julius Peddinghaus** [2], **Norman Heimes** [2], **Ferdi Caner Bayram** [3], **Burak Bal** [3], **Johanna Uhe** [2], **Bernd-Arno Behrens** [2], **Hans Jürgen Maier** [1] **and Christian Klose** [1]

[1] Institut für Werkstoffkunde (Institute of Materials Science), Leibniz University Hannover, 30823 Garbsen, Germany; maier@iw.uni-hannover.de (H.J.M.); klose@iw.uni-hannover.de (C.K.)

[2] Institut für Umformtechnik und Umformmaschinen (Institute of Forming Technology and Machines), Leibniz University Hannover, 30823 Garbsen, Germany; peddinghaus@ifum.uni-hannover.de (J.P.); heimes@ifum.uni-hannover.de (N.H.); uhe@ifum.uni-hannover.de (J.U.); behrens@ifum.uni-hannover.de (B.-A.B.)

[3] Department of Mechanical Engineering, Abdullah Gül University, 38080 Kayseri, Turkey; ferdicaner.bayram@agu.edu.tr (F.C.B.); burakbal@agu.edu.tr (B.B.)

* Correspondence: thuerer@iw.uni-hannover.de; Tel.: +49-511-762-18227

Received: 26 June 2020; Accepted: 26 August 2020; Published: 28 August 2020

Abstract: A novel co-extrusion process for the production of coaxially reinforced hollow profiles has been developed. Using this process, hybrid hollow profiles made of the aluminum alloy EN AW-6082 and the case-hardening steel 20MnCr5 (AISI 5120) were produced, which can be forged into hybrid bearing bushings by subsequent die forging. For the purpose of co-extrusion, a modular tooling concept was developed where steel tubes made of 20MnCr5 are fed laterally into the tool. This LACE (lateral angular co-extrusion) process allows for a variation of the volume fraction of the reinforcement by using steel tubes with different wall thicknesses, which enabled the production of compound profiles having reinforcement contents of either 14 vol.% or 34 vol.%. The shear strength of the bonding area of these samples was determined in push-out tests. Additionally, mechanical testing of segments of the hybrid profiles using shear compression tests was employed to provide information about the influence of different bonding mechanisms on the strength of the composite zone.

Keywords: tailored forming; lateral angular co-extrusion; mechanical behavior; hybrid metal components

---

1. Introduction

The realization of lightweight constructions to increase resource efficiency and reduce CO_2 emissions is of paramount interest to the automotive and aviation industries [1]. In particular, the use of light metals such as aluminum is attractive due to the high specific strength of the respective alloys. Merklein et al. reported that one promising approach to integrate aluminum in automotive designs is the use of hybrid components [2]. In this context, the integration of Tailored Blanks in sheet metal forming has become state of the art in the automotive industry. These allow meeting conflicting design challenges by providing sheet metal components with locally adapted properties. As well as reducing weight by combining sheets of different material grades, thicknesses, etc., Tailored Blanks can also offer improved crash performance [2].

However, the concept of hybrid semi-finished products is not yet widely used in bulk forming of metals. Innovative processing technologies are required for the production of hybrid bulk metal components made of different metals such as aluminum alloys and steel. As part of the novel concept of Tailored Forming, process chains are being developed in which the various bulk materials are first joined before a subsequent forming process such as die forging or impact extrusion is applied.

As discussed by Herbst et al., this differs significantly from conventional process chains in which the individual parts of the components are joined together after the forming step or at the end of the process chain [3].

In contrast to sheet metal forming processes used for the production of components made of Tailored Blanks, the Collaborative Research Center 1153 (CRC 1153) "Tailored Forming" aims at developing suitable processes for the production of three-dimensional solid components with locally adapted properties. The use of aluminum instead of steel in the bulk metal component can result in a reduction of mass of the component. Specifically, only the functional surface, which must be wear-resistant in the solid component, consists of material like hardened steel. A key feature in the present approach is the subsequent joint forming step. The advantage of this process combination lies in the positive influence of the subsequent forming on the local microstructure in the joining zone, and thus further the mechanical properties. This is a key aspect for materials that are difficult to form such as aluminum and steel. An exemplary process chain is shown in Figure 1 for the manufacture of a hybrid bearing bushing made of aluminum and steel that is investigated as a demonstrator part.

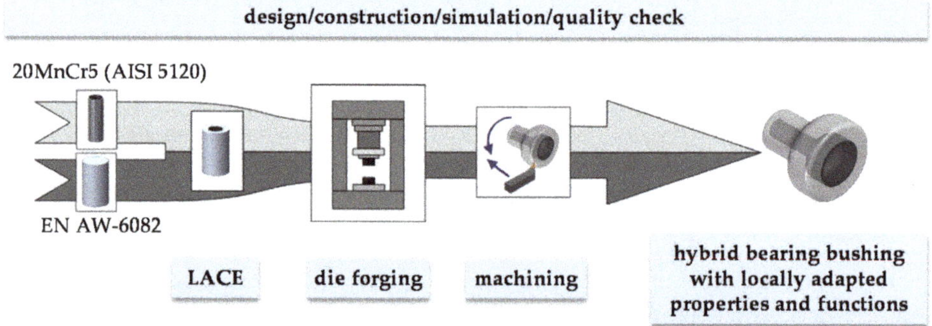

Figure 1. Tailored forming process chain for the production of hybrid bearing bushings.

Similar to Tailored Blanks, the production of hybrid solid semi-finished products can be realized by welding. Pressure welding is one of the processes that enables the formation of bonds in which the intermetallic phase seam is sufficiently small enough to have no negative impact on properties such as tensile strength [4]. In the context of continuous hybrid profiles, a promising approach for joining different materials is co-extrusion, which was used in the present study to manufacture the semi-finished products for the bearing bushing. Co-extrusion enables the production of composite profiles consisting of at least two materials [5]. Co-extrusion can be assigned to joining by forming according to DIN (Deutsches Institut für Normung e. V.—German Institute for Standardization) 8593-5, in which the parts to be joined are formed locally or completely. In principle, co-extrusion can be divided into two different process variants:

- Co-extrusion of modified billets: The reinforcing element is contained in the billet and passes through the entire extrusion process. This variant includes the co-extrusion of metal matrix composites in which reinforcing particles such as Al_2O_3 were introduced in an aluminum billet by powder metallurgy [6] with the intention to achieve an even distribution of the reinforcing elements in the composite profile. A further variant is the local reinforcement of a billet, e.g., by inserting a round rod of a material of higher strength like titanium grade 2 into an aluminum billet and then extruding the materials simultaneously [7].
- Co-extrusion of conventional billets: In this case, the reinforcing element is introduced into the forming zone from outside the tool but is not plastically formed itself [8]. This process variant was investigated especially for the reinforcement of extruded profiles with steel or copper wires. By employing modified chamber tools [9], the wires were introduced into the forming zone via

the support arms of the mandrel. Hence, the wire reinforcement was present in the longitudinal weld seams of the profiles only [10].

With the LACE process (Lateral Angular Co-Extrusion) developed by Grittner et al., reinforcing elements such as titanium sheets and flat profiles that are already relatively rigid can be fed laterally, and thus continuously into the extrusion process [11]. A laboratory-scale LACE process has already been developed within the CRC 1153, which provided a round rod made of 20MnCr5 steel with an aluminum cladding made of EN AW-6082. However, due to the design of the tool the resulting geometry of the compound profile showed significant deviations from the desired coaxial arrangement [12]. For subsequent die forging, the steel reinforcement must be embedded coaxially within the aluminum matrix. In the present study, this issue was addressed by a novel extrusion tool design, which also featured an industry relevant scale. The mechanical properties of the bonding zones of the compound profiles produced with the new tool were characterized in push-out tests. In addition, the influence of the bonding mechanisms, e.g., material closure or form-closure, on the composite strength were determined by shear compression tests on sample segments.

2. Materials and Methods

2.1. LACE Process

A schematic section through the developed modular tool is shown in Figure 2a. Since the LACE process involves the feeding of a rigid reinforcing element instead of a wire, a mandrel part was designed, which is supported by three support arms in the tool housing. In this concept, the aluminum alloy is divided into two metal streams by the portholes in the middle of the symmetrically designed entry. Both metal strands are then directed into pockets milled into each half of the tool cavity. This is intended to change the material flow in such a way that the aluminum alloy evenly envelops the reinforcing element and displacement and/or distortion of the compound profile is avoided. The aluminum alloy then flows around the mandrel part inside the tool. The reinforcing element, in this case a steel tube, is inserted into the tool orthogonally to the movement direction of the extrusion punch and guided through the clamping cover and the mandrel part. This also ensures a coaxial position of the tube in the compound profile.

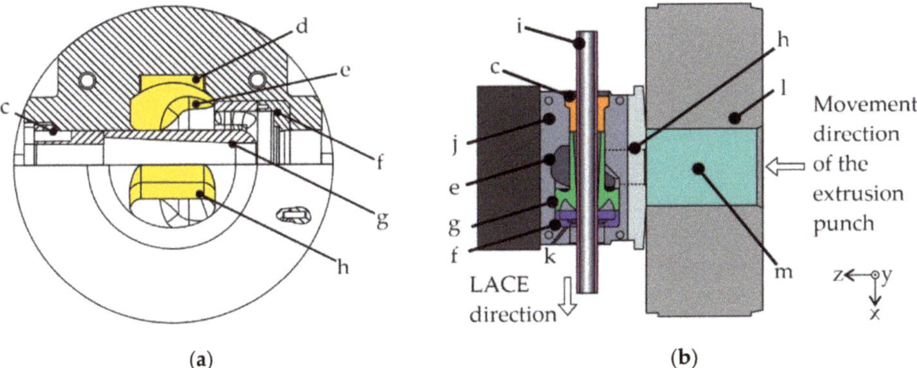

Figure 2. (**a**) Schematic illustration with sectional plane lengthwise through the tool with c: clamping cover, d: pocket, e: deflection, f: die, g: mandrel part with three support arms, h: inlet (the geometry of the inlet and deflection are highlighted); (**b**) schematic illustration of the concept for the 10 MN extrusion press in longitudinal section with i: reinforcing element, j: die housing, k: thermocouple bore, l: container, m: aluminum billet.

The process is shown in Figure 2b as a schematic sectional view including the reinforcing element in the tool and the billet inside the container. The LACE direction is orthogonal to the direction of the movement of the ram.

The LACE experiments were performed on a 10 MN extrusion press (SMS Meer GmbH, Düsseldorf, Germany). A non-heated tool holder specially modified for this project was used, which allowed the reinforcing element to be fed laterally. Aluminum EN AW-6082 billets as well as tubes consisting of the case-hardening steel 20MnCr5 were used as joining partners. To keep the process chain as short as possible and to avoid additional drilling of the hybrid semi-finished products, reinforcing elements with the desired inner diameter were used. Furthermore, the reinforcement content was varied by using steel tubes with different wall thickness. With an inner diameter of the container of 146 mm, an opening diameter of 62.7 mm for the die and 38 mm or 44.5 mm for the outer diameter of the steel tubes, the extrusion ratio equaled 9:1 and 11:1, respectively. This corresponds to a reinforcement content of 14 vol.% for the extrusion ratio of 9:1 and 34 vol.% for the extrusion ratio of 11:1. Since the objective was to achieve a metallurgical bond between the joining partners, the reinforcing element was ground with 40 grit paper and cleaned with ethanol prior to the extrusion process. Previous numerical studies have shown that a bond by material closure can be achieved by employing relatively high temperatures together with long contact times of the joining partners [13]. This translates to high process temperatures and low extrusion speed. Thus, the billets were preheated to 530 °C for 4.5 h, whereas the steel tube had room temperature at the beginning of each experiment. The die was preheated to 490 °C and the container had a temperature of 440 °C. A ram speed of 1.5 mm/s was used at the beginning of the experiment during the upsetting of the billet in the container and the filling of the tool, and it was reduced stepwise until the ram speed reached 0.3 mm/s. In this way, the cooling of the aluminum billet during filling of the tool was kept to a minimum.

2.2. Metallographic Characterization

Cross-sections were extracted from the front and the back ends of the compound profiles to determine deviations of the positions of the reinforcing elements relative to their ideal center positions, as well as to characterize the microstructures of the extruded matrix material. The front end of the actual compound profile was defined as the location where all longitudinal weld seams appeared to be macroscopically closed. The position of the back end was dependent on the particular LACE experiment. If the extrusion was stopped before the entire reinforcing element was jacketed with aluminum, the area closest to the tool was examined in cross-section. For the macro- and microstructural examination of the compound profiles, the samples were prepared metallographically and treated with an etching solution consisting of HF and H_2SO_4 to contrast the secondary precipitates of the aluminum alloy.

2.3. Push-Out Test and Shear Compression Test

The mechanical properties of the bonding area of the compound profiles were measured with push-out tests and shear compression tests. For this purpose, samples were taken over the entire length of the composite profile. Due to the different coefficients of thermal expansion of aluminum and steel, the coaxially reinforced semi-finished products are assumed to have a force closure connection resulting from shrinking of the matrix material onto the reinforcing element [14]. Shear compression tests of sample segments served to determine whether the effective bond mechanism is mainly material closure or force and/or form closure. For the tests, samples were taken in an alternating order from one compound profile per extrusion ratio (Figure 3). The compound profile was divided into slices, each of which had a plane-parallel height of 10 mm after machining on both sides. The first sample was taken 25 mm behind the position, where all four longitudinal weld seams were macroscopically closed. The longitudinal weld seam that was used as the starting point of the sampling was the one closed last during extrusion and is referred to in the schematic illustration in Figure 3 as the relevant longitudinal weld seam.

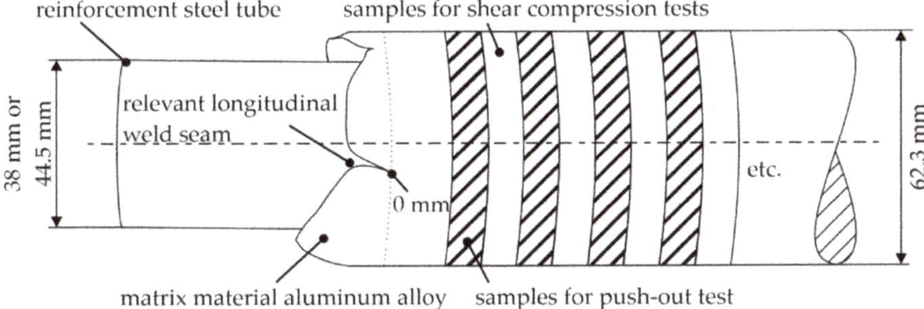

Figure 3. Schematic illustration of a compound profile with alternating sampling for push-out tests and shear compression tests with planned geometry of the compound profile.

Starting at this location, samples were taken from the profile with a thickness of 15 mm each, taking into account the saw cut and the allowance for facing. These samples were used as full samples for the push-out tests or divided into several segments and then used for the shear compression tests. The measured shear strength of the sample segments was then compared with the de-bonding shear strengths from the push-out tests of the adjacent samples in order to be able to determine the contribution of the material bond over the compound profile length.

The push-out tests were carried out using a universal testing machine with a maximum force of 250 kN (type Z250, Zwick, Ulm, Germany) and the test setup is shown schematically in Figure 4. Centering of the sample was realized by a step in the punch. The compound specimens were positioned on a steel ring so that the contact surface with the aluminum alloy was as large as possible. By lowering the punch of the testing machine vertically, the reinforcing element was pressed out and the force-displacement curve was recorded.

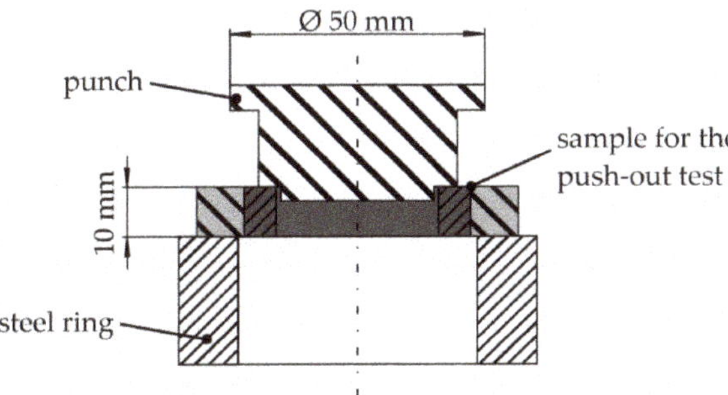

Figure 4. Schematic illustration of the experimental setup of the push-out tests using a sample taken from a hollow compound profile.

The specimen discs for the shear compression test, which had a plane-parallel height of 10 mm, were separated by wire cutting into the segments with an angle of about 65° as shown in Figure 5a. This procedure resulted in two sample segments that did not contain a longitudinal weld seam and a sample that contained two longitudinal weld seams. The latter was taken so that the longitudinal weld seams no. 1 and no. 2 (cf. Figure 5a) were at the edges of the sample segment, and thus did not affect the test results significantly. In order to determine the actual test area, the bonding lengths of all sample segments were measured by using a laser microscope (type VK 9700, Keyence, Neu-Isenburg, Germany). From these data, the actual bonding areas were calculated.

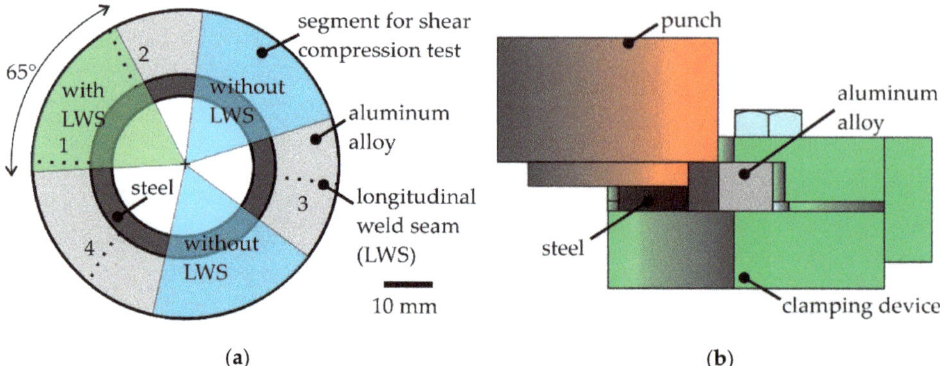

Figure 5. Schematic illustration of (**a**) sample segment extraction from the cross-section of a compound profile with one segment having two longitudinal weld seams (LWS, highlighted by dotted lines) at the edges and two segments each without weld seams, (**b**) test setup used for the shear compression test.

For the characterization of the mechanical properties, the specimens were clamped in the test setup shown in Figure 5b and the steel portion of the specimens was pressed out with a universal testing machine (type Z250, Zwick, Ulm). As with the push-out tests on the entire specimen cross-sections, a test speed of 2 mm min^{-1} was used. A drop in force of 80% was used as the break-off criterion for the push-out tests of the sample segments.

3. Results

3.1. LACE Process

During the LACE experiments, the relevant process parameters such as ram force and ram speed were recorded. Figure 6 shows an exemplary diagram of the ram force and the ram speed vs. process time for a typical LACE experiment with an extrusion ratio of 11:1. At the onset of the test, a ram speed of 1.5 mm s^{-1} was used. This fast ram speed was selected for initial filling of the tool in order to counteract cooling of the tool, whereas the actual extrusion process was performed at lower ram speed. In Figure 6, a rapid increase in ram force to a force plateau of 2.6 MN is seen. The speed of the ram was reduced to a value of 0.5 mm s^{-1} after this plateau. The ram force increased further as the filling process progressed. After reaching 5 MN, the ram speed was reduced to the desired value of 0.3 mm s^{-1} for the LACE extrusion test, which resulted in a slight drop in force. The ram force then increased continuously up to the maximum value of ≈8 MN before termination of the LACE process.

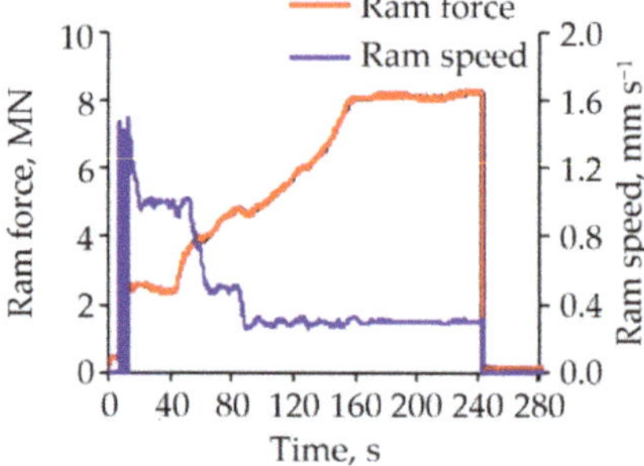

Figure 6. Evolution of ram force during the course of a LACE process with an extrusion ratio of 11:1 and adjustments of ram speed.

3.2. Metallographic Characterization

In Figure 7, representative cross-sections extracted from the front and the back ends of a compound profile are shown. Here, a steel tube with an outer diameter of $\varnothing_a = 38$ mm was used, which resulted in an extrusion ratio of 9:1, and thus a volume fraction of the reinforcement of 14 vol.%. The outer aluminum matrix of the compound profile had a slightly elliptical cross-section at the start of the compound profile. There was a slight material overlap on the side facing the recipient (recipient side), which is interspersed with oxide lines (Figure 7a, left-hand area). The overlap on aluminum extended over a circumference of 55 mm for a total circumference of the cross-section of 202 mm. This overlap resulted from incidental clamping of the steel tube, which in turn resulted in temporarily faulty material flow. This illustrates that is of paramount importance to accurately control the local material flow in the die. For the determination of the lengths of the main and secondary axes of the aluminum jacket, this section was not taken into account. Thus, the main axis had a length l_y of 63.3 mm at the start of the profile and the secondary axis a length l_z of 63.6 mm. The outer contour was thus 1.5% larger in the y-direction and 2.0% larger in the z-direction than the theoretical diameter of the aluminum jacket. No bond was formed between the matrix material and the reinforcing element on the recipient side; instead, there was a 0.5 mm wide gap. The longitudinal weld seam on the side facing away from the recipient (rear side) also showed a gap in the bonding area between the aluminum alloy and the steel tube. By contrast, the cross-section taken from the end of the compound profile showed an almost ideal circular contour without any material overlap. The main axis had a length of 63.2 mm (deviation +1.4%) and the length of the secondary axis was 61.9 mm (deviation 0.7%).

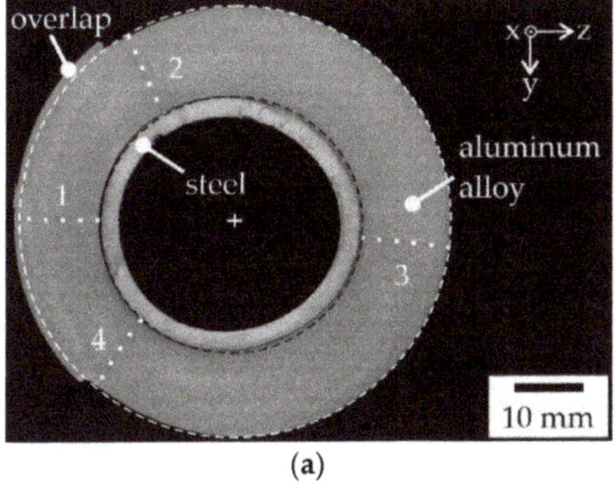

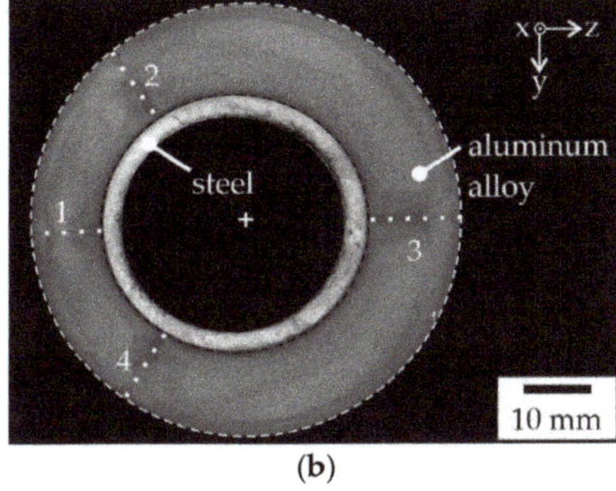

(a)   (b)

Figure 7. Cross-section of a sample taken from (**a**) the start of the compound profile, i.e., after a macroscopically closed longitudinal weld seam had formed and (**b**) the end of the compound profile, which was produced with an extrusion ratio of 9:1; etching: HF/H_2SO_4 mixture; the position of the longitudinal weld seams is highlighted by dotted lines; the interpolated outer contour of the steel tube is highlighted with black dashed lines; the (in case of (**a**) interpolated) outer contour of the aluminum is highlighted with white dashed lines.

In the compound profile shown here, the reinforcing element was not truly embedded coaxially in the matrix. The aluminum metal stream inside the die flowed to the side facing away from the recipient preferentially. This is evident in the greater wall thickness of the matrix material on the right-hand side of the cross-sections shown in Figure 7. In the initial area of the compound profile, this led to a slight deformation of the reinforcing element, which can also be observed in Figure 7a. The reinforcing elements used in these LACE experiments were deep-hole drilled tubes with an uneven wall thickness over the tube circumference. In the compound profile shown here, the wall thickness deviated from the intended 3 mm by up to 0.5 mm, i.e., the wall thickness was between 2.7 mm and 3.5 mm. For the shown cross-section from the start of the compound profile with elliptical reinforcing element, the outer contour of the steel tube was interpolated using the theoretical outer diameter $\varnothing_a$ of 38 mm. The offset of the steel tube at the start of the profile was thus 0.4 mm or 0.6% in the y-direction and 0.8 mm or 1.2% in the negative z-direction. The cross-section from the end piece of the compound profile showed no geometrical deviation of the reinforcing element caused by the LACE process, despite variations in the wall thickness. The offset in the y-direction was 0.4 mm or 0.6% at the start of the profile. In the negative z-direction, the steel tube was shifted by 1.7 mm or 2.7%.

Metallographic etching was used to contrast the secondary precipitates and make the longitudinal weld seams visible. The two longitudinal weld seams that are running horizontally in the metallographic image were caused by the material flowing into the portholes of the tool entry and subsequent welding after flowing around the mandrel part or reinforcing element. Two additional longitudinal weld seams are expected on the side facing the recipient, each of which should be located at an angle of 120° to each other and to the longitudinal weld seam on the rear side. As seen in Figure 7, the weld seams appear close to the expected positions.

The material combination EN AW-6082 and 20MnCr5 was also extruded to a compound profile with an outer diameter of 44.5 mm for the steel tube, and thus an extrusion ratio of 11:1. The cross-sections taken from the start and end of the compound profile are shown in Figure 8a,b. Both cross-sections had almost the desired circular cross-section and did not show any deviations in the wall thickness of the steel tube, which could be attributed to deep hole drilling. The outer contour of the aluminum jacket had a length of 63.4 mm of the main axis at both the start and end of the profile (which had a length of 215 mm), which is 1.6% greater than the expected outer diameter. For the secondary axis, a length of 62.1 mm (−0.4%) could be determined at the start and 62.0 mm (0.5%) at the end of the compound profile. Residual oxides could still be detected inside the matrix material at the front end, but no longer at the back end of the compound profile. In addition, there was no complete bond between the aluminum alloy and the steel tube in the initial area, which became apparent in form of a gap with a width of 15 μm (detail in Figure 8c). At the end of the compound profile, this gap was no longer so pronounced (detail in Figure 8d).

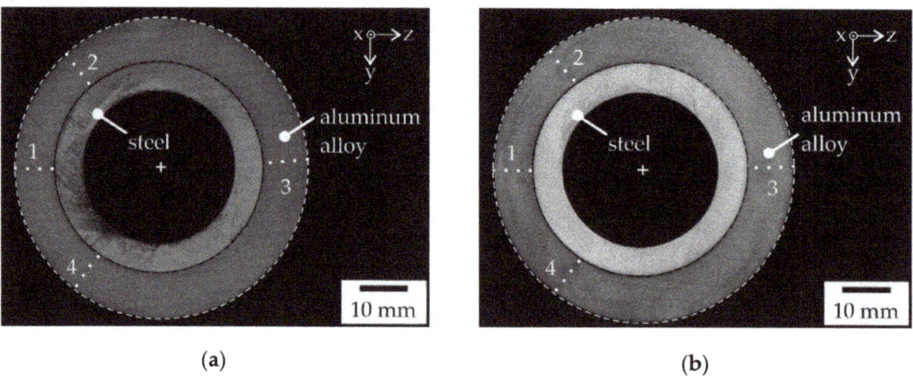

(a)   (b)

Figure 8. *Cont.*

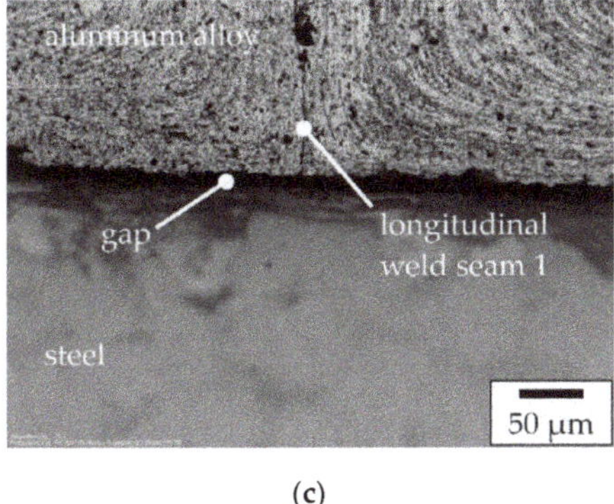

(**c**)   (**d**)

Figure 8. Cross-sections of (**a**) start and (**b**) end of a compound profile made of EN AW 6082 and 20MnCr5 with an extrusion ratio of 11:1; the position of the longitudinal weld seams is highlighted by dotted lines; the outer contour of the steel is highlighted with black dashed lines; the outer contour of the aluminum is highlighted with white dashed lines; with (**c**) detailed image of the bonding area with a gap between the joining partners from the front end of the compound profile; with (**d**) detailed image of the bonding area from the back end of the compound profile; etching: (**a–c**) HF/H$_2$SO$_4$ mixture, (**d**) HNO$_3$/Ethanol mixture.

The position of the reinforcing element remained unchanged over the entire profile length and showed an offset of 0.4 mm or 0.6% in the negative y-direction and an offset of 0.6 mm or 0.9% in the negative z-direction. The position of the longitudinal weld seams did not yet correspond perfectly to the expected position at the start of the compound profile. On the one hand, the longitudinal weld seam on the recipient side was not in a perfectly horizontal position but offset in the negative y-direction. On the other hand, the two longitudinal weld seams, which were formed by the support arms of the mandrel part on the side facing the recipient, had a smaller angle to each other than expected. At the end of the compound profile, the longitudinal weld seams, which are formed horizontally due to the splitting of the matrix material at the portholes and subsequently by passing by one of the support arms of the mandrel part, were on the expected horizontal plane. The angle between the two longitudinal weld seams no. 2 and 4 remained unchanged.

3.3. Mechanical Properties

The strength of the bonding area was determined for the compound profiles with different reinforcement content using push-out tests. Figure 9 shows an exemplary force-path graph from a push-out test on a representative sample that was produced with an extrusion ratio of 11:1. At the beginning the measured force F increases almost linearly until the curve flattens out slightly towards the end and finally reaches its maximum F_{max}. After the maximum, the force decreases rapidly and runs out in a plateau. Based on these data, the shear strength was calculated as [15]

$$\tau_{max} = \frac{F_{max}}{\pi\,d\,h}$$

(1)

where d is the diameter of the reinforcement and h the height of the sample.

Figure 10 shows the de-bonding shear strength calculated by using measured data from all the push-out tests executed over the profile length of both compound profiles. The de-bonding shear strength of the profile with the lower reinforcement content of 14 vol.% was determined over a profile length of 250 mm and varied between 29 MPa and 55 MPa with an average shear strength of 42 MPa ± 7 MPa. In the case of the profile having a reinforcement content of 34 vol.%, the shear strength determined in the push-out test varied between 45 MPa and 63 MPa over the entire profile length of 445 mm with an average shear strength of 54 MPa ± 5 MPa.

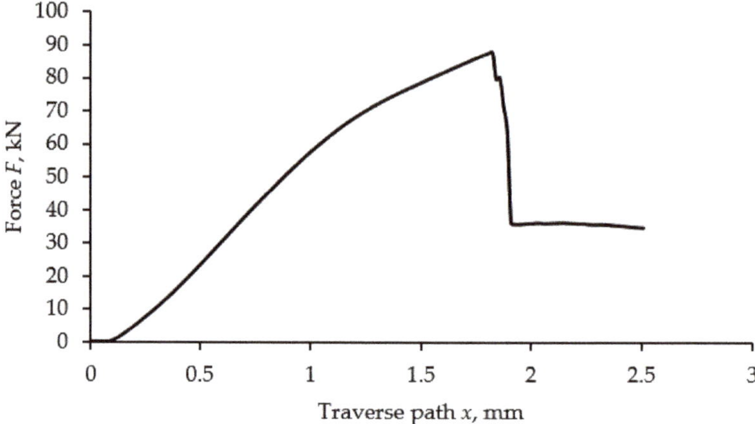

Figure 9. Force-displacement diagram of a push-out tests on a sample sectioned from the front end of a compound profile produced with an extrusion ratio of 11:1.

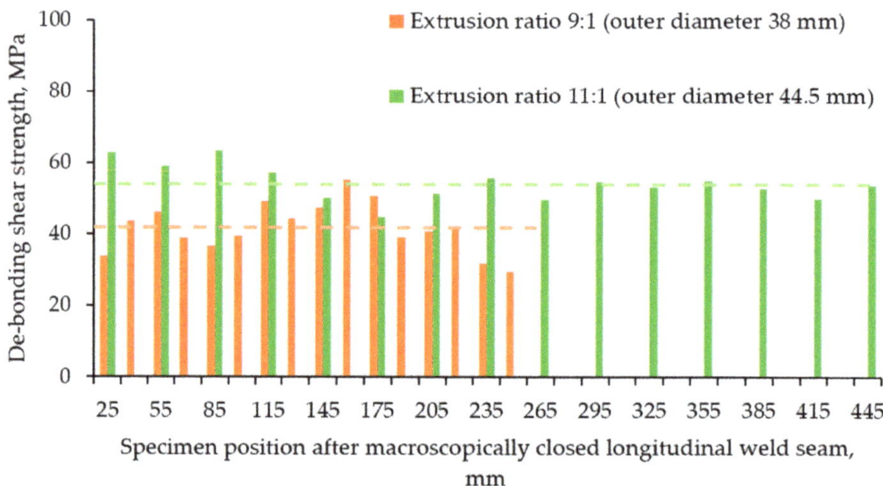

Figure 10. De-bonding shear strength for compound profiles produced via LACE with an extrusion ratio of 9:1 or 11:1, respectively. The respective mean values are indicated by dashed lines, which were determined over the entire length of the profile with macroscopically intact longitudinal weld seams.

For the shear compression test, two segments without longitudinal weld seams and one segment containing two longitudinal weld seams were available for each specimen cross-section. The segments without longitudinal weld seams showed a similar shear strength curve progression as the de-bonding shear strength curves determined by the push-out tests, as it can be seen in Figure 11. The values determined in the shear compression test fluctuated between 47 MPa and 69 MPa. The tested segments, which had two longitudinal weld seams, showed similar behavior over the profile length. However, there were two outliers at 100 mm and 220 mm, which, at 92 MPa and 83 MPa, respectively, had the highest strength values for the compound profiles made of the material combination EN AW-6082 and 20MnCr5. In general, the shear strength of the segment with longitudinal weld seams was below that of the segments without longitudinal weld seams from 250 mm onwards. However, the average shear strength determined in the shear compression test was 56 MPa ± 6 MPa for the segments without longitudinal weld seams and 58 MPa ± 15 MPa for the segments with longitudinal weld seams.

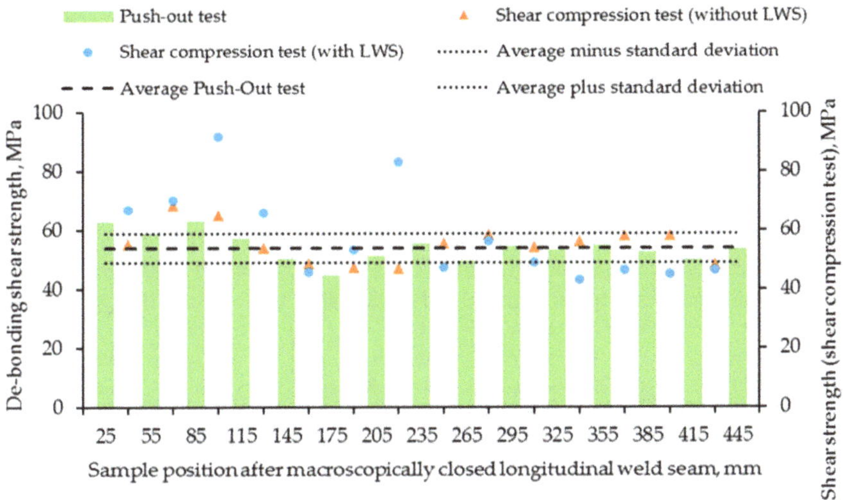

Figure 11. Determined shear strength of the compound profile with a reinforcement content of 34 vol.% starting 25 mm after macroscopically closed longitudinal weld seam with de-bonding shear strength averaged over the entire compound profile length (dashed line); dotted lines represent one standard deviation both for sample segments with and without longitudinal weld seams.

Figure 12 shows one of the two outliers in the shear compression test with longitudinal weld seams, extracted 100 mm behind the location where the longitudinal weld seam was considered to be macroscopically intact. The arrows and dotted lines mark the position of the longitudinal weld seams no. 1 and 2. At the position of the longitudinal weld seam no. 1, which is on the left-hand side in Figure 12, however, a gap between the joining partners can be seen. At the location of the longitudinal weld seam no. 2, which is on the right-hand side in Figure 12, it can be seen that the aluminum still adheres to the reinforcing element after the test. This demonstrates that the separation of the materials did not take place in the bonding zone.

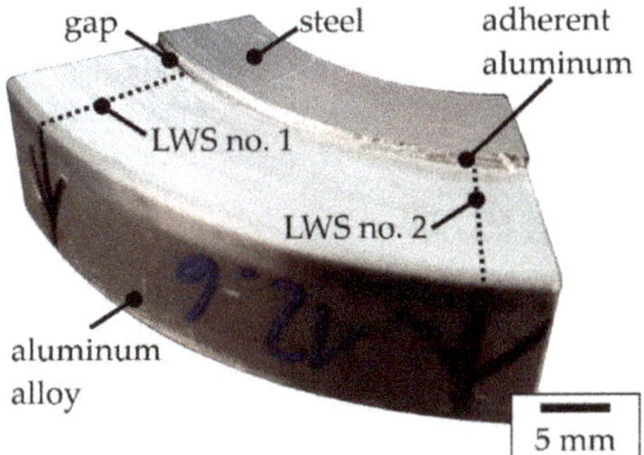

Figure 12. Sample after the shear compression test with adhering aluminum at the level of a longitudinal weld seam (LWS; position marked with arrows and highlighted with dotted lines).

4. Discussion

In order to comply with the requirements of the intended subsequent die forging process, specifications for the compound profiles with regard to the internal geometry of the reinforcing element as well as the position of the joining partners in relation to each other were defined prior to toolset and process development. Thus, the co-extruded semi-finished products were required to have an internal diameter $\varnothing_i$ of 32 mm as well as an aluminum cladding with a uniform wall thickness so that the hybrid semi-finished products can be heated by means of an internal inductor prior to die-forging [16]. Furthermore, a maximum deviation in the coaxial arrangement of the joining partners of 0.2 mm (0.3%) was aimed for with a height of the semi-finished product of 85 mm. This was not fully achieved, as in some cases there was an offset of up to 0.6 mm (0.9%). By turning the semi-finished product and correspondingly increasing the length, the mold filling could still be realized in the subsequent die forging process.

In the present study, coaxially reinforced hollow profiles were produced, with a reinforcement content of 14 vol.% and 34 vol.%. With the extrusion ratio of 9:1, it was possible to envelope a steel tube having a wall thickness of 3 mm in an outer aluminum cladding without the reinforcing element collapsing due to the pressure acting on the tube during the LACE process. At the front end of the compound profile, a slight deviation of the steel tube from the desired circular geometry was detected. This deformation was no longer present at the end of the profile. A similar problem did not occur with the higher reinforcement content due to the higher wall thickness of the steel tube. Compared with previous works [12], in which a steel rod was fed into the process, a significantly improved coaxial positioning of the reinforcing element was achieved using the new guiding mandrel part and the resulting, more uniform material flow inside the novel tool. It should be noted that in the original LACE process with a titanium sheet used as external reinforcement there is no offset, since the compound profile was manufactured asymmetrically for process-related reasons [11]. However, in the case of co-extrusion processes that use wire reinforcements, deviations from the expected position of the reinforcing elements could also be detected [17]. The slight deviations observed in the present study can be explained in terms of the position on the feeding of the reinforcement via a mandrel part. For the application in the process chain of the CRC 1153, the offset could be compensated by further adjustment of the tool or using turned semi-finished products.

The metallographic characterization has revealed that the gaps between the matrix material and the reinforcing element were present on the recipient side in the initial parts of the compound profile. However, these gaps between aluminum and steel were no longer detected in any tests at the end of the compound profile. Thus, these gaps will not be relevant in a Tailored Forming process chain for the production of hybrid bearing bushings once steady-state conditions are attained.

At 42 MPa ± 7 MPa, the average shear strength of the profile produced with the lower extrusion ratio was slightly lower than that of the profile with the higher extrusion ratio, i.e., the one with the higher volume fraction of reinforcement. For the latter, an average shear strength of 48 MPa ± 9 MPa was measured. These values are below the composite strengths that were determined for a LACE process conducted on a laboratory scale [18]. However, since the compound profiles manufactured in these earlier experiments showed a clear warping, their higher bond strengths are attributed to the more pronounced force and form closure.

Compared with the shear strengths achieved in push-out tests on compound forged bearing bushings investigated by Behrens et al. [16], the shear strength achieved with semi-finished products manufactured by LACE was about 20 MPa lower. The bond formation in the compound forging was thus somewhat better, which can be substantiated by the joint forming, and thus by the more extensive formation of new surfaces in the process [14]. In the LACE process, on the other hand, only the aluminum is formed and the relative movement of the materials to each other [11] forms the joint between the partners. The strength of the joining zones is nevertheless promising to withstand the subsequent die forging process.

Due to the coaxial arrangement of the joining partners inside the LACE tool, it can be assumed that the higher thermal expansion coefficient of aluminum causes shrinking of the outer EN AW-6082 cladding onto the steel tube, which results in a form closure as well as a force closure [14]. However, a firm connection that will withstand subsequent bulk forming processes such as forging, requires material closure, as this connects the joining partners by means of physical or chemical bonds, so that they function as one body [19]. Whether the desired material closure is present in the LACE samples, was therefore tested using shear compression tests on segments taken from the sample cross-sections. The existing form and force closures were released by cutting out 65° segments from the sample cross-section.

With respect to the subsequent forging operation, the interface properties are of paramount importance. Ideally, a firm metallurgical bond should be formed. Depending on the process conditions, brittle intermetallic phase can growth at the interface between the joining partners. Herbst et al. reported that intermetallic phase seams narrower than 1 μm have no negative effect on the strength of the composite [3]. In the present study, no intermetallic phase seams were detected metallographically. In addition, due to the coaxial arrangement of the joining partners inside the LACE tool, it can be assumed that the higher thermal expansion coefficient of aluminum causes shrinking of the outer EN AW-6082 cladding onto the steel tube. This will result in a form closure as well as a force closure [14]. However, subsequent bulk-forming processes, such as forging, require material closure, i.e., a firm connection by means of physical or chemical bonds, so that they function as one body [19]. Whether the desired material closure was realized in the LACE samples, was therefore tested using shear compression tests on segments taken from the sample cross-sections. The existing form and force closures were released by cutting out 65° segments from the sample cross-section.

The LACE profile with a steel tube made of 20MnCr5 had a shear strength of 54 MPa ± 5 MPa over the entire profile length, which was determined in the push-out test. The segments without longitudinal weld seams showed almost identical shear strengths with values of 56 MPa ± 6 MPa and the samples with two longitudinal weld seams showed no change over the averaged total profile length, resulting in a value of 58 MPa ± 15 MPa. The strength of the specimen with longitudinal weld seams was both above and below the strength of the segments without longitudinal weld seams. Only the increased deviation in the values shows a slight influence of the longitudinal weld seams on the shear strength. It can therefore be assumed that the proportion of form-fit or frictional connection in the samples manufactured using LACE is low. The shear compression test of the specimens with longitudinal weld seams has shown that the longitudinal weld seams can have a positive influence on the shear strength. The aluminum adhered clearly to the steel near a longitudinal weld seam, which was formed by splitting by a support arm of the mandrel part. The splitting of the aluminum flow and re-welding in the welding chamber thus produced juvenile metal surfaces, which had a positive influence on the formation of the bonding area. According to Weidenmann et al., material closure is also assumed if the reinforcing element is covered by residues of the matrix material after a shear compression test [20], which was the case here for most of the tested samples.

5. Conclusions and Outlook

It could be shown that quasi-continuous hybrid hollow profiles made of EN AW-6082 and 20MnCr5 can be produced on an industrial-relevant scale by employing a lateral angular co-extrusion process together with a new tool concept. The modular design allows the extrusion ratio to be increased, e.g., from 9:1 to 11:1, and thus the reinforcement content could be varied between 14 vol.% and 34 vol.%. The placement of the steel tube inside the aluminum cladding deviated slightly from the desired ideal coaxial position. The de-bonding shear strengths determined by push-out tests were between 42 MPa and 47 MPa. The shear compression tests on sample segments showed that not only form-fit and force-fit is present between the aluminum alloy and the steel tube. The aluminum residues adhering to the steel after the shear compression tests also indicate a material-locking connection, which was observed especially in the areas next to the longitudinal weld seams. The bonding area of these samples needs be investigated more closely in the future to be able to fully exploit the potential of the LACE process.

Author Contributions: Conceptualization, H.J.M., C.K. and B.-A.B.; methodology, S.E.T., J.P. and F.C.B.; validation, S.E.T., N.H. and J.U.; investigation, S.E.T.; writing—original draft preparation, S.E.T.; writing—review and editing, H.J.M., J.U., B.B. and C.K.; visualization, S.E.T.; supervision, H.J.M., C.K. and B.-A.B.; project administration, H.J.M., C.K. and B.-A.B.; funding acquisition, H.J.M., C.K. and B.-A.B. All authors have read and agreed to the published version of the manuscript.

Funding: This study was funded by the Deutsche Forschungsgemeinschaft (DFG, German Research Foundation)—CRC 1153, subproject A1—252662854. The authors thank the DFG for financial support.

Acknowledgments: The results presented were obtained within the subproject A1 "Influence of local microstructure on the formability of extruded composite profiles" of the Collaborative Research Centre 1153 "Process chain to produce hybrid high performance components by Tailored Forming".

Conflicts of Interest: The authors declare no conflict of interest.

References

1. Hirsch, J.; Hirsch, J. Aluminium in Innovative Light-Weight Car Design. *Mater. Trans.* **2011**, *52*, 818–824. [CrossRef]

2. Merklein, M.; Johannes, M.; Lechner, M.; Kuppert, A. A review on tailored blanks–Production, applications and evaluation. *J. Mater. Process. Technol.* **2014**, *241*, 151–164. [CrossRef]

3. Herbst, S.; Maier, H.J.; Nürnberger, F. Strategies for the Heat Treatment of Steel-Aluminium Hybrid Components. *HTM J. Heat Treat. Mater.* **2018**, *73*, 268–282. [CrossRef]

4. Agudo, L.; Jank, N.; Wagner, J.; Weber, S.; Schmaranzer, C.; Arenholz, E.; Bruckner, J.; Hackl, H.; Pyzalla, A. Investigation of microstructure and mechanical properties of steel-aluminium joints produced by metal arc joining. *Steel Res. Int.* **2008**, *79*, 530–535. [CrossRef]

5. Bauser, M.; Sauer, G.; Siegert, K. *Strangpressen*; Aluminium-Verlag: Düsseldorf, German, 2001; ISBN 3-87017-249-5.

6. Foydl, A.; Haase, M.; Ben Khalifa, N.; Tekkaya, A. Co-extrusion of discontinuously, non-centric steel-reinforced aluminum. In Proceedings of the the 14th International Esaform Conference on Material Forming, Esaform 2011, Menary, Gary, 27–29 April 2011; pp. 443–448.

7. Dietrich, D.; Grittner, N.; Mehner, T.; Nickel, D.; Schaper, M.; Maier, H.J.; Lampke, T. Microstructural evolution in the bonding zones of co-extruded aluminium/titanium. *J. Mater. Sci.* **2013**, *49*, 2442–2455. [CrossRef]

8. Kleiner, M.; Schomäcker, M.; Schikorra, M.; Klaus, A. Herstellung verbundverstärkter Aluminiumprofile für ultraleichte Tragwerke durch Strangpressen. *Mater. Werkst.* **2004**, *35*, 431–439. [CrossRef]

9. Pietzka, D.; Ben Khalifa, N.; Gerke, S.; Tekkaya, A. Composite extrusion of thin aluminum profiles with high reinforcing volume. *Key Eng. Mater.* **2013**, *554*, 801–808. [CrossRef]

10. Weidenmann, K.A. *Verbundstrangpressen Mit Modifizierten Kammerwerkzeugen: Werkstofftechnik, Fertigungstechnik, Simulation. Karlsruhe*; KIT Scientific Publishing: Karlsruhe, Germany, 2012; ISBN 978-3-86644-890-2.

11. Grittner, N.; Striewe, B.; Von Hehl, A.; Engelhardt, M.; Klose, C.; Nürnberger, F. Characterization of the interface of co-extruded asymmetric aluminum-titanium composite profiles. *Mater. Werkst.* **2014**, *45*, 1054–1060. [CrossRef]

12. Thürer, S.E.; Uhe, J.; Golovko, O.; Bonk, C.; Bouguecha, A.; Klose, C.; Behrens, B.-A.; Maier, H.J. Co-extrusion of semi-finished aluminium-steel compounds. In Proceedings of the 20th International Esaform Conference on Material Forming: Esaform 2017, Dublin, Ireland, 26–28 April 2017; p. 140002.

13. Behrens, B.-A.; Klose, C.; Thürer, S.E.; Heimes, N.; Uhe, J. Numerical modeling of the development of intermetallic layers between aluminium and steel during co-extrusion. In Proceedings of the 22nd International Esaform Conference on Material Forming: Esaform 2019, Vitoria-Gasteiz, Spain, 8–10 May 2019; p. 040029.

14. Behrens, B.-A.; Kosch, K.-G. Development of the heating and forming strategy in compound forging of hybrid steel-aluminum parts. *Mater. Werkst.* **2011**, *42*, 973–978. [CrossRef]

15. Weidenmann, K.; Kerscher, E.; Schulze, V.; Löhe, D. Characterization of the interfacial properties of compound-extruded lightweight profiles using the push-out-technique. *Mater. Sci. Eng. A* **2006**, *424*, 205–211. [CrossRef]

16. Behrens, B.-A.; Sokolinskaja, V.; Chugreeva, A.; Diefenbach, J.; Thürer, S.; Bohr, D. Investigation into the bond strength of the joining zone of compound forged hybrid aluminium-steel bearing bushing. In Proceedings of the 22nd International Esaform Conference on Material Forming: Esaform 2019, Vitoria-Gasteiz, Spain, 8–10 May 2019; p. 040028.
17. Pietzka, D. *Erweiterung des Verbundstrangpressens zu Höheren Verstärkungsanteilen und Funktionalen Verbunden*; Kleiner, M., Ed.; Shaker Verlag: Düren/Maastricht, Germany, 2014.
18. Thürer, S.E.; Uhe, J.; Golovko, O.; Bonk, C.; Bouguecha, A.; Behrens, B.-A.; Klose, C. Mechanical properties of co-extruded aluminium-steel compounds. *Key Eng. Mater.* **2017**, *742*, 512–519. [CrossRef]
19. VDI-Fachbereich Produktentwicklung und Mechatronik. *Methodische Auswahl Fester Verbindungen-Systematik, Konstruktionskataloge, Arbeitshilfen*; VDI-Gesellschaft Produkt- und Prozessgestaltung: Düsseldorf, Germany, 2004.
20. Weidenmann, K.A.; Kerscher, E.; Schulze, V.; Löhe, D. Grenzflächen in verbundstrangpressprofilen auf aluminiumbasis mit verschiedenen verstärkungselementen. In *Praktische Metallographie Sonderband 37*; Göken, M., Ed.; KIT Scientific Publishing: Karlsruhe, Germany, 2005; pp. 131–136.

Article

Characterization and Modeling of Intermetallic Phase Formation during the Joining of Aluminum and Steel in Analogy to Co-Extrusion

Bernd-Arno Behrens [1], Hans Jürgen Maier [2], Christian Klose [2], Hendrik Wester [1], Susanne Elisabeth Thürer [2], Norman Heimes [1] and Johanna Uhe [1,*]

[1] Institut für Umformtechnik und Umformmaschinen (Forming Technology and Machines), Leibniz Universität Hannover, 30823 Garbsen, Germany; behrens@ifum.uni-hannover.de (B.-A.B.); wester@ifum.uni-hannover.de (H.W.); heimes@ifum.uni-hannover.de (N.H.)

[2] Institut für Werkstoffkunde (Materials Science), Leibniz Universität Hannover, 30823 Garbsen, Germany; maier@iw.uni-hannover.de (H.J.M.); klose@iw.uni-hannover.de (C.K.); thuerer@iw.uni-hannover.de (S.E.T.)

* Correspondence: uhe@ifum.uni-hannover.de; Tel.: +49-511-762-2427

Received: 30 October 2020; Accepted: 24 November 2020; Published: 26 November 2020

Abstract: The reinforcement of light metal components with steel allows to increase the strength of the part while keeping the weight comparatively low. Lateral angular co-extrusion (LACE) offers the possibility to produce hybrid coaxial profiles consisting of steel and aluminum. In the present study, the effect of the process parameters temperature, contact pressure and time on the metallurgical bonding process and the development of intermetallic phases was investigated. Therefore, an analogy experiment was developed to reproduce the process conditions during co-extrusion using a forming dilatometer. Based on scanning electron microscopy analysis of the specimens, the intermetallic phase seam thickness was measured to calculate the resulting diffusion coefficients. Nanoindentation and energy dispersive X-ray spectroscopy measurements were carried out to determine the element distribution and estimate properties within the joining zone. The proposed numerical model for the calculation of the resulting intermetallic phase seam width was implemented into a finite element (FE) software using a user-subroutine and validated by experimental results. Using the subroutine, a numerical prediction of the resulting intermetallic phase thicknesses is possible during the tool design, which can be exploited to avoid the weakening of the component strength due to formation of wide intermetallic phase seams.

Keywords: aluminum-steel compound; intermetallic phases; co-extrusion; tailored forming; nanoindentation

1. Introduction

The industrial efforts to reduce the mass of vehicles in order to save fuel and reduce CO_2 emissions result, inter alia, in the use of hybrid components and thus in the demand for new joining techniques of dissimilar materials. In order to achieve a reduction in mass at low cost, the combination of aluminum and steel has lately received substantial attention. The joining of 6xxx series aluminum alloys and steel has extensively been investigated using several joining processes, such as laser welding [1], friction stir welding [2], friction welding [3], compound forging [4] or co-extrusion [5]. The occurrence of intermetallic phases presents a challenge for both fusion welding and solid-state joining processes, as these phases are very hard and brittle and can reduce the strength of the hybrid component. Control of the resulting phase seam width is therefore essential to achieve reliable compounds [6]. The growth of intermetallic phases is diffusion controlled, and thus strongly dependent on the prevailing temperature and time [7].

Intermetallic phases typically exhibit low crystal symmetry, which curtails dislocation movements. Due to the low mobility of the dislocations, intermetallic phases are generally characterized by high hardness values and a particularly brittle material behavior [8]. For this reason, the thickness of the intermetallic phase seam is an indispensable aspect in assessing the strength of hybrid components. Intermetallic phase seams with a given width often consist of different intermetallic phases, which in the case of the Fe-Al system, are Fe_3Al, $FeAl$, $FeAl_2$, Fe_2Al_5 and $FeAl_3$. When joining aluminum and steel in the solid state, the phase Fe_2Al_5 is mainly formed [9].

The effect of intermetallic phases on the mechanical properties of a joint has been evaluated by several authors. Yamamoto et al. report a linear decrease in the joint strength with an increase in the thickness of the intermetallic layer [10]. Kimapong and Watanabe state that the joint strength increases exponentially with a decrease in the intermetallic seam thickness [11]. Yilmaz et al. determined that the highest strength can be achieved by the thinnest possible intermetallic phase in friction welding [12]. According to Fukuora, even with a thickness of the intermetallic layer less than 1 μm, the joint demonstrated premature fracture at the interface in friction bonding of high-strength Al alloys to steels [13].

Clearly, it is crucial for these hybrid materials to control the thickness of the intermetallic phase seam that forms at the interface during bonding and to characterize its properties, especially the mechanical ones. Nanoindentation enables to probe the local hardness at the nanometer scale, and was used by several authors to investigate the mechanical properties of intermetallic phases. Ogura et al. determined the nano hardness of different Fe-Al intermetallic phases. They stated that the nano hardness of intermetallic phases of type Fe_xAl_y increases with increasing proportion of aluminum, with the exception of the $FeAl_3$ phase, which is less hard than $FeAl_2$ and Fe_2Al_5. The increase in hardness can be explained by the increasing complexity of the lattice structures [6].

Within the framework of the Collaborative Research Centre 1153, co-extrusion is used to produce coaxial hybrid profiles of aluminum and steel. In the further course of the process chain, these profiles are used as joined hybrid semi-finished workpieces for the die forging of bearing bushings. The use of already joined semi-finished workpieces allows a geometrical and thermomechanical tailoring of the joining zone, resulting in improved mechanical properties. For a sufficient formability of these hybrid semi-finished products, the intermetallic phase seam must not exceed a certain size after co-extrusion. In order to consider the resulting phase seam thickness already in the numerical process design, a phenomenological model was developed that can predict the phase seam width during the post-processing of a commercial finite element (FE) system. In the present study, the influence of the process parameters temperature, time and force on the resulting intermetallic phase seam thickness were investigated using analogy experiments and subsequent scanning electron microscopy (SEM) analysis. In the following, the numerical model being developed and its implementation into the FE software are presented. The parameters required to describe the development of the intermetallic phase seam thickness were determined from the analogy experiment. To correlate the properties of the joining zone with the formed intermetallic phases, additional nanoindentation and energy dispersive X-ray measurements (EDS) were carried out.

2. Materials and Methods

2.1. Experimental Procedure

An experimental setup for analogy experiments was developed in order to simulate the boundary conditions physically during co-extrusion on a laboratory scale and to be able to set them independently, see Figure 1a. The specimens, consisting of two steel cylinders (20MnCr5, AISI 5120), an aluminum cylinder (AlMgSi1, EN AW-6082) and a steel sleeve, were placed on a forming dilatometer DIL 805 A/D + T (BÄHR Thermoanalyse GmbH, Hüllhorst, Germany) between two deformation punches to which the forming force was applied. The steel cylinders and the aluminum cylinder had a diameter of 5 mm and 3 mm, respectively. The diameter of the aluminum specimen was chosen to be smaller in order to break up the surface layers during forming, analogous to the co-extrusion process. A steel sleeve was used to prevent the aluminum specimen from being displaced between the steel specimens and to obtain higher stresses than the yield stress of aluminum. An external induction coil realized the heating of the specimen. The temperature was controlled by a thermocouple (TC, type K) welded to the outside of the sleeve. To minimize temperature gradients and avoid oxidation effects, the experiments were carried out in a vacuum of 3.5×10^{-3} Pa. The vacuum was generated by means of a vacuum pump integrated in the dilatometer.

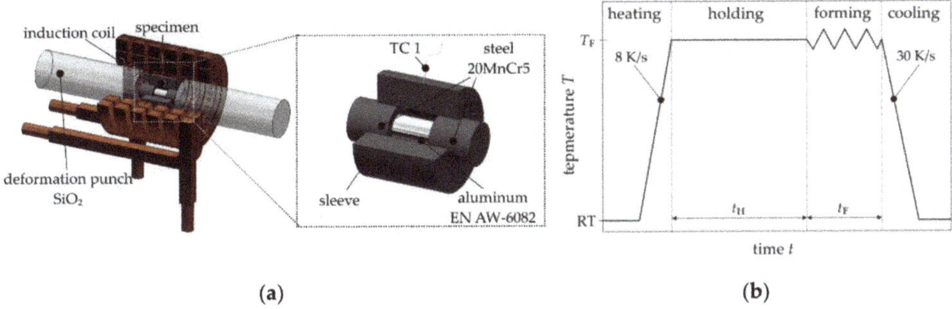

Figure 1. (a) Schematic test setup of the analogy experiments on a forming dilatometer, (b) temperature–time curve used in the experiments.

The temperature–time course employed during the experiments is shown schematically in Figure 1b. The jagged line indicates the actual forming. In the first step, the specimens were heated to the respective forming temperature T_F with a heating rate of 8 K/s and kept at T_F for 240 s to ensure a near homogeneous temperature distribution. The subsequent forming was force-controlled by a pre-defined forming force F. The force was applied and maintained over the given time period t_F. After forming, the specimens were cooled to room temperature by nitrogen with a rate of 30 K/s.

For the reference specimen, a temperature T_F of 560 °C, a forming force F of 5890 N, and a forming time t_F of 120 s were used. Based on this parameter configuration, time, force and temperature were varied to determine their influence on the resulting thickness of the intermetallic phase seam. The parameter matrix employed is shown in Table 1. The temperature, time and force values were chosen based on the numerical investigation of the co-extrusion process described in [5]. For each parameter combination, three repetitions were carried out to estimate variability in the data.

Table 1. Test matrix used for the dilatometer experiments; the values deviating from the reference configuration are highlighted in italics.

Test Series	Temperature T_F in °C	Forming Time t_F in s	Force F in N
Reference variant	560	120	5890
Temperature variation	*450, 505, 560, 590*	120	5890
Time variation 1	560	*30, 60, 120, 240, 480*	5890
Time variation 2	*590*	*60, 120, 240, 480*	5890
Force variation	560	120	*1500, 2797, 5890*
Validation	*575*	120	5890

The chemical compositions of the aluminum alloy EN AW-6082 and the steel 20MnCr5 used in the present study are listed in Tables 2 and 3. The aluminum and steel specimens—produced by wire cutting—were ground and polished shortly before the tests to remove excess oxide layers from the surfaces.

Table 2. Chemical composition of the aluminum alloy EN AW-6082 in wt.%, the balance is Al.

Material	Si	Fe	Cu	Mn	Mg	Cr	Zn	Ti
EN AW-6082	1.11 ± 0.0295	0.19 ± 0.0349	0.0349 ± 0.001	0.438 ± 0.006	0.656 ± 0.027	0.0352 ± 0.001	0.0169 ± 0.002	0.0186 ± 0.002

Table 3. Chemical composition of the steel alloy 20MnCr5 in wt.%, the balance is Fe.

Material	C	Si	Mn	Cr	S
20MnCr5	0.22 ± 0.02	0.21 ± 0.01	1.10 ± 0.02	1.01 ± 0.0109	0.0131 ± 0.0007

After the analogy experiments, the hybrid specimens were embedded and then cut perpendicular to the bonded surfaces. The specimens were ground and polished down to 1 µm. The interface was analyzed by SEM with EDS using a Supra 40VP (Zeiss, Oberkochen, Germany). Images were recorded using secondary (SE) and backscattered electrons (BSE). SEM analysis was performed at both joining zones at several locations. Figure 2 schematically shows a formed specimen where the areas probed in the SEM and nanoindentation analysis are highlighted. The width of the intermetallic phase seam was measured by evaluating the recorded BSE images using a MATLAB (R2020a, The MathWorks, Natick, MA, USA) script descripted in detail in by Herbst et al. [14]. In each specimen, at least two images were taken of each side of the joining zone.

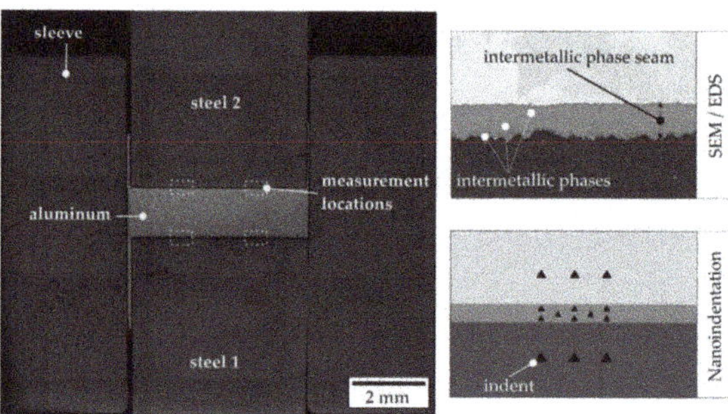

Figure 2. Schematic representation of a formed specimen with highlighted positions of the scanning electron microscopy (SEM) and nanoindentation measurements.

Nanoindentation measurements were carried out on a Hysitron TI 980 Nanoindenter (Bruker, Billerica, MA, USA) to characterize the mechanical properties of the individual detected phases in the intermetallic phase seam. Due to their small size, the visual localization of the intermetallic phases was challenging and scanning probe microscopy (SPM) integrated in the nanoindenter was used for this purpose. The joining zone was scanned tactilely with a scanning force of 2 µN. Due to the different material hardness, the specimen preparation by grinding and polishing results in different surface roughness of the different materials. The nano hardness measurements were carried out with a Berkovich diamond tip. The tip geometry was calibrated for a minimum penetration depth of 25 nm on a fused quartz probe. The hardness was evaluated based on the method proposed by Oliver and Pharr [15]. A trapezium function with a duration of 12 s was used as the load function to eliminate dynamic effects [16]. For the nanoindentation measurements, a test force of 3000 µN was used. Thus, the minimum penetration depth of 25 nm in the intermetallic phase could be achieved and at the same time, the remaining indentation was small enough to position several indents across the phase width. In the longitudinal direction, a distance of about 2 µm was maintained. In both steel and aluminum, three indentations each were created with a distance of 7.5 µm from the intermetallic phase and in longitudinal direction with a distance of 7.5 µm using a programmed grid.

In addition, the joining zone was characterized using accelerated mechanical property mapping (XPM). A 10 × 10 indentation grid was stretched out applying a test force of 1000 µN. The distance between the indents was set to 0.7 µm to obtain a high lateral resolution of the hardness in the joining zone. The quantitative measurement of steel and aluminum was carried out separately near the joining zone.

2.2. Setup of Numerical Model and Subroutine

To validate the subroutine, a numerical model of the analogy experiments was built up in the first step. For the FE analysis of the analogy experiments, the commercial software FORGE NxT 3.0 (Transvalor S. A., Mugins, France) was used. The 3D FE model of the analogy experiments is shown in Figure 3a. The temperatures from the experimental measurements were assigned to the components as homogeneous temperature. A hydraulic press with a constant speed of 0.19 mm/s was assigned to steel cylinder 1, which was determined from the experimental data. In this way, steel cylinder 1 was moved in a positive z-direction until the defined maximum force was reached, after which the force was kept for a given period of time. The contact between the steel cylinders and the aluminum work piece was modeled as bilateral sticking, following the contact modeling from the lateral angular co-extrusion (LACE) process. Between the steel sleeve and aluminum specimen a friction factor m of 0.95 was chosen in accordance to the co-extrusion process, as determined numerically in previous studies [5]. Flow curves of aluminum were recorded by cylindrical upsetting tests and implemented in the FE software. Details about the material data used are given in [5].

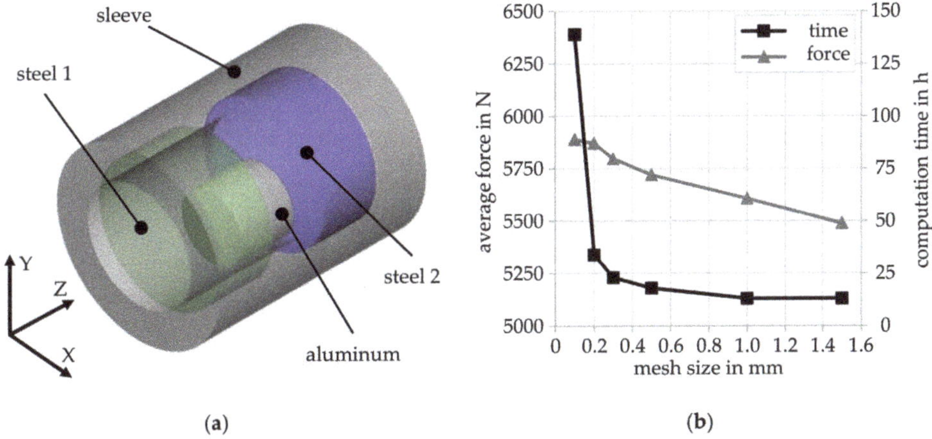

Figure 3. (**a**) 3D finite element (FE) model of the analogy experiment, (**b**) influence of the minimal element length on the calculated force and the computation time.

The model comprised approximately 44,408 volume elements (tetrahedral) of linear interpolation type typically used for metal forming simulations. The steel specimens and the sleeve were modeled as rigid bodies to limit the computation time. In addition, the effect of the element size was analyzed, since both the accuracy of the calculated results and the computation time strongly depend on the selected minimal element size. In Figure 3b, the results of the mesh study are shown. For a minimum element size of 0.2 mm a sufficient accuracy, as well as an adequate computation time were achieved and therefore used for the subsequent analysis. Remeshing was applied, combining two remeshing criteria, a periodic initiated remeshing criterion and an automatic size criterion. Thus, remeshing followed a fixed incremental step of 20. In addition, the automatic size criterion was activated to refine the mesh of the work piece in the contact zone with the steel cylinders. The time step equaled 0.2 s, which resulted in 681 time steps in total.

To describe the growth of intermetallic phases numerically, the Einstein–Smoluchowski equation

$$p = \sqrt{2Dt} \tag{1}$$

was employed, where p is the intermetallic phase seam thickness after time t and D is the diffusion coefficient [17]. The phase seam thicknesses p determined via SEM images was then used to calculate the apparent diffusion coefficient D with the Einstein–Smoluchowski equation.

To enable the usage of a temperature-dependent diffusion coefficient in the numerical simulation, a functional correlation was derived from the measured data, which describes the dependence of the diffusion coefficient on temperature. Thus, the subroutine can be used to describe the growth of the intermetallic phase seam as a function of contact time and contact temperature. The intermetallic phase seam was calculated only where the element nodes of the aluminum and the element nodes of the steel cylinders came into contact. To check if the requirements were fulfilled, the contact modeling was queried in each increment for each node of the aluminum. If the conditions were fulfilled, the subroutine was continued. Within the subroutine the temperature was requested from the solver for the considered node. If the nodal temperature was equal to or between 450 °C and 590 °C, the routine was continued. Otherwise, the temperature is too low for development of an intermetallic phase and the subroutine is terminated. The temperatures were chosen according to the upper and lower limits of the test matrix and the developed subroutine is only valid in this range. The diffusion coefficients, which are calculated with temperatures outside the temperature range are extrapolated and not interpolated values and therefore, are not permitted. If the node temperature is in the permitted range, the diffusion coefficient is calculated for the considered node, depending on the node temperature. The phase growth in the current increment is calculated, taking into account the phase width of the previous increment. Finally, the calculated phase thickness is stored and is available in the next increment as an already existing phase thickness and is included again in the calculation of the next increment. The sequence of the subroutine is shown in Figure 4 in a flow chart. In each increment the result variable *IMP_Layer* is updated. The subroutine was programmed in Fortran 90 and implemented by compiling the dynamically linked library in FORGE NxT 3.0.

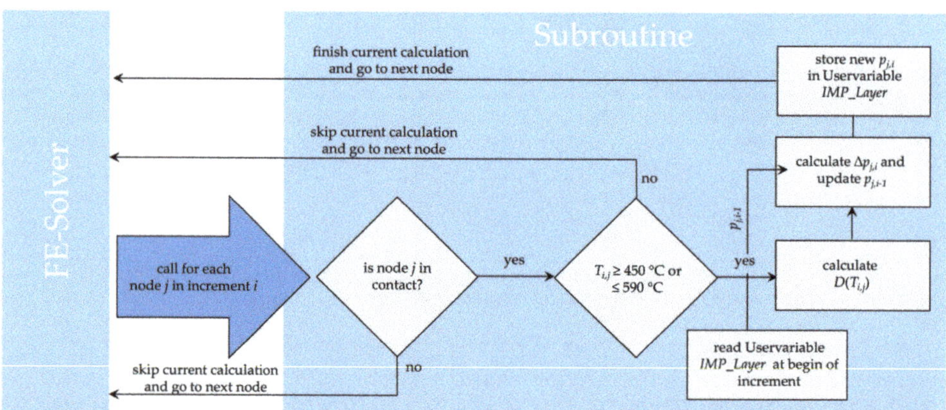

Figure 4. Flow chart for the developed subroutine that is executed for each node of the aluminum in each increment.

3. Results and Discussion

3.1. Evolution of the Intermetallic Phase Seam Thickness

The growth of intermetallic phases can be divided in different steps as described by Ryabov et al. [7]. At first, small areas of an intermetallic phase are created by diffusion processes at the joining zone and continue to grow along the joining zone with advancing time. The different phases subsequently grow together and start to grow transversely to the joining zone. In Figure 5, SEM images of the resulting joining zone from analogy experiments at a forming temperature of 560 °C, a force of 5890 N and different forming times are shown. After 30 s intermetallic phases are present, which exist as isolated islands along the joining zone, cf. Figure 5a. After 60 s, the individual phases have already partially grown together and have begun to grow across the joint zone, as shown in Figure 5b. A larger band of intermetallic phases is shown in Figure 5c which was detected for experiments with a forming time of 120 s. The intermetallic phase fringe continues to grow into the aluminum with a further increase in time to 240 s. As seen in Figure 5d, the different grey tones indicate the presence of different types of intermetallic phases within the joining zone.

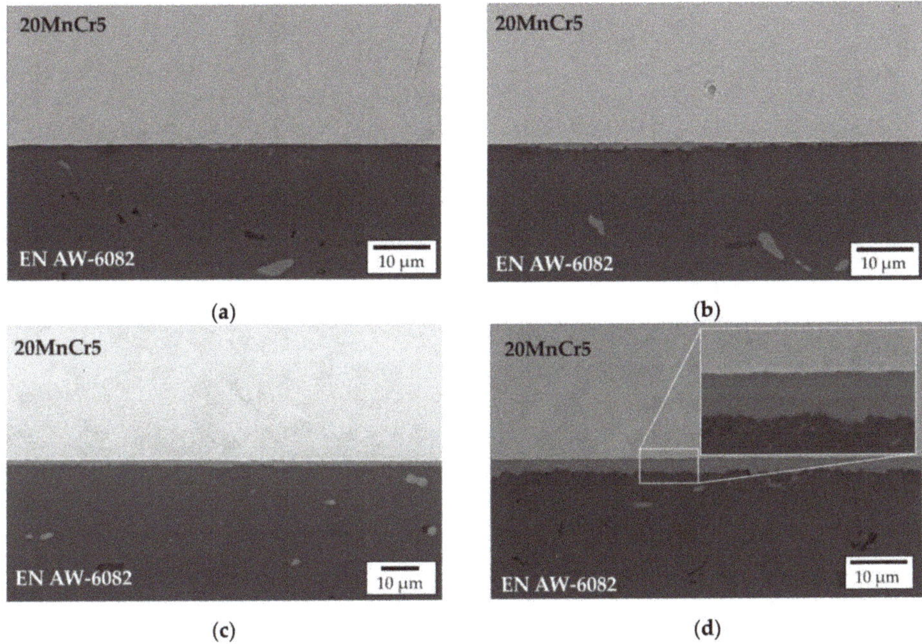

Figure 5. SEM images of the resulting joining zones formed in the analogy experiments at a forming temperature of 560 °C, a force of 5890 N and a forming time of (**a**) 30 s, (**b**) 60 s, (**c**) 120 s and (**d**) 240 s.

Based on the SEM images, the average thickness of the intermetallic phases was determined depending on the parameters forming time, forming force and forming temperature, using the described MATLAB script. The graphic representation of the dependence of the intermetallic phase seam thickness on time is shown in Figure 6a. An increase in the intermetallic phase seam width with time can be seen for both forming temperatures. As expected, the comparison of the results for different forming temperatures demonstrates an increasing growth rate with increasing temperature. The variation of force shows no significant influence on the resulting phase seam thickness, as shown in Figure 6b. Theoretically, the phase growth should decrease with increasing force, as diffusion is hindered. Rummel et al. [18] were able to show this effect in investigations on a single crystal, but this effect only occurred at contact pressures being significantly higher than those employed in the analogy experiments conducted in the present study.

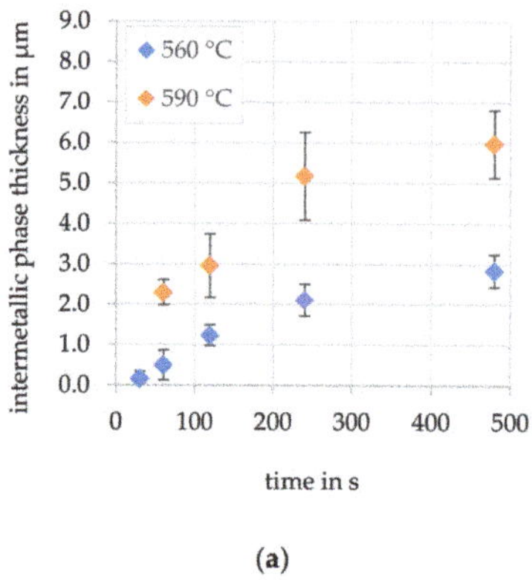
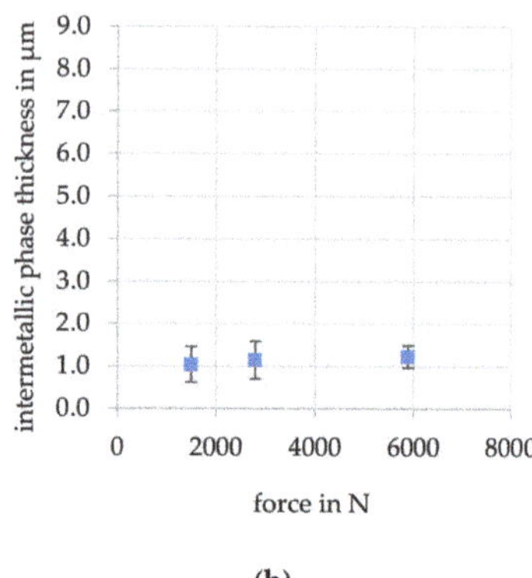

(a)

(b)

Figure 6. Development of the intermetallic phase seam thickness with indication of the standard deviation as a function of (**a**) forming time and forming temperature for a force of 5890 N, (**b**) force for a temperature of 560 °C and a forming time of 120 s.

Depending on the temperature, a parabolic increase in the intermetallic phase seam width can be observed, as shown in Figure 7a, according to literature [9]. The standard deviation of the measured values increases with increasing temperature, which can be explained by the locally irregular and stem-like growth of the layers, especially on the aluminum side.

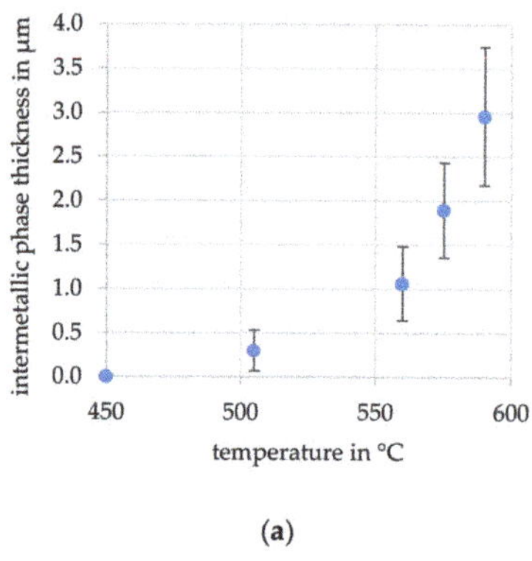
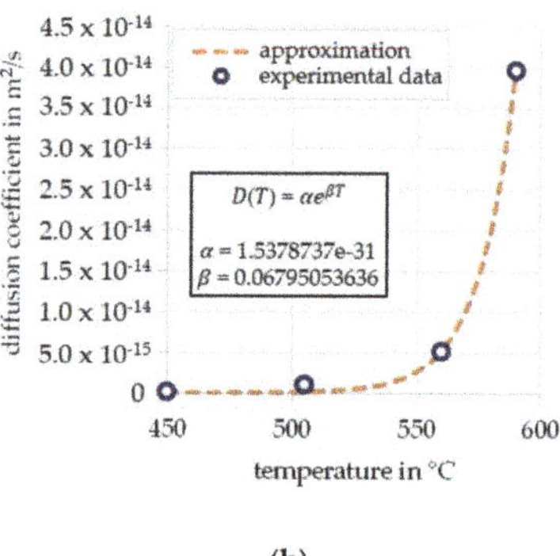

(a)

(b)

Figure 7. (**a**) Development of intermetallic phase seam thickness as a function of forming temperature with indication of the standard deviation for a forming time of 120 s and force of 5890 N, (**b**) apparent diffusion coefficient as a function of forming temperature with representation of the approximation function.

As no significant influence of the applied force on the resulting phase thickness was detected, a force-dependent calculation of the intermetallic phase seam width was not carried out in further investigations. Since a decisive influence on the formation of time and temperature on the intermetallic phases was observed, time- and temperature-dependent modeling was chosen. For temperature-dependent modeling, the corresponding diffusion coefficients were first calculated from the measured intermetallic phase seam widths using the Einstein–Smoluchowski equation. Then, an approximation function was determined to describe the diffusion coefficient as a function of temperature, which is represented in Figure 7b. An exponential function with the two constants α and β was used. The constants were determined by fitting the measured data and are valid in the temperature range from 450 °C to 590 °C. The value determined at 575 °C was not used for the approximation but later for the validation of the subroutine. The exponential increase in the intermetallic phase seam thickness and diffusion coefficient with increasing temperature is as expected for a thermally-activated process [19].

EDS investigations of the joining zones were carried out to determine the chemical composition of the phases formed in the joining zones. Of particular interest were the joining zones, in which up to three different phases could be identified based on line scans carried out perpendicular to the joining zone. In Figure 8, the distribution of the elements over the measured distance for a specimen formed at 575 °C compared to a specimen formed at 590 °C is presented, whereby the first shows two and the second shows three intermetallic phases with different thicknesses. A rapid transition of iron and aluminum content can be recognized in both line scans. However, the specimens formed at the lower temperature of 575 °C (Figure 8a) show an aluminum content of approximately 60 to 75 at.% at the interface. In contrast, for the specimen formed at the higher temperature of 590 °C (Figure 8b) the line scan shows individual plateaus, which indicate individual phases. The aluminum content in the joining zone increases from approximately 60 at.% in the first plateau to up to 70 at.% in the second plateau. A final plateau with an aluminum content of approximately 80 at.% is present in the area of the third phase at a distance of 6 to 8.5 μm to the interface. The iron content behaves reciprocally to the development of the aluminum content. In the steel, a weak signal of chromium and manganese was also detected, which is caused by the alloying elements of the mono material. Based on the determined ratios of aluminum to iron, the presence of the Al-rich phases $FeAl_2$, Fe_2Al_5 and $FeAl_3$ would be possible. The aluminum contents of the phases $FeAl_2$, Fe_2Al_5 and $FeAl_3$ are approximately 66–67 at.%, 70–73 at.% and 74–76 at.% according to the Fe-Al phase diagram [20]. A statement regarding the stoichiometry solely based on the EDS measurements is, however, not possible due to the interaction volume. For this reason, nanoindentation measurements were carried out to obtain further information on the composition and properties of the different intermetallic phases.

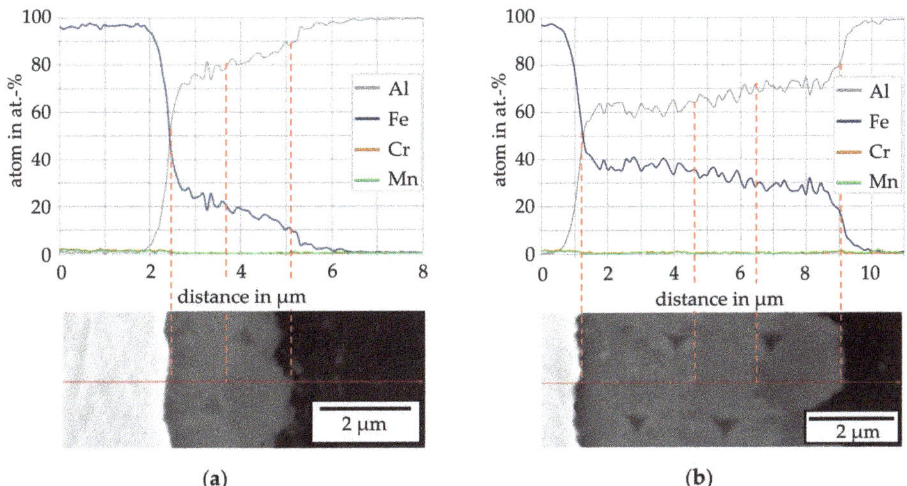

Figure 8. EDS and SEM analysis of the joining zone of a specimen for (**a**) 575 °C, 120 s, 5890 N and (**b**) 590 °C, 480 s, 5890 N.

3.2. Nanoindentation Measurement of the Intermetallic Phase Seam

A scanning probe microscopy (SPM) image of the nanoindentation measurement of a specimen, which was formed at a temperature of 575 °C, a forming time of 120 s and a force of 5890 N is shown in Figure 9a, illustrating the different topographies of aluminum, the joining zone and steel caused by the specimen preparation. In Figure 9b, the corresponding BSE image is shown. The contour of the intermetallic phase is very well represented in the SPM image, as seen, for example, in the V-shaped notch between the two right-hand indents in the intermetallic phase. The different shades of grey within the intermetallic phase seam that can be seen in the BSE image are not reproduced in the SPM image. To clearly assign the measured hardness to a specific phase, it was therefore necessary to examine the measuring points by SEM after indentation.

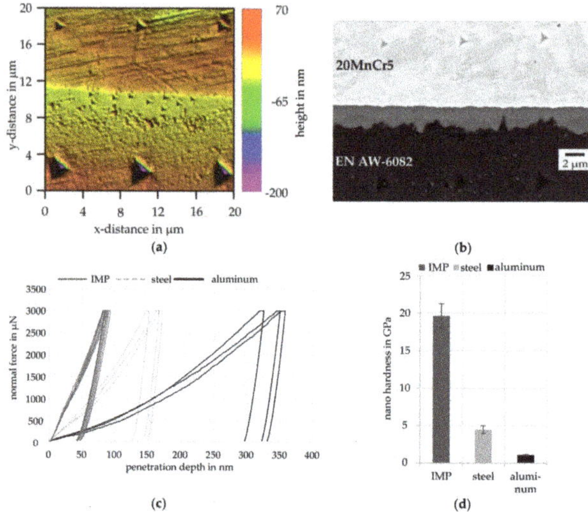

Figure 9. Examination of a joining zone formed at 575 °C, 120 s and 5890 N with nanoindentation measurements with a test force of 3000 µN: (**a**) scanning probe microscopy (SPM) image, (**b**) SEM image, (**c**) force-penetration curves, (**d**) nano hardness of aluminum, steel and the intermetallic phase seam.

The indents shown in the BSE image in Figure 9b were all indented with a test force of 3000 µN. It can be recognized that the indents of the intermetallic phase are significantly smaller than the indents in both base materials. In Figure 9c, the force-penetration curves of the indents are shown. The penetration depth of the indent correlates with the amount of plastic deformation. The average penetration depth of the indent of the intermetallic phase is 75 nm, and thus, is well above the minimum penetration depth of 25 nm. In Figure 9d, the nano hardness of aluminum, steel and the intermetallic phase is compared. The average hardness of aluminum and steel is 1 GPa and 4.85 GPa, respectively. In comparison, the hardness values of the intermetallic phase are about 20 GPa. In addition, the measured values for the intermetallic phase show more scatter, which is due to the different phases present within the intermetallic phase seam.

To correlate the measured nano hardness values to the different phases that are present in the joining zone, the measuring points were assigned to the different phases based on the BSE images, as exemplarily shown in Figure 10a. The marked hardness values show that the hardness in intermetallic phases seam increases towards the steel side. Nevertheless, it is evident that it was not always possible to place the hardness indentations in the middle of the phases present, making it sometimes difficult to assign the hardness unambiguously. This was most pronounced for the phase formed next to aluminum. In particular, this was the case for specimens produced at lower temperatures or times, some of which featured significantly thinner phases than those shown in Figure 10a. For this reason, three lines of hardness measurements were made on all specimens, as shown in Figure 10b.

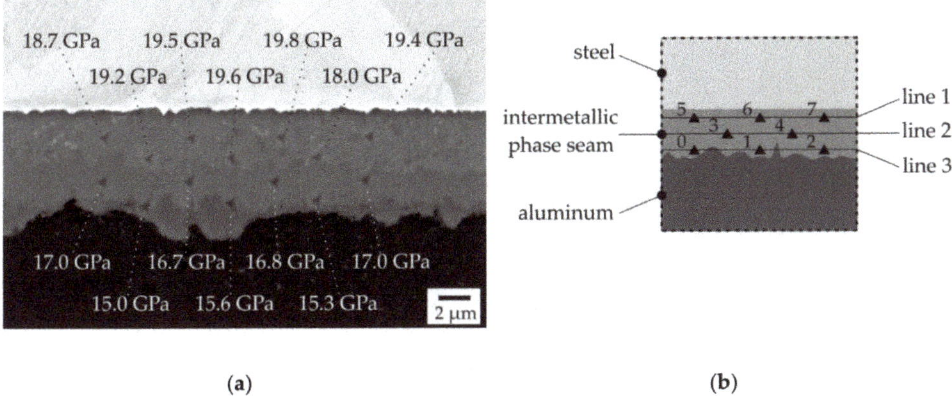

(a)   **(b)**

Figure 10. (**a**) SEM image and resulting nano hardness of a specimen, which was formed at a temperature of 590 °C, a forming time of 480 s and a force of 5890 N, (**b**) Schematic representation of the approach used during the nano hardness measurements, see main text for details.

The evaluation of the three series of measurements in Figure 11 shows that the hardness decreases continuously from line one to line three for all investigated specimens. As shown in Figure 11a, the contact time does not seem to have a significant influence on the resulting nano hardness. When time is varied, the hardness of the investigated specimens is at a similar level for each line, although the phase seam width increases with time. The larger scatter in line three is caused by the fact that the transition of the intermetallic phase to the aluminum formed an irregular jagged pattern. Therefore, the positioning of the indent is more critical. If the indentations are placed too close to the jagged transition, they will partially slide off into the much softer aluminum, which affects the hardness value determined. A further explanation for the strong scattering of the nano hardness in line three might be that if the indent lies optically within the intermetallic phase in the SEM image, no statement can be made regarding the geometrical shape of the intermetallic phase in the depth direction. Contrary to the observation of the development of the nano hardness as a function of time, the nano hardness increases slightly over all phases with increasing temperature, see Figure 11b. The high hardness values measured in the intermetallic phase seam indicate, in accordance with the results of the EDS measurements and the hardness data from literature, the presence of the Al-rich intermetallic phases $FeAl_2$, Fe_2Al_5 or $FeAl_3$. Ogura et al. stated that the hardness of the Al-rich phases is higher than the hardness of the Fe-rich phases, which is due to their lattice structure and a smaller number of slip systems. They also observed a decrease in the hardness of the intermetallic phase from steel to aluminum [6].

To investigate the properties of the joining zone with a high later resolution, XPM measurements were performed. In Figure 12a the resulting indentations in the SEM image of a joining zone that was formed at a temperature of 590 °C, a holding time of 120 s and a force of 5890 N are exemplarily shown. In the intermetallic phase, the indents are significantly smaller than in the base material due to its high hardness. In the base material, the hardness values can only be assessed qualitatively due to the small distance between the indents. In the case of steel and aluminum, the distance between the indentations of 0.7 μm at a test load of 1000 μN is too small to be able to evaluate the hardness values quantitatively. The indents influence each other so that the hardness values increase, as can be seen in the hardness map in Figure 12b. Here, the nano hardness for steel is 5.75 GPa is higher compared to 4.85 GPa in Figure 9d. However, the objective of the XPM measurements was to characterize the hardness development over the intermetallic phase with high resolution. The indents of the intermetallic phase are difficult to see in the SEM image because the penetration depth was about 35 nm and therefore, not clearly visible in the BSE images. The indent was more clearly visible using the in-lens detection, where a gap between the intermetallic phase and the aluminum could also be detected, which is shown in the image here as a black spot in the lower left corner of the measuring field. It is not clear whether the crack was caused by the preparation of the specimen.

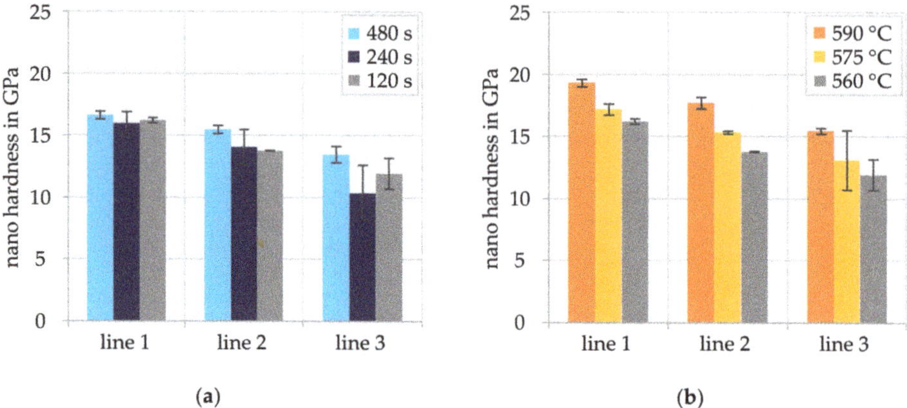

Figure 11. Nano hardness in the joining zone. (**a**) For a forming temperature of 560 °C, a force of 5890 N and different forming times. (**b**) For a forming time of 120 s, a force of 5890 N and different forming temperatures.

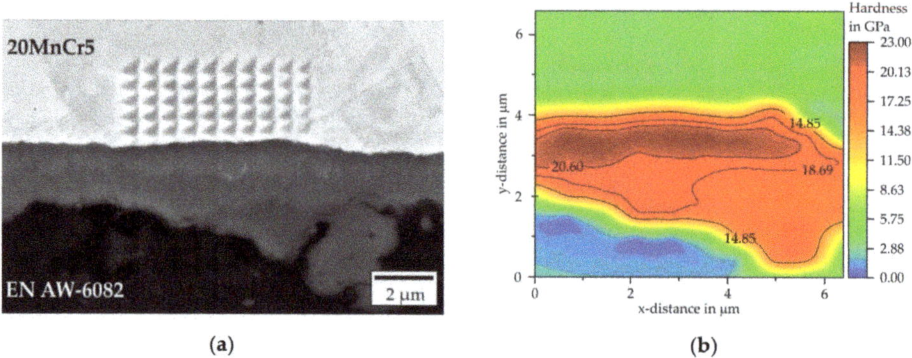

Figure 12. XPM measurement of a joining zone formed at 590 °C, 120 s, 5890 N: (**a**) SEM image of the measuring site; (**b**) mapping of the nano hardness.

Due to the lower penetration depth of the indent within the intermetallic phase, the indents do not influence each other and the minimum penetration depth of 25 nm was also exceeded, so that the recorded hardness values could also be evaluated quantitatively. The different phases, which are visible in the SEM picture, are also displayed in the hardness map. Isolines were included in the hardness map to represent the transitions of the individual phases. Compared to the hardness values of line one at 590 °C (see Figure 10a), it is shown that the hardness in the hardness map is 3 GPa higher in the area of line one. With a higher test load of 3000 µN the indent must be moved further away from the transition, so that no mixed hardness is produced due to the larger indentation. By means of the XPM measurement, it was possible to make measurements closer to the transition of steel due to the low-test load. This allows the individual phases to be visualized by ISO lines in the hardness map.

3.3. Validation of the Subroutine

A first verification of the functionality of the developed subroutine for calculating the intermetallic phase seam width was presented in [21]. For the final validation of the subroutine within the present study, further experiments with parameter sets that differ from the sets used for parameterization of the diffusion coefficient were performed. The following results refer to an experiment with a temperature of 575 °C, a forming time of 120 s and a forming force of 5890 N. In Figure 13a, the comparison of the resulting numerical and experimental force–time and temperature–time curves is shown. The temperature was kept constantly at 575 °C in the simulation and experiment. In the experiment, a slight drop in temperature can be seen at the beginning of the forming step, which is due to the increasing contact of the aluminum with the sleeve and the resulting change in heat flow, which cannot be compensated instantly by the temperature controller due to the inertia of the heating coil. Shortly after reaching the maximum force of 5890 N, the temperature became stable again. The force–time curves of the simulation and the experiment show a very good agreement just like the temperature progression. Deviations in the force–time curve are most pronounced at the beginning of the forming process, where the force in the experiment shows a small drop. This drop can be attributed to the alignment of the faces of the specimens in the experimental setup. In the further course of the test, the force increases exponentially in both courses due to the upsetting of the aluminum cylinder and the filling of the sleeve.

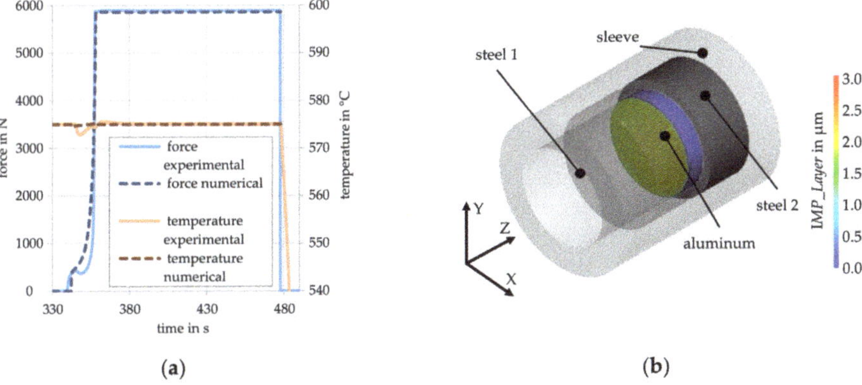

(**a**) (**b**)

Figure 13. Numerical simulation of the analogy experiment at a temperature of 575 °C, a forming time of 120 s, and a force of 5890 N: (**a**) comparison of force–time and temperature–time curves, (**b**) simulated thickness of the intermetallic phase seam.

The calculated intermetallic phase seam width for a test temperature of 575 °C is shown in the corresponding FE model, see Figure 13b. The comparison of the experimentally determined phase seam widths and the simulated phase seam widths is shown in Figure 14a. The simulated phase seam widths of the temperatures, which were used to create the function in Figure 7b, show a very high degree of agreement. At 575 °C, the median of the experimental phase width is 1.96 µm, at the simulated one the phase width is on average 2.05 µm in the contact area. This comparison shows a very good agreement, even if the difference in values is larger compared to the other temperatures. Figure 14b shows the corresponding result of the simulations. At 575 °C, it is clear to see that the nodes, which were in contact with the steel cylinders at the beginning of the forming process, have a larger phase seam width than the nodes which reached the contact area only after the forming process. The different phase seam widths are due to the different contact times.

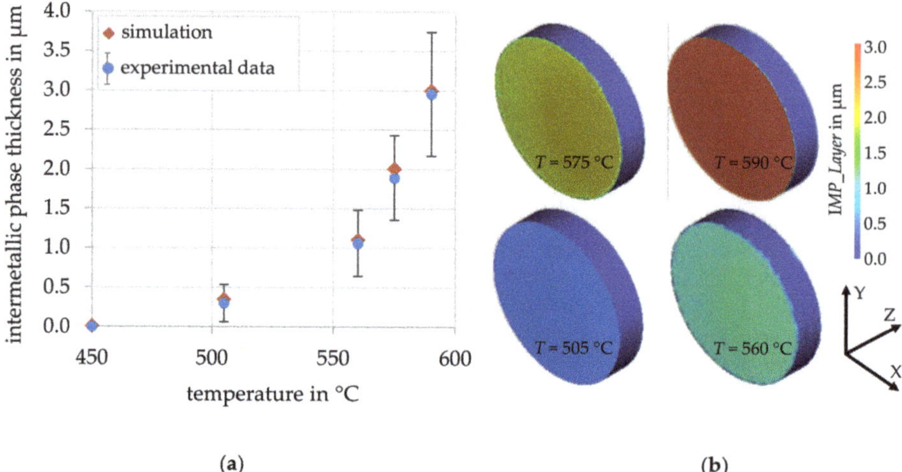

(a)   (b)

Figure 14. (**a**) Comparison of the experimentally determined intermetallic phase seam thickness with the simulated phase seam thickness of the respective forming temperature for a forming time of 120 s and a force of 5890 N, (**b**) simulated thickness of the intermetallic phase seam.

4. Conclusions and Outlook

Within the present study a model for the calculation of the intermetallic phase seam thickness was developed, which is based on the Einstein–Smoluchowski equation and implements a temperature dependent diffusion coefficient. An experimental setup was designed for analogy experiments on a forming dilatometer to analyze the influence of process conditions and determine the input parameters. Specifically, experiments were performed at different temperatures, forces and forming times to investigate the influence of the boundary conditions on the resulting phase seam width. The thickness of the resulting intermetallic phases was subsequently measured by SEM examinations and optical image analysis. Nanoindentation and EDS measurements were carried out to determine the element distribution and properties of the joining zone. Based on experimental findings, the diffusion coefficient within the developed model was determined and implemented into the commercial FE software Forge NxT 3.0 by means of a user-subroutine. The subroutine was successfully validated by comparison with the experimental results of the analogy experiments, and a high correlation between the experimental and simulated phase thicknesses was achieved.

Based on the presented results, the following conclusions can be drawn:

- As expected, the growth of intermetallic phases in the joint zone of EN AW-6082 and 20MnCr5 is time-dependent. By calculating the intermetallic phase seam width using the Einstein–Smoluchowski equation, a good agreement with the experimental data was achieved.
- With rising temperature, the phase seam width and the apparent diffusion coefficient increase.
- The force showed no significant effect on the phase formation in the investigated range.
- The SEM images indicated the presence of up to three different intermetallic phases in the joining zone. This was confirmed by EDS and nanoindentation measurements. At temperatures above 560 °C or times above 120 s, the formation of different intermetallic phases was observed.
- Due to the high aluminum content and the hardness determined in the EDS and nanoindentation analysis, the presence of the phases $FeAl_2$, Fe_2Al_5 or $FeAl_3$ would be possible, but could not be clearly proven. To confirm this assumption, further electron probe micro analysis (EPMA) and X-ray diffraction (XRD) investigations are to be carried out. In future work, the usability of the subroutine for the co-extrusion process will be tested and validated based on experimental investigations.

Author Contributions: Conceptualization, B.-A.B., C.K. and H.J.M.; methodology, J.U., N.H. and H.W.; software, J.U. and N.H.; validation, J.U., N.H. and S.E.T.; investigation, J.U. and N.H.; writing—original draft preparation, J.U. and N.H.; writing—review and editing, S.E.T., H.W., H.J.M., and B.-A.B.; visualization, J.U. and N.H.; supervision, B.-A.B., C.K. and H.J.M.; project administration, B.-A.B., C.K. and H.J.M.; funding acquisition, B.-A.B., C.K. and H.J.M. All authors have read and agreed to the published version of the manuscript.

Funding: This research was funded by the Deutsche Forschungsgemeinschaft (DFG, German Research Foundation) grant number 252662854. The APC was funded by the Deutsche Forschungsgemeinschaft (DFG, German Research Foundation).

Acknowledgments: The results presented in this paper were obtained within the Collaborative Research Center 1153 "Process chain to produce hybrid high performance components by Tailored Forming" in the subproject A01, funded by the Deutsche Forschungsgemeinschaft (DFG, German Research Foundation)—252662854. The authors thank the German Research Foundation (DFG) for their financial support of this project.

Conflicts of Interest: The authors declare no conflict of interest.

References

1. Sierra, G.; Peyre, P.; Deschaux-Beaume, F.; Stuart, D.; Fras, G. Steel to aluminium key-hole laser welding. *Mater. Sci. Eng. A* **2007**, *447*, 197–208. [CrossRef]

2. Watanabe, T.; Takayama, H.; Yanagisawa, A. Joining of aluminum to steel by friction stir welding. *J. Mater. Process. Technol.* **2006**, 342–349. [CrossRef]

3. Ashfaq, M.; Sajja, N.; Khalid Rafi, H.; Prasad Rao, K. Improving strength of stainless steel/aluminum alloy friction welds by modifying faying surface design. *JMEPEG* **2013**, *22*, 376–383. [CrossRef]

4. Behrens, B.-A.; Kosch, K.-G. Development of the heating and forming strategy in compound forging of hybrid steel-aluminum parts. *Mat. Werkstofftech.* **2011**, *42*, 973–978. [CrossRef]

5. Behrens, B.-A.; Klose, C.; Chugreev, A.; Heimes, N.; Thürer, S.E.; Uhe, J. A numerical study on co-extrusion to produce coaxial aluminum-steel compounds with longitudinal weld seams. *Metals* **2018**, *8*, 717. [CrossRef]

6. Ogura, T.; Ueda, K.; Saito, Y.; Hirose, A. Nanoindentation measurement of interfacial reaction layers in 6000 series aluminum alloys and steel dissimilar metal joints with alloying elements. *Mater. Trans.* **2011**, *52*, 979–984. [CrossRef]

7. Ryabov, V.R. Welding of aluminum alloys to steels. *Weld. Surf. Rev.* **1998**, *9*, 1–139.

8. Agudo, L.; Jank, N.; Wagner, J.; Schmaranzer, C.; Arenholz, E.; Bruckner, J.; Hackl, H.; Pyzalla, A. Investigation of microstructure and mechanical properties of steel-aluminium joints produced by metal arc joining. *Steel Res. Int.* **2008**, *79*, 530–535. [CrossRef]

9. Springer, H.; Kostka, A.; Payton, E.J.; Raabe, D.; Kaysser-Pyzalla, A.; Eggeler, G. On the formation and growth of intermetallic phases during interdiffusion between low-carbon steel and aluminum alloys. *Acta Mater.* **2011**, *59*, 1586–1600. [CrossRef]

10. Yamamoto, N.; Tkahashi, M.; Aritoshi, M.; Ikeuchi, K. Effect of intermetallic compound layer on bond strength of friction-welded interface of commercially pure aluminum to mild steel. *Q. J. Jpn. Weld. Soc.* **2005**, *23*, 622–627. [CrossRef]

11. Kimapong, K.; Watanabe, T. Lap joint of A5083 aluminum alloy and SS400 steel by friction stir welding. *Mater. Trans.* **2005**, *46*, 835–841. [CrossRef]

12. Yilmaz, M.; Çöl, M.; Acet, M. Interface properties of aluminum/steel friction-welded components. *Mater. Charact.* **2003**, *49*, 421–429. [CrossRef]

13. Fukumoto, S.; Tsubakino, H.; Okita, K.; Aritoshi, M.; Tomita, T. Static joint strength of friction welded joint between aluminium alloys and stainless steel. *Weld. Int.* **2000**, *14*, 89–93. [CrossRef]

14. Herbst, S.; Dovletoglou, C.N.; Nürnberger, F. Method for semi-automated measurement and statistical evaluation of iron aluminum intermetallic compound layer thickness and morphology. *Metallogr. Microstruct. Anal.* **2017**, *6*, 367–374. [CrossRef]

15. Oliver, W.C.; Pharr, G.M. An improved technique for determining hardness and elastic modulus using load and displacement sensing indentation experiments. *J. Mater. Res.* **1992**, *7*, 1564–1583. [CrossRef]

16. Fischer-Cripps, A.-C. *Nanoindentation*, 3rd ed.; Springer: New York, NY, USA, 2011; pp. 21–37.

17. Heumann, T. *Diffusion in Metallen. Grundlagen, Theorie, Vorgänge in Reinmetallen und Legierungen*, 1st ed.; Springer: Berlin/Heidelberg, Germany, 1992; pp. 4–13.

Metals **2020**, *10*, 1582

18. Rummel, G.; Zumkley, T.; Eggersmann, M.; Freitag, K.; Mehrer, H. Diffusion of implanted 3d-transition elements in aluminium part II: Pressure dependence. diffusion implantierter 3d-Übergangselemente in aluminium. Teil II: Druckabhängigkeit. *Z. Met.* **1995**, *95b*, 131–140.

19. Rummel, G.; Zumkley, T.; Eggersmann, M.; Freitag, K.; Mehrer, H. Diffusion of implanted 3d-transition elements in aluminium part I: Temperature dependence. diffusion implantierter 3d-Übergangselemente in aluminium. Teil I: Temperaturabhängigkeit. *Z. Met.* **1995**, *95a*, 122–130.

20. Massalski, T.B.; Okamoto, H. Binary alloy phase diagrams. In *American Society for Metals*; ASM International: Novelty, OH, USA, 1990.

21. Behrens, B.-A.; Klose, C.; Thürer, S.E.; Heimes, N.; Uhe, J. Numerical modeling of the development of intermetallic layers between aluminium and steel during co-extrusion. *AIP Conf. Proc.* **2019**, *2113*, 040029.

Publisher's Note: MDPI stays neutral with regard to jurisdictional claims in published maps and institutional affiliations.

Article

Influence of Ultrasound on Pore and Crack Formation in Laser Beam Welding of Nickel-Base Alloy Round Bars

Jan Grajczak [1,*], Christian Nowroth [2], Sarah Nothdurft [1], Jörg Hermsdorf [1], Jens Twiefel [2], Jörg Wallaschek [2] and Stefan Kaierle [1]

[1]   Laser Zentrum Hannover e.V., Hollerithallee 8, 30419 Hannover, Germany; s.nothdurft@lzh.de (S.N.); j.hermsdorf@lzh.de (J.H.); s.kaierle@lzh.de (S.K.)

[2]   Institute of Dynamics and Vibration Research, Leibniz University Hannover, An der Universität 1, 30823 Garbsen, Germany; nowroth@ids.uni-hannover.de (C.N.); twiefel@ids.uni-hannover.de (J.T.); wallaschek@ids.uni-hannover.de (J.W.)

*   Correspondence: j.grajczak@lzh.de; Tel.: +49-511-2788-391

Received: 31 August 2020; Accepted: 25 September 2020; Published: 29 September 2020

Abstract: Welding by laser beam is a method for creating deep and narrow welds with low influence on the surrounding material. Nevertheless, the microstructure and mechanical properties change, and highly alloyed materials are prone to segregation. A new promising approach for minimizing segregation and its effects like hot cracks is introducing ultrasonic excitation into the specimen. The following investigations are about the effects of different ultrasonic amplitudes (2/4/6 μm) and different positions of the weld pool in the resonant vibration distribution (antinode, centered, and node position) for bead on plate welds on 2.4856 nickel alloy round bars (30 mm diameter) with a laser beam power of 6 kW. The weld is evaluated by visual inspection and metallographic cross sections. The experiments reveal specific mechanisms of interaction between melt and different positions regarding to the vibration shape, which influence weld shape, microstructure, segregation, cracks and pores. Welding with ultrasonic excitation in antinode position improves the welding results.

Keywords: ultrasound; laser beam welding; excitation methods; melt pool dynamics; nickel base alloy 2.4856

1. Introduction

Nickel alloys with an addition of about 20% chromium show high corrosion resistance and high strength at elevated temperatures, so they are commonly used for high-temperature applications and in corrosive environments, as example for fan blades in turbines or for parts in chemical plants [1]. There are several trade names for this material group like Inconel, Chronin or Nicrofer.

Welding of highly alloyed nickel leads to similar problems as in welding stainless steel. Segregation is likely to occur due to a high amount of alloying elements. Those can form lately solidifying phases between the primary grains, and effect hot cracking [1]. In addition, segregated alloying elements are missing in the residual material, which changes its properties like corrosion resistance in the case of chromium [2].

Ultrasonic excitation of solidifying melt leads to mixing of the melt [1]. Hence, it can prevent segregation including its consequences and foster a finer grain structure, because dendrites break through mixing, which in turn creates more nucleation sites [3]. A finer grain structure improves strength as well as ductility and hardness [4]. In addition, porosity and hardness in the heat affected zone can be decreased and penetration depth can be increased for aluminium welds [5,6]. Another effect of ultrasonic excitation at high amplitudes is acoustic cavitation, which occurs in fluids and can be divided into gas bubble cavitation and vapour bubble cavitation. In vapour bubble cavitation the fluid turns to gas when its pressure becomes low enough. This gas bubble implodes immediately, when the pressure rises sufficiently high for condensation. In result, liquid is drawn into the generated vacuum and a pressure shock of around 1010 Pa with temperatures of about 10,000 K occurs [7]. Close to solid interfaces a liquid micro jet, coursed by an imploded bubble, generates a stress peak on the interface and may lead to damage. In ultrasonically assisted welding, cavitation may happen in the melt pool and for a stationary wave cavitation is unlikely to occur in a melt pool at nodal position. Gas bubble cavitation occurs when gas is present in a liquid. The gas itself can form bigger bubbles when the pressure changes periodically by ultrasonic excitation or the bubbles can grow at the surface of particles. Such big bubbles slowly rise up by buoyancy, even if they shrink slightly when the ultrasonic pressure increases again. In ultrasonic laser beam welding the keyhole interaction with cavitation effect has to be examined [7].

The melt dynamics are also influenced by ultrasonic excitation due to the effect of acoustic streaming. Besides an induced oscillating motion, a time-dependent unidirectional flow can be induced [8], which suggests the possibility of effecting asymmetrical welds. Wu [8] gave a mathematical overview of how a fluid flow can be generated by ultrasound. He assumed that the flow force in a standing wave originates at the wave antinode. If the molten pool lies in this position, it is loaded equally, resulting in a symmetrical seam. The same applies if the molten pool is located in the wave node. Due to the same distance to the left and right wave antinode, cf. Figure 1, a symmetrical influence is also created. If, however, welding is carried out in an intermediate position, the distance to one wave antinode is smaller than to the other wave antinode. Wang et al. [9] investigated the influence of acoustic streaming on the solidification process of an Al_2Cu alloy. They describe that the force F_n emanating from acoustic streaming depends, among other things, on the distance to the antinode. It leads to the assumption that the asymmetrical weld seam is a result of the force difference due to the unequal distances which causes a material flow to one side.

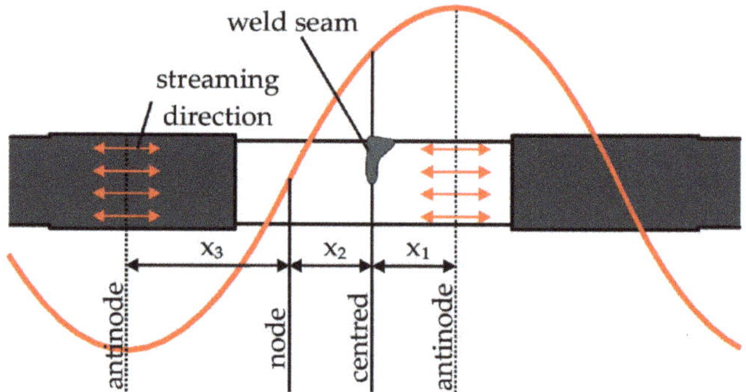

Figure 1. Influence of the wave-position on the weld seam.

Porosity can be influenced by the melt solidifaction morphology: Exogenous solidification originates from the melt border with growing columnar crystals and endogenous solidification happens in the melt volume with growing equiaxed crystals. At high supercooling of exogenous solidification also equiaxed crystals form in the melt volume. Equiaxed crystals can as well be promoted by small particles for heterogeneous nucleation. Gas bubbles can move through a mushy melt with equiaxially solidifying crystals, but in a columnar dendrite network they are trapped [10].

Ultrasonic excitation was applied in many cases only for welding steel and aluminium alloys. Zhou et al. [1] conducted pulsed laser welding with ultrasonic excitation on dissimilar welds of nickel-based Hastelloy C-276 with austenite stainless steel 304 in overlap configuration. No grain refinement occurred, but segregation was strongly reduced.

In previous simulations [11] and experiments on stainless steel and aluminium alloy suitable ultrasonic frequency range and vibration amplitudes were found, which will be tested on a nickel alloy. The effect of antinodal ultrasonic excitation on a liquid pool surrounded by solid material at its sides and air at its top was investigated. The liquid is pushed up at the solid wall sides and can be ejected at very high excitation levels. In result, a V-shaped weld seam collapse forms [11]. Experiments with similar setups showed that with laser beam welding sagging appears in the weld and the keyhole's bottom part can be closed by ultrasonic excitation [4,12]. The experiments in our contribution investigate the ultrasonic excitation effects on partial penetration laser welded nickel alloy round bars, see Table 1, to improve the weld properties and minimize pore and crack formation. The welds will be evaluated by visual inspection and metallographic cross sections.

Table 1. Chemical composition of 2.4856 nickel alloy [13].

Element in wt.%	Ni	Cr	Fe	C	Mn	Si	Co	Al	Ti	P	S	Mo	Nb + Ta
Minimum	58	21	-	-	-	-	-	-	-	-	-	8.00	3.20
Maximum	71	23	5.00	0.03	0.50	0.40	1.00	0.40	0.40	0.01	0.01	10.00	3.80

2. Experimental Setup

2.1. Laser Beam Welding Setup

For the welding tests, a diode-pumped solid state disk laser system (TruDisk 16002, Trumpf, Ditzingen, Germany) was used with specifications according to Table 2.

Table 2. Laser beam welding system specifications.

Model	Trumpf TruDisk 16002
Wavelength in nm	1030
Optical fiber diameter in μm	200
Collimation length in mm	200
Focal length in mm	400
Focal spot diameter in μm	400

The laser head is held in one single position by a robot system (Kuka, Augsburg, Germany) and the welding speed is provided by rotating the specimen for the ultrasonic assisted laser beam welding process, see Figure 2. The complete setup is described in detail in [11]. Only adapters and specimens have been changed.

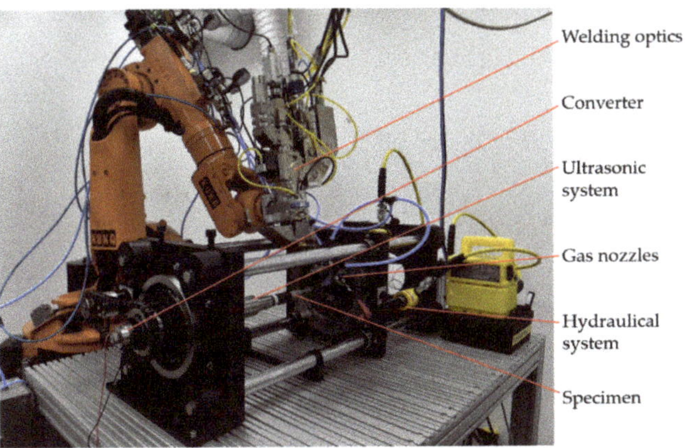

Figure 2. Experimental setup for ultrasonic assisted laser beam welding.

2.2. Ultrasonic System

The specimens, as part of the ultrasonic system, are excited to resonant longitudinal oscillations in the ultrasonic range during the welding process. For this purpose, the specimens are pre-stressed in the ultrasonic system between two adapters, see Figure 3a. Beside the specimens the system consist of adapters as well as boosters and bearing boosters both amplifying the vibration amplitude of the ultrasonic transducer up to the joining point. The vibration amplitude is adjusted between 0 µm and 10 µm in the antinode (closest one to the weld), depending on the current amplitude set. Since at resonance the current amplitude is proportional to the vibration amplitude. The ultrasonic transducer (built by IDS, Institute of Dynamics and Vibration Research, Garbsen, Germany) is driven in the systems resonance frequency (7. Longitudinal mode) at approximately 20 kHz. The control of the phase between current and voltage of the transducer guaranties the stable operation at resonance as well as the control of the current amplitude keeps the vibration amplitude, at the closest anti-node to the weld, steady. Both is carried out utilizing a control unit, the DPC 500/100 [14]. With different adapters, the joint can be placed at any point of the amplitude distribution. Three significant positions are defined for the investigations. In the antinode position, the vibration amplitude has the highest value, see Figure 3b, (this value is always used to describe the amplitude) and the mechanical stress amplitude is minimal. If the joint is placed within the node position, the vibration amplitude becomes zero and the mechanical stress amplitude has its local maximum. In addition, the joint is placed exactly between the antinode and the node of the wave in the centered position. Both vibration amplitude and mechanical stress have significant amplitudes here. The system is preloaded hydraulically. This preloading force is selected to be high enough, that the components cannot clatter during the excited longitudinal vibration. It is selected to be 120 kN, based on a specimen diameter of 30 mm.

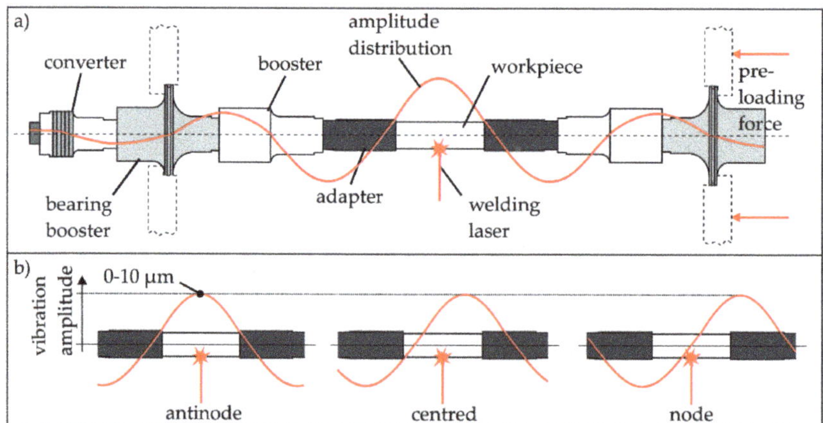

Figure 3. Ultrasonic system for laser beam welding of round bars with vibration distribution (**a**) and selected positions for welding tests (**b**).

3. Experimental Procedure

The specimen are round bars with a diameter of 30 mm and length of 30 mm made of the nickel alloy 2.4856. Those are used for all bead on plate welds. For joining, the optics are angled by 20°. The laser focus point is adjusted to be 4 mm in the sample and 6 mm backwards over the specimen's surface from the angular point. This orientation of the laser spot avoids the formation of melt drops on the specimen's surface resulting from the weight of the melt while rotating the round bars. A laser beam power of 6 kW and a welding speed of 1.00 m/min are used. The focal lense (welding optics: BEO D70, Trumpf, Ditzingen, Germany) is protected by a crossjet and two flat nozzles provide argon as shielding gas with a pressure of 6 bar and a flow rate of 60 L/min at an angle of 45°. The flat nozzles are positioned with 50 mm distance to each other and to the specimen. One nozzle is aiming at the specimen bottom and the other one is aiming above the specimen. According to the test plan (see Table 3), for an ultrasonic amplitude of 6 μm only one welding test per wave position is conducted for testing the effects of excessive excitation, which effects spatter. The applied parameters were determined by previous experiments [4,12].

Table 3. Test plan for ultrasound assisted laser beam welding of nickel-base alloy round bars.

Number of Specimens		Amplitude in μm			
		0	2	4	6
Wave position of weld	Node		3	3	1
	Centred	3	3	3	1
	Antinode		3	3	1

Metallographic cross sections of each specimen are prepared. The etching is conducted with Adler's etchant until the microstructure appears. The weld width, weld depth, weld metal area and pore area are identified and evaluated. In addition, scanning electron microscopy (SEM)-investigations including energy dispersive X-ray spectroscopy (EDX)-analysis (Seifert ISOVOLT 320, 80 kV, 20 mA, 3.6 min, Rich. Seifert & Co., Ahrensburg, Germany) are used for analyzing cracks and chemical composition.

4. Experimental Results

4.1. Weld Appearance

The outer appearance of weld seams in the nickel alloy 2.4856 (see Figure 4), depends strongly on position according to both the vibration distribution and vibration amplitude. With ultrasonic excitation and weld in node position the weld surface becomes wavy starting with an amplitude of 4 μm.

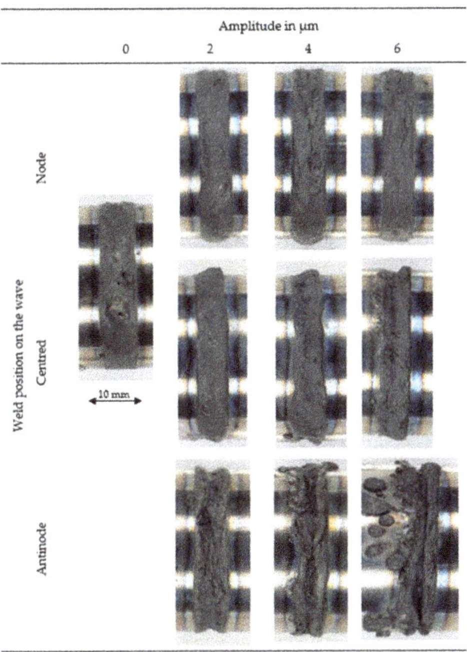

Figure 4. Visual inspection of 2.4856 weld seams, depicted side opposite to welding start/end point.

For the weld in the centred position the ultrasound leads to an unsymmetrical material distribution. This effect increases with increasing amplitudes. In addition, sagging appears starting with 4 μm amplitude. For the weld in antinode position sagging appears with ultrasound. At an amplitude of 4 μm the sagging becomes narrower and the edges touching each other partially. At an amplitude of 6 μm strong spatter occur additionally where the edges touch each other. This shows the melt pool dynamics similar to our simulation in [11].

Evaluating the specimens' cross-sections, see Figure 5, reveals more effects due to the ultrasonic excitation. Pores appear mostly in the weld root area. The weld seam without ultrasonic excitation has one small pore in the root area with a diameter of about 0.5 mm. For the welds in the nodal position, the pores increase up to a diameter of 1.3 mm and the weld surface becomes wavier. For the weld in centred position at an amplitude of 2 μm a big pore with a diameter of about 1.0 mm is located in the root area and the upper weld area is not symmetrical anymore. At amplitudes of 4 μm and 6 μm big pores with a diameter of about 1.4 mm appear and the weld symmetry decreases further until one weld side is straight from the specimen surface to its root. On this side, a reinforcement appears on the specimen surface. In the antinode position, sagging appears in the middle of the weld surface. The welds with amplitudes of 2 μm and 4 μm show show no or little pores with a diameter of about 0.2 mm in the root area, whereas an amplitude of 6 μm shows a big pore with a diameter of about 1.0 mm in the root and a little pore in the upper area. The weld shapes in centred and antinode position are similar for amplitudes of 4 μm and higher, although symmetrical conditions are expected for antinode position.

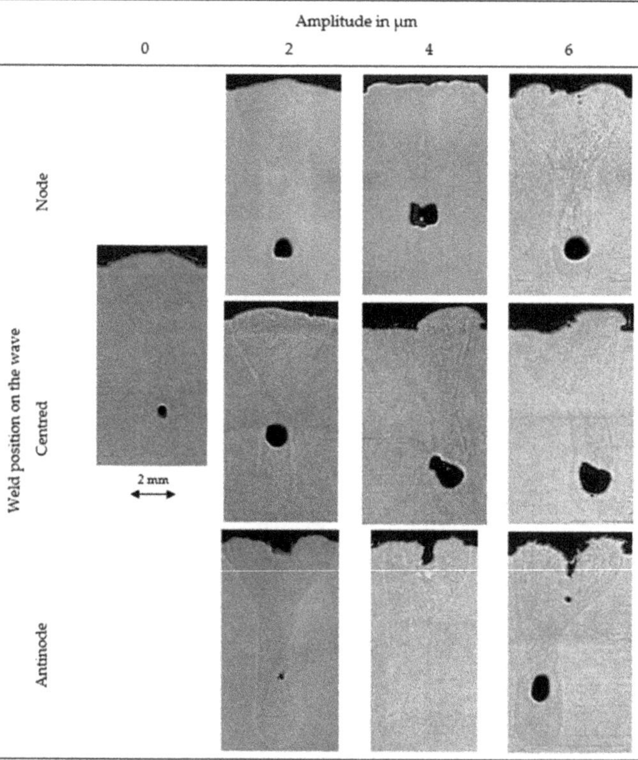

Figure 5. Exemplary micrographs of metallographic cross sections of 2.4856 weld seams.

It is important to consider the laser spot position on the weld seam. On those specimens where ultrasound was applied and asymmetric welds and sagging has been found, the laser spot aims onto one edge of the welding seam. Also considering the cross sections, it becomes clear that the laser beam still creates a keyhole, which penetrates the specimen rectangular to its surface, but the upper weld area shifts towards one side. The reason is likely a directed melt flow, which removes solid metal from one side. The directed melt flow may result from changed weld pool dynamics by acoustic streaming [8,9]. Another reason for asymmetric welds could be the dynamics of the weld during the oscillation, what is simulated in [11]. Effects such as non-parallel interfaces, caused by the V-shape of the melt, between the melt pool and the environment can promote the ejection of the melt towards one side. In order to validate the influence of acoustic streaming on the weld seam, further test series are planned for the future. Special attention will be paid to the ultrasonic amplitude distribution directly at the weld pool in order to investigate the additional influence on the weld shape due to the different coupling between solid and viscous liquid material.

4.2. Pore Formation

The pore areas in Figure 5 show large differences. Without ultrasonic excitation the pore area is low, with diameters of about 0.5 mm. Welds at the centred and nodal positions have similar sized pore areas with diameters of about 1.3 µm at all ultrasonic amplitudes. In contrast, welds at antinode position have very small pore areas with diameters of about 0.2 mm at an amplitude of 2 µm and no pores at an amplitude of 4 µm. However, at an amplitude of 6 µm the pore area increases strongly to a diameter of about 1.0 mm. Commonly pores appear in the middle or bottom weld area, see Figure 5. In general, the pores have round shapes, because they contain gas. This gas can be residual keyhole-gas. Pore formation is modified by an ultrasonically modified melt flow. According to [11], welding at antinode position can course an ejection of melt and a V-shaped opening of the melt due to the strong dynamics. The opening supports the keyhole and promotes gas escapement from the melt. In contrast, during welding at the centred position the melt is moved to one single direction, as described in Section 4.1, crossing and disturbing the keyhole, which effects turbulences. Hence, the melt absorbs keyhole-gas resulting in many times bigger pores. Welding at nodal position neither foster a V-shaped weld seam collapse nor turbulences, but increases the pore area due to compressing the melt, which either closes the keyhole's middle part before the bottom keyhole gas can escape or promotes combining of gas bubbles and holding the gas in the metal resulting in pores. Acoustic cavitation could as well influence this process by inducing growth and shrinkage, which promotes bubble merging. Welds at nodal and centred position show similar pore areas, which can mean that the keyhole is disturbed only little and that acoustic cavitation has a big influence on increasing pores by merging little pores. Another possibility is that keyhole-disturbance and node-effects are of similar influence. For welds at antinode position there should be an optimal vibration amplitude depending on the welding speed to support the keyhole with no keyhole-disturbance by flow direction. At an amplitude of 6 µm big pores form due to a previously described strong melt flow directing mechanism for welds at antinode position.

The porosity shape changes with the position on the wave, see Figure 6. In the weld at nodal position at an amplitude of 4 µm a slightly rectangular-, compressed and collapsed-looking pores has been found due to node position compression. In the centred positioned weld with an amplitude of 4 µm elongated asymmetric pores are found due to the directed melt flow in centred position. At antinode position, the pores are rounder and smaller.

For further investigation of pore formation, the amount of equiaxed microstructure is evaluated visually regarding very low and high amount, see Figure 7. Without ultrasonic excitation there is a very low amount of equiaxed microstructure. With ultrasonic excitation the equiaxed microstructural amount is high for the welds at node and centred position In the micrographs of the welds at antinode position, the equiaxed microstructural amount is very low until an amplitude of 4 µm and becomes high with an amplitude of 6 µm.

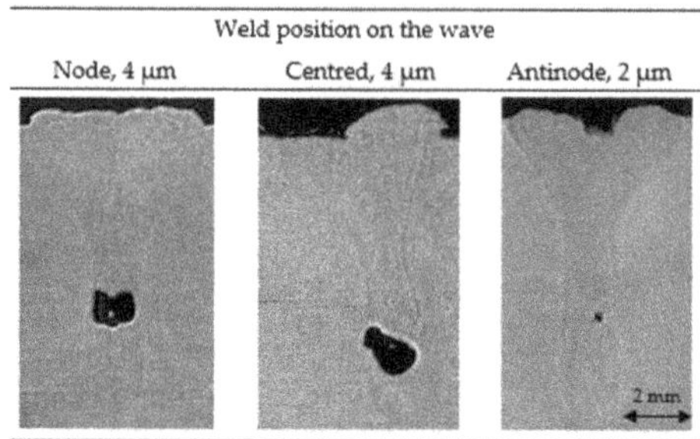

Figure 6. Exemplary micrographs of metallographic cross sections of welding seams (Material 2.4856) regarding porosity shape.

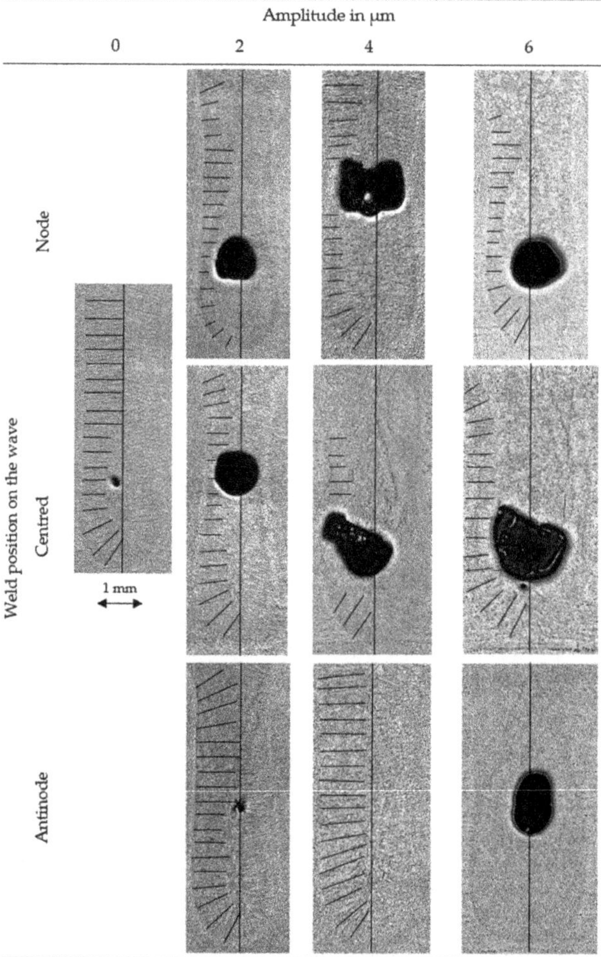

Figure 7. Exemplary micrographs of metallographic cross sections of welding seams (Material 2.4856), the columnar solidification is indicated by lines on one half of the weld with indicated columnar solidification areas, root area.

The comparison between pore area and equiaxed microstructural amount, see Figures 6 and 7, reveals a relation. Equiaxially solidifying melt promotes mobility, combining and growing of gas bubbles. Therefore, gas bubbles are located in the equiaxed crystal core zone in the middle of the weld, see Figure 8.

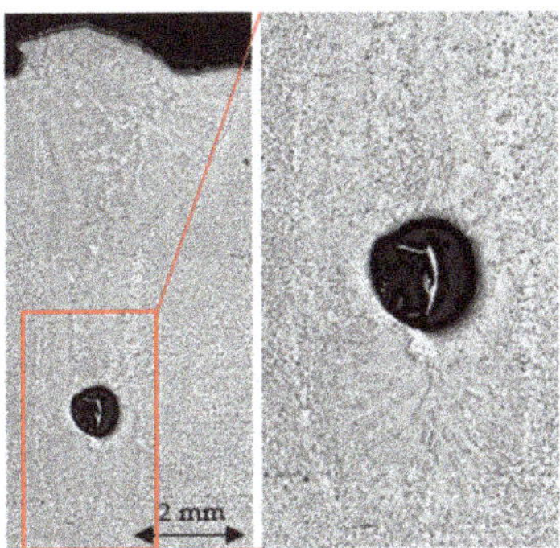

Figure 8. Exemplary microstructural surrounding of pores.

Following mechanisms contribute to an equiaxial microstructure. For welds in centred position the induced melt flow crushes dendrites, which promotes heterogeneous nucleation. For welds at nodal position dendrites are crushed by nodal pressure, which again promotes heterogeneous nucleation. For welds at antinode position with amplitudes below 6 µm equiaxial solidification is reduced to a similar amount as without excitation because no dendrites are crushed and hence no heterogeneous nucleation is fostered. At an amplitude of 6 µm the equiaxed amount rises due to strongly enhanced directed melt flow, which crushes dendrites and supports heterogeneous nucleation. Additionally, welding at antinode position eliminates porosity by promoting the absorption of gas into the keyhole by the previously described V-shaped weld seam collapse.

4.3. Crack Formation

Cracks appear in some specimens with welds in centred and nodal position beginning with amplitudes of 4 µm, see Figure 9. The cracks appear around the grain boundaries, which indicates, that those are hot cracks. They appear in the porosity containing equiaxed crystal core zone in the middle of the weld, because they originate from the segregation of alloying elements, which are mainly pushed ahead by the columnar crystal solidification front. As the equiaxed crystal core zone solidifies endogenously most segregated elements remain at the columnar-equiaxed interface and initiate hot cracks.

Cracking does not happen in welds in antinode position, because the amount of equiaxed microstructure is not increased for amplitude lower or equal than 4 µm. The low number of specimens including cracks does only allow assumptions, but cracks in welds in nodal position appear finer than those in welds in centred position and in general cracks become finer with increasing amplitude, because the melt is pressed together periodically and the segregations are pressed between the solidifying phases. With increasing amplitude at the antinode there is an increasing pressure at the node and the segregations are distributed over larger areas. In result, extremely fine cracks or detachments form, which result in severely weakened welds.

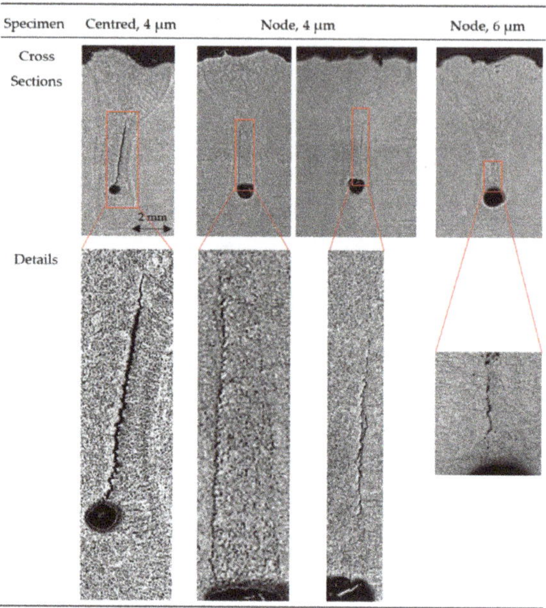

Figure 9. Micrographs of metallographic cross sections of specimens with cracks.

Scanning electron microscope (SEM)-investigations, see Figure 10, validate the hot crack assumption, since the grains are intact without sharp edges and the crack surfaces are coated by segregated melt. Energy dispersive X-ray spectroscopy (EDX)-analysis show only small differences in the chemical compositions of weld metal and crack surface because of the small segregation layer thickness. The niobium- and tantalum-contents increase at the crack surface from about 4 wt.% to 7 wt.% for niobium and from about 2 wt.% to 4 wt.% for tantalum; the nickel-content decreases from about 60 wt.% to 55 wt.%.

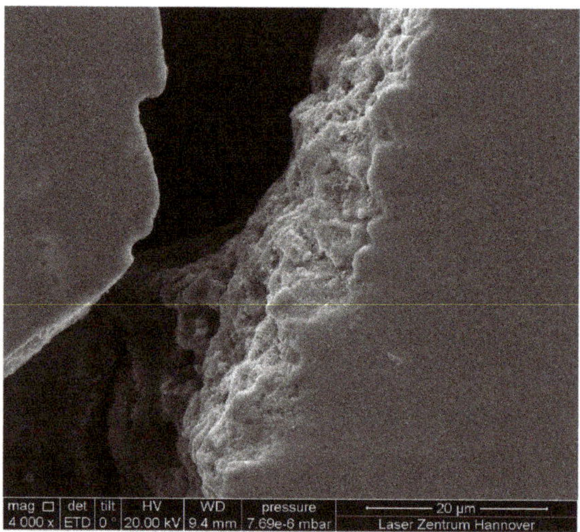

Figure 10. SEM-picture of a crack surface, centred position, ultrasonic amplitude: 4 µm.

5. Conclusions

The different welding positions (regarding the vibration distribution) show remarkably different effects compared to samples without ultrasonic excitation, see Table 4. No clear cavitation effects were observed, but cavitation may support the porosity-related effects. In terms of the few incidents of crack formation, segregation is subordinated to the change of microstructure caused by nodal and centred position.

Table 4. Effects of different ultrasonic excitation positions.

| Position | Weld Reinforcement | Results | | Equiaxial Structure Amount |
		Cracks	Pores	
Without excitation	standard	no	yes	medium
Node	wavy	yes	yes	high
Centred	unidirectional	yes	yes	high
Antinode	V-shaped	no	no	low

For welds in nodal position the melt underlies pressure, which crushes the dendrites and effects a fine microstructure. Still a small area at the weld margin solidifies columnar and alloying elements segregate at the columnar-equiaxial interface, which can result in hot cracks. Pores form due to enhanced gas mobility in the equiaxially solidifying melt area.

In welds at centred position cracks and pores form due to the enhanced equiaxial solidification. It may result from an ultrasonically modified directed melt flow by acoustic streaming. This would crush dendrites and foster equiaxial solidification. In addition, a typical weld shape results, where the top area is shifted towards one side.

For welds at antinode-position a V-shaped weld seam collapse is promoted, cf. [11], which supports the keyhole and therefore gas escapement from the melt. The equiaxial microstructural amount is reduced, because directed melt flow is enhanced only by a small extent, which only reduces supercooling. Concluding, in antinode position porosity is eliminated and neither segregations nor hot cracks are formed.

Ultrasonic excitation in antinode position appears to be the most suitable excitation option for the welding of 2.4856 nickel-base alloy round bars due to the absence of cracks and due to the suppression of pore formation. It correlates with sagging on the welded round bar's surface. To avoid sagging, positions close to antinode position with suitable keyhole-support and without sagging as well as pores and cracks should be investigated.

6. Outlook

Pore formation in welds in nodal and centred position probably results from keyhole-disturbance, which will be investigated by optical coherence tomography subsequently. For exact conclusions regarding porosity distribution scanning acoustic microscopy will be used in future and for statistical coverage more experiments have to be conducted.

The temperature distribution over the specimen and the whole ultrasonic system will be investigated because ultrasonic excitation brings in much heat depending on the position in the amplitude distribution. This affects solidification.

As mentioned in Section 5, welding positions close to the antinode position should be investigated. Welds at centred position shows stronger melt shifting effects than antinode position and no V-collapse effect so that it could be used to segregate the effects of V-collapse including gas escapement through the keyhole in antinode position from the melt shifting effect, which also effects antinode position. Finally, the exact mechanism of flow directing will be investigated by simulations regarding the ultrasonic wave behaviour during passing the weld.

Author Contributions: Conceptualization, J.G., C.N., S.N., J.H., J.T., J.W. and S.K.; methodology, J.G., C.N. and S.N.; software, C.N.; validation, J.G., C.N. and S.N.; formal analysis, J.G., C.N. and S.N.; investigation, J.G., C.N. and S.N.; resources, J.G., C.N. and S.N.; data curation, J.G., C.N. and S.N.; writing—original draft preparation, J.G., C.N. and S.N.; writing—review and editing, J.G., C.N., S.N., J.H., J.T., J.W. and S.K.; visualization, J.G. and C.N.; supervision, J.W. and S.K.; project administration, S.N., J.T. and J.H.; funding acquisition, J.W. and S.K. All authors have read and agreed to the published version of the manuscript.

Funding: Funded by the Deutsche Forschungsgemeinschaft (DFG, German Research Foundation)-CRC 1153, subproject A3–252662854.

Acknowledgments: The authors would like to thank the German Research Foundation (DFG) for the financial and organisational support of this project.

Conflicts of Interest: The authors declare no conflict of interest.

References

1. Zhou, S.; Ma, G.; Dongjiang, W.; Chai, D.; Lei, M. Ultrasonic vibration assisted laser welding of nickel-based alloy and Austenite stainless steel. *J. Manuf. Process.* **2018**, *31*, 759–767. [CrossRef]
2. Lee, H.T.; Wu, J.L. Intergranular corrosion resistance of nickel-based alloy 690 weldments. *Corros. Sci.* **2010**, *52*, 1545–1550. [CrossRef]
3. Dommaschk, C.D.; Hübler, J.H. Auswirkungen einer Vibrationsbehandlung auf das Erstarrungs- und Speisungsverhalten von Gusswerkstoffen. *Gießerei-Prax.* **2003**, *12*, 505–512.
4. Nothdurft, S.; Ohrdes, H.; Twiefel, J.; Wallaschek, J.; Mildebrath, M.; Maier, H.J.; Hassel, T.; Overmeyer, L.; Kaierle, S. Influence of ultrasonic amplitude and position in the vibration distribution on the microstructure of a laser beam welded aluminum alloy. *J. Laser Appl.* **2019**, *31*, 022402. [CrossRef]
5. Krajewski, A.; Włosiński, W.; Chmielewski, T.; Kołodziejczak, P. Ultrasonic-vibration assisted arc-welding of aluminium alloys. *Bull. Pol. Acad. Sci. Tech. Sci.* **2012**, *60*, 841–852.
6. Kim, J.S.; Watanabe, T.; Yoshida, Y. Ultrasonic vibration aided laser welding of Al alloys: Improvement of laser welding-quality. *J. Laser Appl.* **1995**, *7*, 38–46. [CrossRef]
7. Deutsche Gesellschaft für Akustik, e.V. DEGA-Empfehlung 101: Akustische Wellen und Felder. Available online: https://www.dega-akustik.de/fileadmin/dega-akustik.de/publikationen/DEGA_Empfehlung_101.pdf (accessed on 14 August 2020).
8. Wu, J. Acoustic Streaming and Its Applications. *Fluids* **2018**, *3*, 108. [CrossRef]
9. Wang, G.; Croaker, P.; Dargusch, M.; McGuckin, D.; StJohn, D. Simulation of convective flow and thermal conditions during ultrasonic treatment of an Al-2Cu alloy. *Comput. Mater. Sci.* **2017**, *134*, 116–125. [CrossRef]
10. Foundry Lexicon: Gas Blister. Available online: www.giessereilexikon.com/en (accessed on 17 August 2020).
11. Ohrdes, H.; Nothdurft, S.; Nowroth, C.; Grajczak, J.; Twiefel, J.; Hermsdorf, J.; Kaierle, S.; Wallaschek, J. Influence of ultrasonic vibration amplitude on the weld shape of EN AW-6082 utilizing a new excitation system for laser beam welding. *Prod. Eng.*. Special Issue under review.
12. Nothdurft, S.; Ohrdes, H.; Twiefel, J.; Wallaschek, J.; Hermsdorf, J.; Overmeyer, L.; Kaierle, S. Investigations on the effect of different ultrasonic amplitudes and positions in the vibration distribution on the microstructure of laser beam welded stainless steel. In *High-Power Laser Materials Processing: Applications, Diagnostics, and Systems 2020, IX*; International Society for Optics and Photonics: Bellingham, WA, USA, 2020; p. 112730J.
13. VDM Metals. Datenblatt Nr. 4118: VDM Alloy 625. Available online: https://www.vdm-metals.com/fileadmin/userupload/Downloads/DataSheets/DatenblattVDMAlloy625.pdf (accessed on 11 August 2020).
14. Ille, I.; Twiefel, J. Model-based Feedback Control of an Ultrasonic Transducer for Ultrasonic Assisted Turning Using a Novel Digital Controller. *Phys. Procedia* **2015**, *70*, 63–67. [CrossRef]

Article

Computer-Aided Engineering Environment for Designing Tailored Forming Components

Tim Brockmöller *, Renan Siqueira, Paul C. Gembarski, Iryna Mozgova and Roland Lachmayer

Institute of Product Development, Leibniz University Hannover, An der Universität 1, 30823 Garbsen, Germany; siqueira@ipeg.uni-hannover.de (R.S.); gembarski@ipeg.uni-hannover.de (P.C.G.); mozgova@ipeg.uni-hannover.de (I.M.); lachmayer@ipeg.uni-hannover.de (R.L.)

* Correspondence: brockmoeller@ipeg.uni-hannover.de; Tel.: +49-0511-762-5340

Received: 29 October 2020; Accepted: 19 November 2020; Published: 27 November 2020

Abstract: The use of multi-material forming components makes it possible to produce components adapted to the respective requirements, which have advantages over mono-material components. The necessary consideration of an additional material increases the possible degrees of freedom in product and manufacturing process development. As a result, development becomes more complex and special expert knowledge is required. To counteract this, computer-aided engineering environments with knowledge-based tools are increasingly used. This article describes a computer-aided engineering environment (CAEE) that can be used to design hybrid forming components that are produced by tailored forming, a process chain developed in the Collaborative Research Center (CRC) 1153. The CAEE consists of a knowledge base, in which the knowledge necessary for the design of tailored forming parts, including manufacturer restrictions, is stored and made available. For the generation of rough and detailed design and for elaboration the following methods are used. The topology optimization method, Interfacial Zone Evolutionary Optimization (IZEO), which determines the material distribution. The design of optimized joining zone geometries, by robust design. The elaboration of the components by means of highly flexible computer-aided design (CAD) models, which are built according to the generative parametric design approach (GPDA).

Keywords: tailored forming; multi-material; IZEO; topology optimization; computer-aided engineering environment; GPDA; manufacturing restrictions

1. Introduction

The progress of manufacturing technologies tends to astonish observers. For example, at Leibniz University Hannover, the Collaborative Research Center (CRC) 1153, which is funded by the German research association, explores process chains for tailored forming [1]. Here, semi-finished hybrid workpieces, consisting of two different materials like steel and aluminum, are processed by forming, heat treatment and cutting technologies to produce high-performance multi-material parts [2,3].

Ashby and Cebon have shown that for special purposes, a multi-material design achieves superior performance than a conventional design [4]. Thinking of structural components like wheel carriers, rocker levers, or even pinion shafts, areas where high stiffness and wear resistance are needed can be made from steel, all other areas are made from aluminum [5]. So, from a design point of view, a new degree of freedom is introduced, which is the material distribution within the multi-material part. What initially appears to be an interesting avenue for leveraging even more efficiency of such parts results in higher complexity for the design since the material distribution generally influences the mechanical properties [6–9].

However, more than this, the complexity rises also from the manufacturing point of view [10]. The tailored forming technology requires production lines where many manufacturing steps follow each other, whose processes are linked and therefore precisely coordinated [11]. Setting up and running in a new product variant result in large efforts, especially when process windows need to be (re-)evaluated and the quality of semi-finished materials varies [12,13]. Thus, it is necessary for the designer to consider the available capabilities of manufacturing as early as possible since they restrict the possible solution space of the part geometry [14]. Examples, therefore, range from simple manufacturing restrictions like maximum traveling distances or hardening depths [15], over appropriate tolerances of dimensions, form and positioning to the consideration of design guidelines, as is discussed today as Design for Excellence (DfX) [16,17]. Here it also must be considered that both materials of a tailored forming part may differ in their processing, i.e., forming temperatures, cutting speed, etc., [18,19].

In order to avoid iterations during design, computer-aided engineering environments (CAEE) support the designer in making the right decisions, checking the design with respect to the solution space and finding the optimum between requirement fulfillment, capabilities and resulting production costs [15,20,21]. On the one hand, they include all necessary synthesis and analysis tools for a design task [5,22–24]. On the other hand, e.g., artificial intelligence technologies offer the possibility to process data from production, find patterns and formalize new manufacturing knowledge automatically [25–27]. Thus, such CAEE serve as a central information hub for all experts that are involved in the according to the design process [28–30].

Within the scope of this work, a CAEE is set up to reduce the uncertainties in the development of tailored forming components and to help ensure that they are adapted as optimally as possible to the respective use case. The CAEE has different tools that are used in different phases of the product development process. The manufacturing and process knowledge needed for the development is provided by the subprojects of the CRC 1153. Accordingly, the CAEE offers the possibility to extend the underlying knowledge base with new insights gained in the CRC 1153 and delivers rough as well as detailed designs of tailored forming components accordingly. Special consideration is given to the material distribution in the component as well as the implementation of the applicable manufacturing restrictions. The article is structured as follows. The second Section deals with the state of the art on the relevant topics of tailored forming, design theory and knowledge-based systems. Section 3.1 presents the methods and tools of the CAEE with which rough and detailed designs are implemented. In the following Section 4 these are implemented by means of corresponding examples. In the last two Sections 5 and 6 the contribution is summarized and a conclusion and an outlook to further research projects are given.

2. Research Background

2.1. Manufacturing Processes for Multi-Material Parts

The hybridization of semi-finished parts is a widely discussed and promising topic for raising the performance of structural components. In order to create a composite of different metallic materials, different methods are researched and applied in the industrial environment.

In explosive welding, two plate-shaped workpieces are joined together by a controlled explosion. The workpieces are welded at the joint surfaces without heat input by applying an abrupt force caused by the pressure wave generated by the detonation of explosives, preferably without filler metal [31,32]. Another family of production processes that is used to create hybrid semi-finished materials is additive manufacturing [33]. Different from this is a relatively new approach that combines two different materials in laser powder bed fusion processes [34]. All of these approaches have in common that an inter-metallic joining zone between both materials occurs.

With the hybrid forging developed by the Leiber company, non-plate shaped semi-finished products of steel and aluminum alloys can also be joined together. This approach does not aim at an inter-metallic bonding of both materials [35]. In e.g., hybrid compound forging, this is different since a material joint is created using a soldering material [36].

In the CRC 1153 of Leibniz University Hannover, various process chains for the production of multi-material and formed solid components are being researched. The materials used are mainly aluminum (EN AW 6082) and steel alloys (20MnCr5, 41Cr4). The general process chain can be seen in Figure 1.

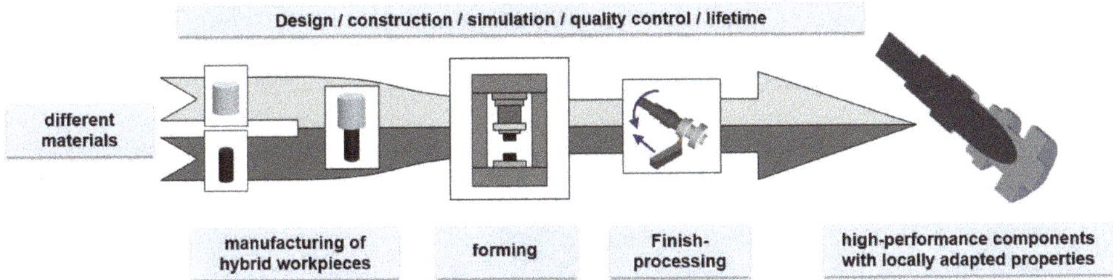

Figure 1. General tailored forming process chain, according to [2].

In the first step, a so-called hybrid semi-finished product is produced. Two mono-material semi-finished products are joined together by friction welding [37,38], ultrasonic-assisted laser beam welding [39], deposition welding [40] or composite rod extrusion (LACE (LACE = Lateral Angular Co-Extrusion) process) [41]. The hybrid semi-finished product is then shaped in a forming process. Here, cross wedge rolling [42], impact extrusion [43] and drop forging [40] are investigated. Both materials are formed during the process step and thus form an intermetallic compound. The process differs here, for example, from hybrid forging, in which the materials are not joined before forming and only one material is formed. In the end, heat treatment processes follow in order to be able to influence the component's mechanical properties [44] and the component is finished by machining [19]. The manufacturing process relevant for this article consists of the process steps friction welding, impact extrusion and machining. As the heat treatment processes only influence the component properties, but not the geometric shape, these are not considered.

Rotary friction welding is suitable for welding different materials that cannot be joined by other welding processes [45]. For friction welding of hybrid components, various investigations have been carried out in the CRC 1153 in which the strengths of steel–steel alloy combinations have been analyzed [46], also in comparison to US laser beam welding [45]. In addition, the strengths of steel–aluminum alloy combinations at different temperatures have been investigated [47]. Furthermore, it has been explored how different geometries (different cone angles in the semi-finished steel product), properties of the surfaces and temperatures also affect the strength of the composite [37].

Impact extrusion is a metal forming process in which a semi-finished workpiece is pressed through a die to obtain a product with a smaller cross-sectional area. These are differentiated according to the direction of material flow and the geometry of the formed product. The process used here is called forward rod extrusion [48]. In connection with extrusion, SFB 1153 has developed heating strategies for inhomogeneous heating, since the required forming temperature is different for each material. In addition, it is investigated how the shape and strength of the joining zone can be influenced by impact extrusion so that the strength of the composite is increased. The first concepts for impact extrusion and inhomogeneous heating by induction can be found in [2,49]. In Goldstein et al. 2017 the simulative results of heating are validated by experimental tests on steel–aluminum semi-finished products (20MnCr5, EN AW 6082) [48]. Based on these results, the joining zone geometries and properties of manufactured components are investigated in [18], which are then optimized in [43] by adjustments in the forming process. In addition, it has been shown that forming can improve the strength of the joining zone of components (41Cr4, C22.8) produced by US laser beam welding [50].

Besides pure shaping, machining is used to manipulate the properties of the surfaces of hybrid components [51,52]. However, these aspects are not yet relevant for the current state of the CAEE. Behrens et al. 2019 illustrate the entire product development and manufacturing process from the creation of the joining zone geometry, joining, forming and heat treatment to the finished machined component [11]. Figure 2 illustrates the manufacturing process.

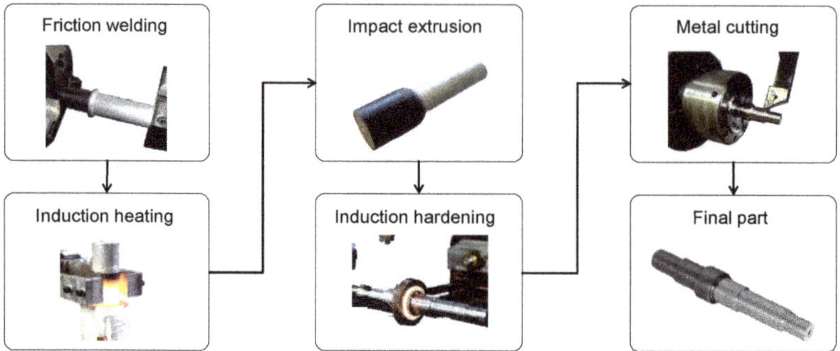

Figure 2. Process chain for the production of a hybrid shaft by tailored forming [11].

2.2. Computer-Aided Engineering Environments

The development of technical products follows well-known process models that are either of a sequential or networked nature [53]. As an example, the process according to Pahl and Beitz divides the development process into four phases which are task clarification, concept determination, embodiment design and detailed design [54]. Another example is Suh's approach of Axiomatic Design where customer requirements are translated into functional requirements, design parameters and process variables for manufacturing [12]. The translation is achieved with design matrices and is thus strictly formalized [14]. Usually, the processes allow iterations and zig-zagging through the phases, as requirements are sharpened and new knowledge is created continuously while the design team converges the solution space against the final design [55].

In modern product development, independently from the process, the application of software tools for synthesis and analysis of design artefacts is state of the art for many disciplines [56]. Beside these, such computer-aided engineering environments (CAEE) comprise product data management and collaboration support systems that allow for coordination of large teams as well as formalizing and communicating knowledge between all relevant stakeholders [15,57]. A very central tool for mechanical engineering is still the computer-aided design (CAD) system for defining e.g., product shape and production information [58,59]. Over time, these CAD systems have developed from tools for 2D line drawing to powerful parametric 3D design systems where a designer is able to modify his parts and assemblies simply by changing values of e.g., dimensions for lengths and adding or deleting features [60]. Hereby, it has to be considered that only a part of the product's characteristics may be modeled directly, like geometry, material, or surface quality. e.g., stress distribution is a resulting property that is influenced by the characteristics and thus modeled indirectly which leads to synthesis-analysis loops during development [61].

Two lines of development stand representative for the progress in CAEE implementation. First, knowledge-based engineering and design systems use formal, explicit knowledge that has been integrated into the according to synthesis and analysis systems [62–66]. As an example, knowledge-based CAD uses dimensioning formulae, design rules, spreadsheet integration and intelligent templates to automate routine design tasks [67,68]. Exemplary works from this line of development describe CAEE for fixture design [30,69,70], automotive and aircraft engineering [67,71] or mechanical and plant engineering [20,72–74]. In Sauthoff 2017, the automatic configuration and optimization of structural components from automotive engineering are proposed, integrating a knowledge-based design system and an evolutionary optimization algorithm [75]. All of these works have in common that a more or less closed solution space of predetermined designs is modeled. The resulting artefact description is usually of high quality and corresponds to detailed design.

As the second line of development, computational design synthesis systems rely on a more informal and implicit formulation of knowledge in order to design an artefact [59]. Their aim is more to capture the laws-of-creation of how a design artefact is developed. The consequence is that computational design systems commonly deliver more abstract artefact descriptions which have to be reengineered e.g. into parametric CAD [76]. An example of this is 3D topology optimization that considers manufacturing restrictions [6]. Other works from this line include the synthesis of additively manufactured parts using object-oriented programming, CAD and parametric optimization [21] or the design and optimization of mechanical engineering parts using CAD and multi-agent systems [27].

3. Computer-Aided Engineering Environment for Tailored Forming Parts

In order to design a tailored forming part, both lines of development make a contribution. The determination of the material distribution is more subject to computational design synthesis as laws-of-creation, therefore, may be formulated, independently from distinct geometry. Especially the design of the joining zone necessitates a formal representation that considers the restrictions of the later manufacturing processes precisely. Thus we propose a CAEE that uses both approaches for the respecting phases of the development of tailored forming parts.

The basic structure of the CAEE is shown in Figure 3. It essentially consists of four different areas. Three of the areas represent the product development process and provide tools for potential determination (1), for the creation of the rough design (2), and for the generation of the embodiment design or elaboration (3). The fourth area is the knowledge base (4), in which the expertise required for development is stored. The focus of this paper is on the areas (2), (3) and partly (4). Further information on area (1) can be found in [5,77] and is not part of this paper.

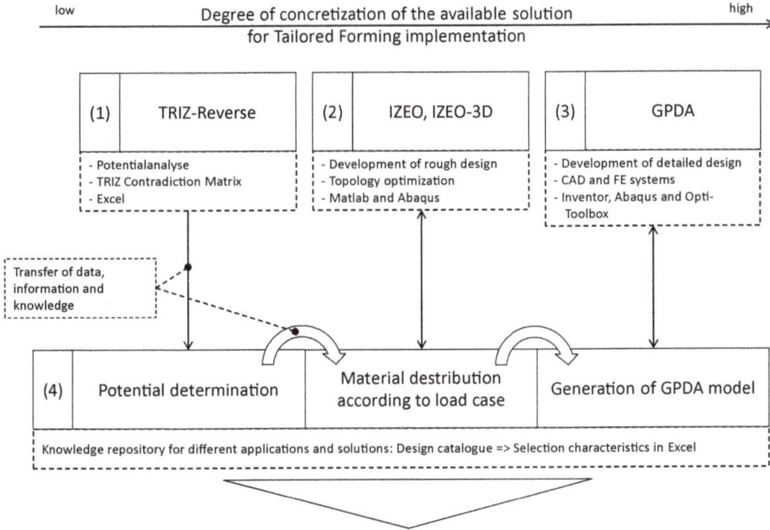

Figure 3. Structure of the computer-aided engineering environment (CAEE) for tailored forming.

3.1. Rough Design by Interfacial Zone Evolutionary Optimization

The Interfacial Zone Evolutionary Optimization (IZEO), developed in [78], is a method able to deal with the specific challenges of the present study since it can solve general multi-material problems that have the presence of strong manufacturing restrictions. As recommended in [79], taking these restrictions into account in an early phase of the design process avoids the loss of the optimized properties when these are applied later.

The working principle of IZEO can be visualized in Figure 4. This method is based on evolutionary optimization algorithms, such as the Bidirectional Evolutionary Structure Optimization (BESO) [80], where the domain is discretized into elements and the material of the elements are changed iteratively, following a sensitivity function. The primary difference in IZEO is how these changes occur, which is limited to the interfacial zone between the different materials.

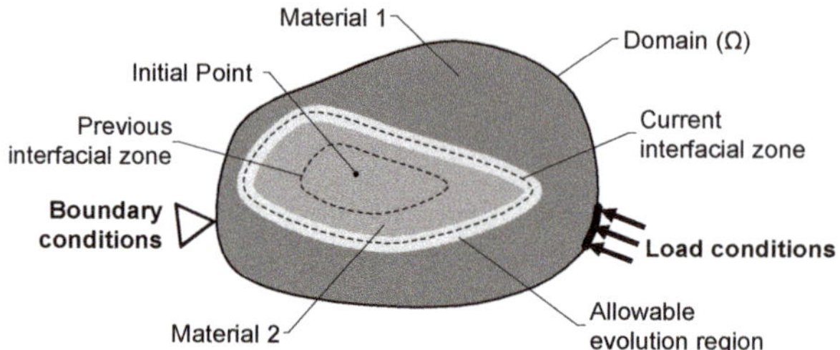

Figure 4. Model representation of the interfacial evolutionary process [81].

This strategy allows the implementation of a variety of manufacturing restrictions [81]. Following the theory proposed in [79], each manufacturing technique can be modeled as a combination of geometric constraints, as shown in Figure 5. IZEO follows the same principle, allowing the designer to apply different constraints at the same time.

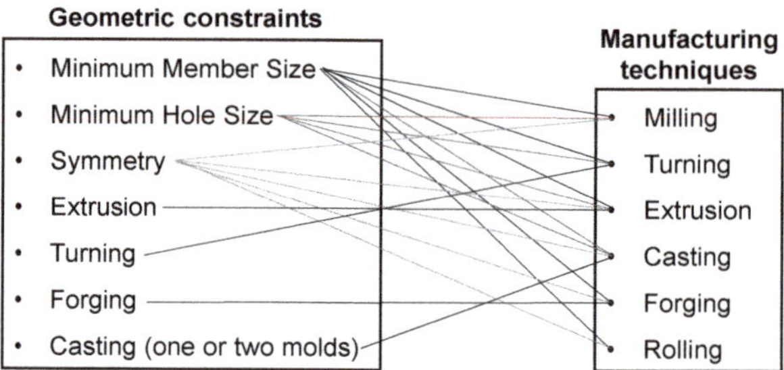

Figure 5. Relationship between geometric constraints and manufacturing techniques [79].

The constraints shown in Figure 5 can be also serialized in the simulation, which works as a prioritization from the first one applied until the last one. This is in accordance with typical manufacturing process-chains, where many restrictions are applied in different stages of the process. For multi-material processes, this also allows different constraints for the connections between the materials and the component body. This way, with the inclusion of all necessary geometric constraints, a general approach can be implemented to attend to the specific challenges inherent to a manufacturing process and generate optimized conceptual designs.

In the current study, the implementation of IZEO was extended for a 3D environment, differently from previous works. This was implemented in the FE-software Abaqus, using its scripting capabilities in Python. Therefore, the full IZEO program was implemented with Python, using the solving capabilities of the FE-software. In this case, the implementation of the manufacturing restrictions described in [81] was made following the same concepts, but considering the third dimension and a higher degree-of-freedom to control them. Table 1 presents the implemented geometric constraints and the respective control parameters.

Table 1. Table with the implemented Interfacial Zone Evolutionary Optimization (IZEO) geometric constraints.

Geometric Constraint	Definition	Control Parameters	Representation
Minimum member size	Level of detail in the manufacturing process	Minimum size	
Uni/Bidirectional growth	Unidirectional access of the manufacturing tools or serial connection of materials	Vector of growth direction	
Extrusion	Extrusion direction in the manufacturing	Vector of extrusion direction	
Planar symmetry	Symmetry imposed by the processes	Point and normal vector to the plane	

It can be observed that with the inclusion of the control parameters, the implementation of geometric constraints adds new degrees of freedom to the generation of optimized solutions. Naturally, these restrictions will be selected according to the chosen manufacturing process. Ideally, the optimization should be performed several times with a variation of these constraints, in order to find the most suitable geometry and manufacturing process at the same time. In this case, not only the control parameters (radius, points and vectors), would be varied, but also different combinations of the constraints, simulating different process chains. Since the current study is focused on tailored forming, only the constraints related to the proposed process are here investigated.

3.2. Detailed Design Using the Generative Parametric Design Approach

A CAD-centric KBE environment was proposed by Sauthoff for the automatic configuration and optimization of structural components in mechanical engineering [75]. It combines a CAD modeling strategy called generative parametric design approach (GDPA) with knowledge integration and an evolutionary optimization algorithm. In order to achieve the necessary flexibility, the CAD model of a structural component is divided into several design zones which are linked by a common skeleton (Figure 6a). For each design zone, independent CAD models are implemented as so-called design elements that reflect parts of the structural component and may be understood as generic parametric templates (Figure 6b). In such a design element, all relevant design knowledge, like dimensioning, design rules or manufacturing restrictions, are stored [65]. The top-level assembly of the component is implemented in such a way that adjacent design zones communicate with each other and exchange interface parameters. The design elements can be replaced with other design elements that are also approved for the design zone, as required. When now a control parameter of the skeleton or general requirements for the structural component change, this is propagated through all design elements that check themselves for consistency, technical correctness and violation of restrictions. The result is that highly flexible models are created which can be rebuilt without errors even after topological changes [82]. If a sufficient library of generic and task-specific design elements exists, a large solution space of structural components like vehicle chassis or bodies, is available [83].

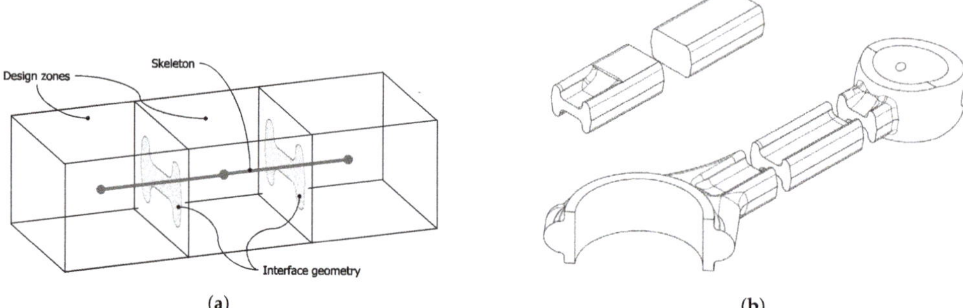

Figure 6. Generative parametric design approach (GPDA) model of a connection rod: (**a**) skeleton and interfaces, (**b**) computer-aided design (CAD) model, according to [84].

Due to the flexible model structure, it is possible to optimize the shape of the GPDA models in automated synthesis-analysis loops. According to Sauthoff, the CAD system is coupled with an FE system via an optimization program, the so-called Opti-Toolbox. The Opti-Toolbox generates several component variants on the basis of e.g., evolutionary algorithms by automatically adjusting the parameters in the GPDA model and exchanging design elements. These are then analyzed in the FE system and the results are evaluated by the Opti-Toolbox. If the requirements are not met, further component variants are generated. This loop is repeated until the requirements are met [75]. Figure 7 shows the schematic structure of the GPDA engineering environment.

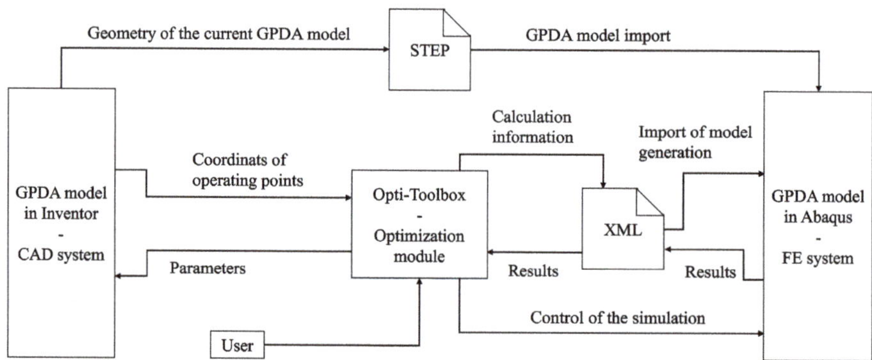

Figure 7. Schematic structure of the GPDA engineering Environment [83].

4. Implementation for Shaft-Like Tailored Forming Parts

A hybrid demonstrator shaft developed in CRC 1153 is used as an application example. The shaft is manufactured by the above-described manufacturing processes of friction welding, impact extrusion and machining. The material combination under consideration and 41Cr4 and EN AW-6082, whose properties are given in Table 2. The objective function is to generate a component that is as light as possible with sufficient strength.

Table 2. Material properties.

Material	Density	Yield Stress	Ultimate Stress
Steel (41Cr4)	7.85 $\frac{g}{cm^3}$	660 MPa	1020 MPa
Aluminum (EN AW-6082)	2.70 $\frac{g}{cm^3}$	280 MPa	385 MPa

Figure 8 shows the load and boundary conditions considered in this example. Furthermore, the represented geometry describes the boundaries of the domain in which the optimization is allowed to take place. The absolute values for force and torque were set to generate a global safety factor of 1 when the shaft is completely made of steel and the proportion between them was set to generate 15% of maximal stress through the bending and the rest through the torsion.

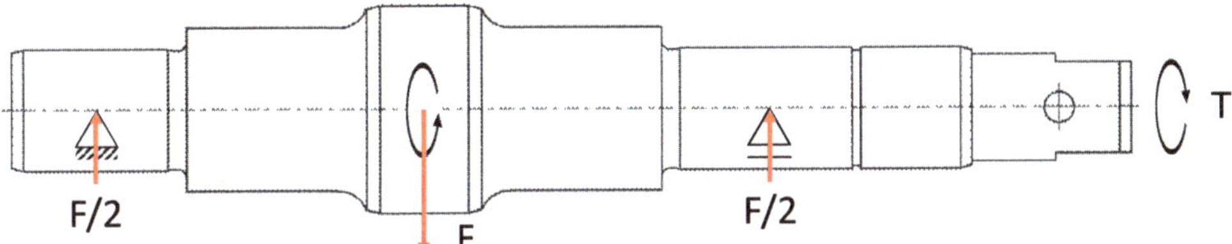

Figure 8. Domain with load and boundary conditions of the shaft for the 3D optimization.

4.1. Expansion of Geometric Constraints

With the geometric constraints described in Table 1 and the idea of a combination of constraints from Figure 5, a great variety of processes can already be simulated. However, for the current application, two constraints were added: rotational symmetry and radial growth.

Rotational symmetry is self-explained, being related to components that are subjected to processes such as rolling or turning. Two control parameters are necessary: initial and final coordinates of the symmetry line. In the case of multi-materials, this constraint can be applied not only to the component body as a whole, but also separately to the connection between the two materials. IZEO allows these possible configurations, as presented in Figure 9. This restriction was implemented using the same principle of planar symmetry presented in [81], where the sensitivity of all elements present in the rotational curve are averaged.

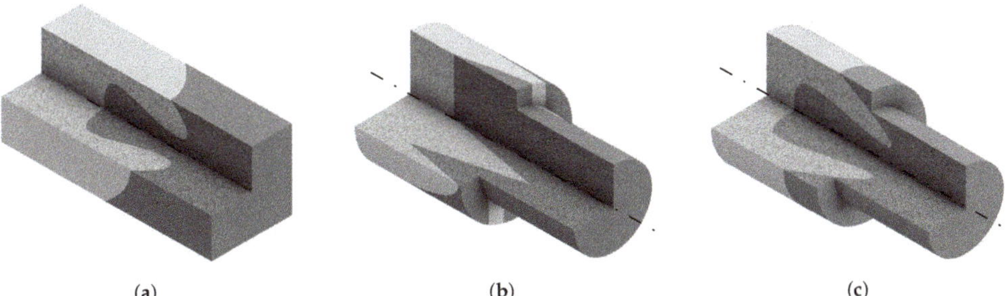

(a)   (b)   (c)

Figure 9. Rotational symmetry applied to: (**a**) joining zone only; (**b**) component body only; (**c**) both joining zone and component body.

Radial growth is a special constraint present in tailored forming. In the manufacturing of rotational symmetric components, the possible processes do not allow the presence of the softer material inside the harder material. Due to thermal properties, the harder material always flows inside the softer material. This translates to the optimization method as a special type of "unidirectional growth" constraint, where the direction is not linear, but radial coming from outside, similar to what is seen in a turning machine (Figure 10). Therefore, the same as rotational symmetry, the initial and final coordinates of the center-line are required as control parameters.

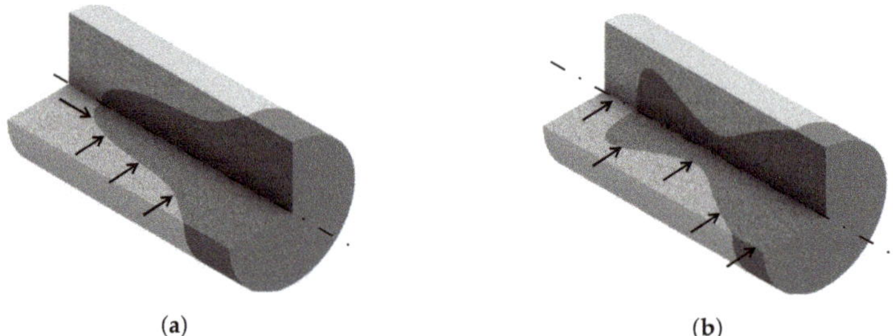

(a)   (b)

Figure 10. Rotational symmetric components with joining zone constrained by: (**a**) radial growth only; (**b**) radial and unidirectional growth.

This radial growth is not only important because of the thermal effects of the multi-material connection, but it also describes the main restriction involved in the manufacture of shafts during turning in a mono-material approach.

4.2. IZEO and Robust Design for Tailored Forming

The model described was submitted to IZEO with the following constraints: minimum member size (3 mm), unidirectional growth (same direction of the aluminum in the friction welding), rotational symmetry and radial growth (aligned to the axis of the shaft). Since the outer geometry of the shaft should remain unchangeable and the addition of aluminum will tend to reduce the strength of the shaft, it was set as the objective function a safety factor of 50% the value for a shaft made entirely of steel. The last interactions are presented in Figure 11.

During implementation, it became clear that design and manufacturing process development need to be aligned towards a common objective. The information exchange between the two fields is commonly of a sequential nature. Thus, an additional information exchange platform for continuous improvement was created to prevent from losing the knowledge acquired in past interactions.

For that purpose, the use of Knowledge-Based Engineering (KBE) tools are necessary for the creation of this common interface between design and manufacturing processes, and for the operationalization of both, as proposed in [85]. Therefore, an adaptation of a case-based reasoning (CBR) cycle was proposed, where the decision-making process is supported by a unified information management system. This method makes use of parametric models to analyze the information generated on both sides, compare them and suggest innovative design solutions based on new specifications and previous experiences. The topology optimization result will serve as the first input in the construction of this parametric model. Thereby, both design and manufacturing research can be performed in parallel, exchanging information in a continuous way and enhancing the system with its use.

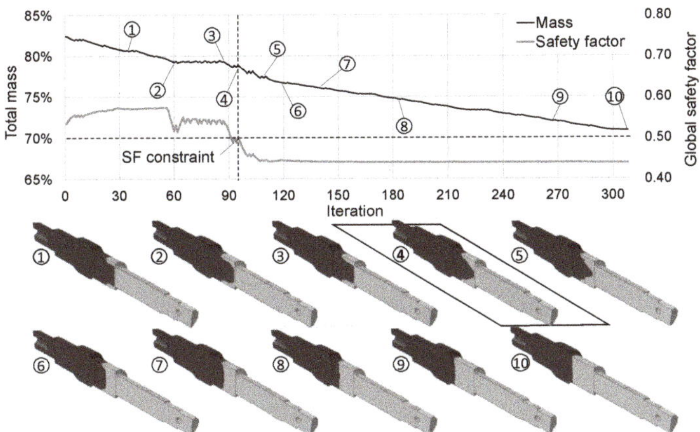

Figure 11. Optimization evolution for the tailored forming shaft model with IZEO.

With the results obtained with IZEO, a parametric model of the joining zone was constructed for the submission in the adapted CBR. With different parametric models and parameters, a large number of variations were simulated. Figure 12 shows a graph where the two objectives are set at both axes and every variation is represented as a point in the space. A Pareto front of optimal solutions can be easily recognized, where the simulations close to this curve are considered optimal solutions.

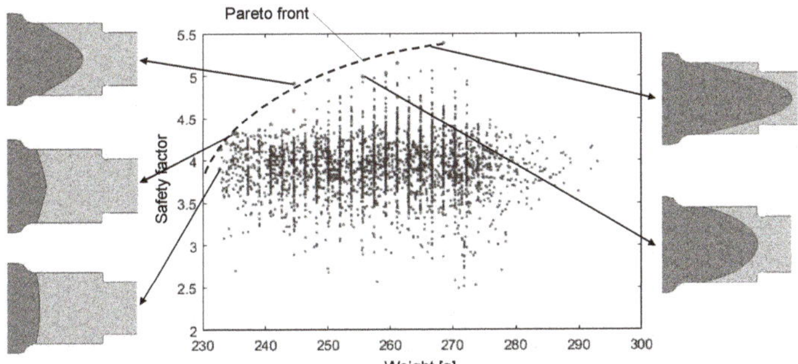

Figure 12. Plot of every parametric simulation over safety factor and weight, where a Pareto front is observed.

With the completion of the CBR cycle, the best candidates for manufacturing can be selected and submitted to the process chain of tailored forming. In this way, the process learns on every cycle while more optimized solutions are being generated.

For validation purposes, various joining zone geometries were examined in test bench trials, e.g., on the torsion test bench, and compared with the simulation results. Subsequently, the parameters were adjusted so that the simulation provides an adequate representation of the manufactured components [6].

4.3. Intermediate Results

The comparison with a mono-material shaft cannot be straightforward executed, since multi-materials are intrinsically connected to more requirements, but it serves to show the potential of the technology for lightweight. This potential, however, is also connected to some of the geometric restrictions imposed, such as the allowable size of the component. Figure 13 makes a comparison between the multi-material design achieved and an equivalent mono-material shaft with the same requirements for strength and wear, considering a life-span of 1 billion cycles.

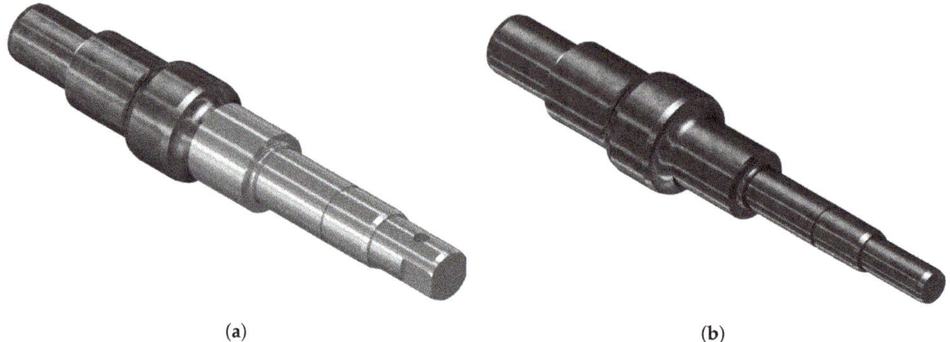

(**a**)   (**b**)

Figure 13. Shaft design for same requirements, where a reduction of 11% in weight is seen for the multi-material shaft (**a**) in comparison to the mono-material one (**b**).

4.4. GPDA for Tailored Forming

In the GPDA implementation for tailored forming, the design elements are carriers of the knowledge that gives the design its shape. In addition, a design catalog [86,87] of the CAEE controls the GPDA models and serves as a superordinate knowledge base. Depending on the application and load case, the knowledge in the catalog determines which skeleton and which design elements should form the basis for the development of the tailored forming component. The more knowledge is available in concrete form, the better the selected starting point and the lower the effort required for subsequent optimization. The design catalog does not consist of a single catalog, but of a general main catalog that refers to concrete detail catalogs. The connection of the catalogs is shown in Figure 14.

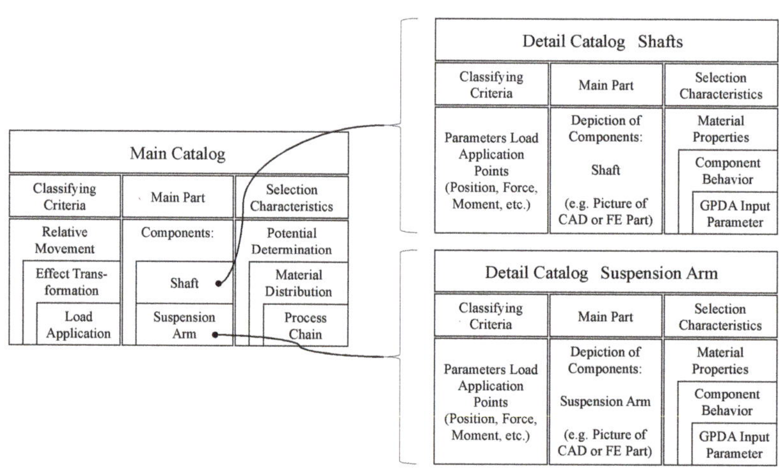

Figure 14. Structure of design catalogs.

Different component types and the corresponding general application and load cases are defined in the main catalog. It shows how a tailored forming implementation for conventional mono-material parts can look like, e.g., by showing the general material distribution according to IZEO. The main catalog also provides the skeleton and thus the basic structure for the GPDA model. For each case in the main catalog, there is a detailed catalog in which concrete characteristics are derived from the general case. Here, concrete values have been assigned to the parameters that describe the load cases and geometry characteristics. In addition, the resulting and relevant component properties such as max. deformation or stress are also stored.

The structure of the GPDA model of the shaft is shown in Figure 15 The skeleton consists of an axis on which the interface geometries are defined. Along the axis, there is a design zone between the interfaces in which the design elements are attached. The design elements are defined in such a way that they represent exactly one shaft step. The leading diameter of each design element is defined by the interface geometry of the skeleton.

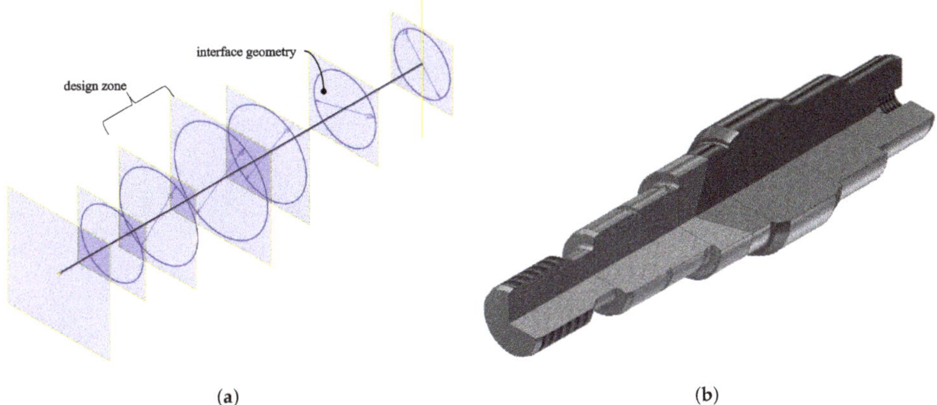

(a)   (b)

Figure 15. Skeleton (**a**) and CAD model (**b**) of tailored foming shafts.

The design elements contain the concrete knowledge of geometry and take into account the manufacturing restrictions and design guidelines. Figure 16 shows, for example, how the relief grooves required on a shaft are implemented in the model. The dimensions of the relief grooves depend directly on the leading diameter of the shaft shoulder and are described according to DIN 509 in Table 3 [88]. Furthermore, the shape of the relief grooves can vary depending on the application. In the case of a relief groove of type F, the definition goes beyond the limits of the design element, so that the geometry in the adjacent element must adapt accordingly. For this case, parameters are already stored in the adjacent design element, which are then filled accordingly via the skeleton. These parameters are suppressed for relief grooves of the type E that do not extend beyond the design zone.

Table 3. Relief groove parameters for shafts according to DIN 509 [88].

R1	f	t1	t2	D1
0.4	2	0.2	0.1	>3...18
0.8	2.5	0.3	0.2	>18...80
1.2	4	0.4	0.3	>80

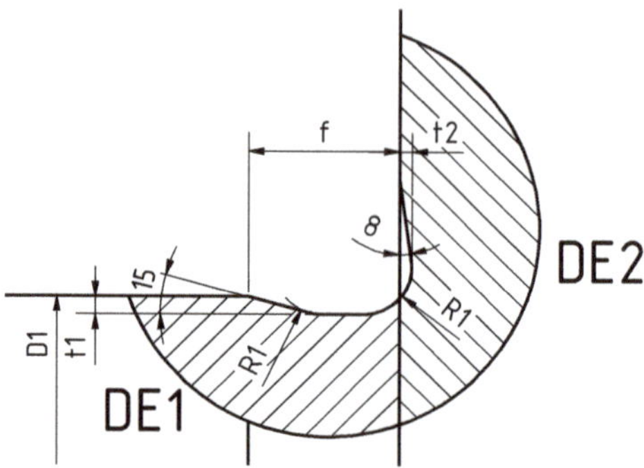

Figure 16. Parameters for a relief groove (type F) that extends over two design elements (DE1 and DE2).

4.5. Application Example of the GPDA: Model Adaptation in Case of Changes in Boundary Conditions

In the GPDA a load case of the shaft is considered as an example, where $F = 5.5$ kN and $T = 40$ Nm. For this load case, the joining zone position from the results of IZEO (Figure 11, result 5) and the shape from the results of CBR are used. In the GPDA model, the joining zone position is the distance from the left shaft end to the center of the joining zone area ($P = 73$ mm; Figure 17a). As can be seen in Figure 17b, the v. Mises stress does not exceed the yield strength of 280 MPa of the aluminum alloy in the relief groove under consideration, with an assumed safety factor of 1.

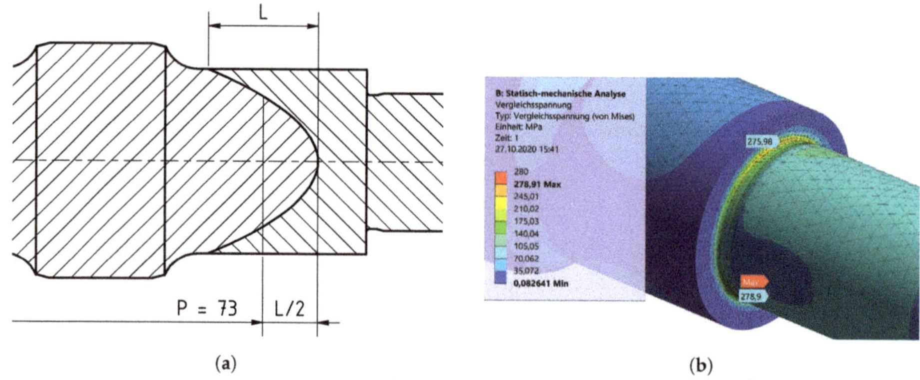

(a)   (b)

Figure 17. (**a**) Joining zone position and (**b**) resulting v. Mises stresses (max. $278.9 \frac{N}{mm^2}$) at $F = 5.5$ kN and $T = 40$ Nm.

If the force is increased at a constant torsional moment, the yield strength is exceeded. Figure 18 shows the case at $F = 8$ kN. To reduce the stresses, the position or geometry of the joining zone must now be adjusted. It is not possible to increase the diameter of the shaft on which the relief groove lies, because the bearing size is determined by the external connection dimensions.

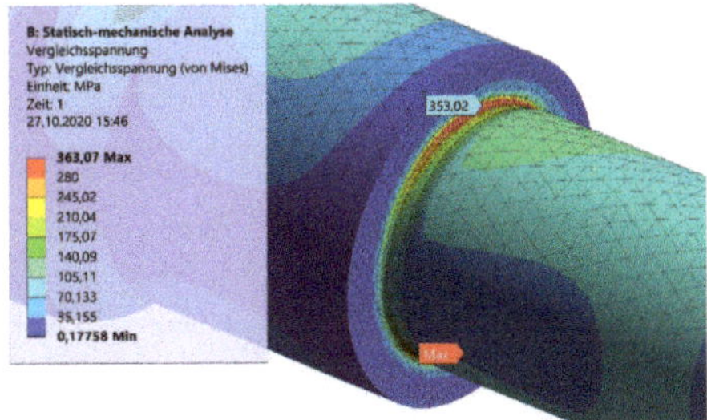

Figure 18. Exceeded yield strength (max. $363.07\frac{N}{mm^2}$) at a joining zone position of $P = 73$ mm at $F = 8$ kN and $T = 40$ Nm.

Therefore, the position of the joining zone is shifted 10 mm to the right to $P = 83$ mm in the following. The yield strength of the aluminum alloy is no longer exceeded in the undercut. Figure 19 shows the new joining zone position (a) and the resulting stresses (b).

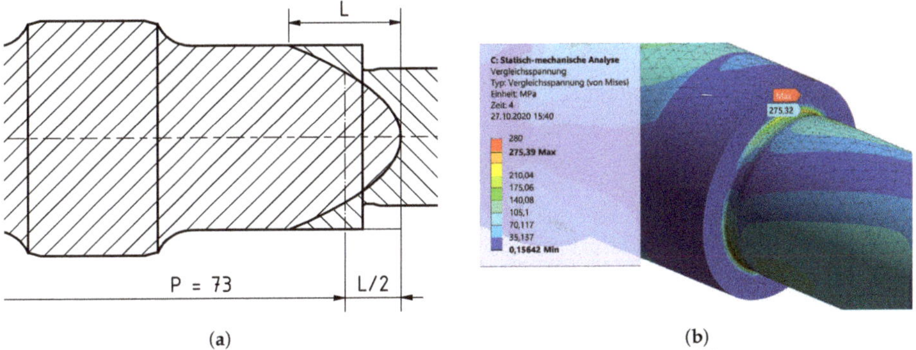

(a)	(b)

Figure 19. (a) Joining zone position (b) and resulting stresses (max. $375.39\frac{N}{mm^2}$) at $F = 8$ kN and $T = 40$ Nm.

Table 4 summarizes the individual results. By increasing the proportion of steel alloy in the component, the weight of the shaft increases from 245.61 g to 264.05 g.

Table 4. v. Mises Stress in the relief groove at different forces.

Position	v. Mises Stress at 5.5 kN	v. Mises Stress at 8 kN	Weight
73 mm	278.9 MPa	353.02 MPa	245.61 g
83 mm	–	275.39 MPa	264.05 g

In this case that the GPDA offers the possibility to move the joining zone over the boundaries of the individual design elements. This increases the proportion of steel and reduces the stresses in the undercut of the aluminum area. Because the model is designed according to the approach of the GPDA, it can be used to develop similar shafts that are exposed to similar load cases. Due to the flexible structure, which is based on the use of the design elements, parametric and topological changes can be made without much effort if they are necessary for another load case under different boundary conditions. The test bench trials required for validation are still pending.

5. Discussion

All in all, it can be said that the computer-aided engineering environment has met the required expectations. On the basis of the given load cases and other boundary conditions like design space, a tailored forming high-performance part was generated. In addition, the restrictions resulting from the manufacturing processes could be fully considered. Furthermore, the CBR system provides a platform for a data-driven development of tailored forming components.

Since the tailored forming process chain is novel, basic research is conducted in CRC 1153. In order to develop controllable manufacturing processes, in the beginning, only simple, rotationally symmetric components were investigated. For these components, the most robust results have been achieved and most knowledge about manufacturing restrictions is known. For these reasons, the shaft presented in this paper is the subject of the investigations on the creation of the CAEE. However, due to the existing load cases, the full tailored forming potential cannot be developed for shafts. Therefore, mirror-symmetric geometries such as rocker arms, which offer a higher tailored forming potential, are currently being investigated in CRC 1153 (Figure 20). Nevertheless, as shown in Section 4.5, there is also tailored forming potential for shafts under certain boundary conditions and these components are therefore also suitable for the development of the CAEE.

Figure 20. Rocker arm (**left**) and derived tailored forming component variants in IZEO (**center**) and CAD (**right**).

Furthermore, it should be noted that all simulations have been carried out with linear-elastic material behavior up to the yield strength, since this describes the limit in which a component can be used in practice. The joining zone is designed as a simple adhesive contact. Within the scope of CRC 1153, special finite elements are being developed that can simulate the material properties of the joining zone [89]. These are currently not yet included in the simulations described here, but will be added in the future.

In addition, test bench trials have been conducted to validate the strength of the joining zone geometries generated by IZEO and Robust Design. An optimized joining zone geometry helps the shaft to withstand higher loads. Analogy tests on simplified shafts have shown that a shaft with optimized joining zone geometry has nearly the same strength as a reference shaft made of the aluminum alloy. With a non-optimized geometry, the shaft fails in the area of the joining zone and the strength is reduced [6].

For future approaches to the development of CAEE, ontology-based approaches are probably more beneficial than the approaches presented in this paper. The ontology would serve as a mediator between the knowledge base and the instantiated CAD model. The result would be a model architecture in which e.g., the design elements could be used much more flexibly. Currently, the parameters of the design elements are hard-coded by the CAD system and are explicitly addressed so that they can practically only be used for a single or similar component.

The challenge with GPDA is that an enormous amount of work is required in advance to generate a functioning model. In order to ensure the modularity of the approach, great care is required in the generation of the skeleton, the interface geometries and the design elements. The creation of a model-free of errors within defined limits requires increased programming effort and a well-planned structure, especially at the beginning. Further degrees of freedom are added in the context of tailored forming by taking the joining zone into account, which must be defined both in the top-level assembly and in each design element. As shown in Figure 19a, the design elements must be controlled by the top-level assembly so that the joining zones form a smooth transition from design element to a design element. However, the work has also shown that the effort for embedding new design elements and new joining zone geometries is reduced the more the GPDA model is built up, since they can be derived from the previously created design elements and can be integrated into the working top-level assembly relatively easily.

Furthermore, there is a significant difference in the programming effort required to implement formal, explicit and informal, implicit knowledge (see Section 2.2). While explicit knowledge can be implemented very easily, e.g., by means of table values and If-Then-Else queries, the translation effort for implicit knowledge is significantly higher and also ties up more computing capacity. However, as IZEO has shown, implementation is quite possible. In summary, it can be said that computer-aided methods can handle explicit knowledge very well, but there is still a need for research on the implementation of implicit knowledge.

6. Summary and Outlook

The desire for components that are always better adapted to external conditions than their predecessors leads to the technological advancement of the components, but also of the processes required for their manufacture. As a result, components and processes are becoming more and more complicated, so that the effort for planning, conception, design and elaboration is increasing. In some cases, components and process chains are already so complicated that the optimal solution is no longer readily apparent. When newer approaches, such as multi-material design, are added, the degrees of freedom to be considered increase even further. Especially in this case, systematic, computer-aided approaches are needed to meet the challenge of finding the best solution from an objective point of view. Therefore, modeling approaches and design methods are needed that take into account the manufacturing processes throughout the entire product development process.

The methodology presented in this work works as a framework to develop the technology of tailored forming further and generate continuously better solutions. As seen, the topology optimization method IZEO was able to handle dynamic manufacturing restrictions while optimizing the use of multi-materials. Additionally, different strategies for solution exploration were presented, such as CBR and GPDA, where the influence of manufacturing is direct. For these reasons, this design methodology is able to support this manufacturing technology to be further developed. This translates into first transfer projects for real industry applications that are being currently performed under the umbrella of the CRC 1153.

All in all, computer-aided engineering environments help to find the optimum shape for a component in order to derive the best possible manufacturing process. The stored knowledge base provides a clear and objective set of rules that can protect companies from undesirable developments. This provides a better starting point for the development of components and processes. Routine processes can also be automated, giving designers more room for creative work.

In further developments, additional non-rotationally symmetric components will be investigated and developed. For this purpose, complicated manufacturing restrictions have to be implemented for IZEO. The GPDA also needs skeletons and design elements with more complicated shapes and extended functionalities. For example, the skeleton will no longer be one-dimensional, but two or three dimensional. The design elements may have more than two neighboring elements. In order to better link product development with process development in the future, a transfer model is currently being worked on within the framework of CRC 1153, which will allow conclusions to be drawn about the upstream production stages. For this purpose, a GPDA model is currently being developed, which, depending on the manufacturing process, can map the individual stages of component production. In this case, the research results from the CRC will also serve as a basis.

Author Contributions: Conceptualization, R.S. and T.B.; methodology, R.S. and T.B.; software, R.S. and T.B.; validation, R.S., T.B. and P.C.G.; formal analysis, R.S. and T.B.; investigation, R.S. and T.B.; resources, R.L.; data curation, R.L.; writing—original draft preparation, R.S., T.B. and P.C.G.; writing—review and editing, T.B. and P.C.G.; visualization, R.S. and T.B.; supervision, P.C.G., I.M. and R.L.; project administration, P.C.G., I.M. and R.L.; funding acquisition, I.M. and R.L. All authors have read and agreed to the published version of the manuscript.

Funding: This research was funded by the Deutsche Forschungsgemeinschaft (DFG, German Research Foundation) grant number 252662854.

Acknowledgments: The results presented in this paper were obtained within the subproject C2 "Configuration and design of hybrid solids" of the Collaborative Research Center 1153 "Process chain to produce hybrid high performance components by Tailored Forming". The authors would like to thank the German Research Foundation (DFG) for the financial and organizational support of this project.

Conflicts of Interest: The authors declare no conflict of interest.

Abbreviations

The following abbreviations are used in this manuscript:

BESO	Bidirectional Evolutionary Structure Optimization
CAD	Computer-aided Design
CAEE	Computer-aided Engineering Environment
CBR	Case-based Reasoning
CPM	Characteristics-Properties Modeling
CRC	Collaborative Research Center
DfX	Design for X
GPDA	Generative Parametric Design Approach
IZEO	Interfacial Zone Evolutionary Optimization
KBE	Knowledge-based Engineering
LACE	Lateral Angular Co-Extrusion
PDD	Property-Driven Development

References

1. Behrens, B.A.; Chugreev, A.; Matthias, T.; Poll, G.; Pape, F.; Coors, T.; Hassel, T.; Maier, H.; Mildebrath, M. Manufacturing and Evaluation of Multi-Material Axial-Bearing Washers by Tailored Forming. *Metals* **2019**, *9*, 232.
2. Behrens, B.A.; Bouguecha, A.; Frischkorn, C.; Huskic, A.; Stakhieva, A.; Duran, D. Tailored Forming Technology for Three Dimensional Components: Approaches to Heating and Forming. In Proceedings of the 5th International Conference on Thermomechanical Processing, Milan, Italy, 6–28 October 2016.
3. Denkena, B.; Bergmann, B.; Witt, M. Automatic process parameter adaption for a hybrid workpiececeduring cylindrical operations. *Int. J. Adv. Manuf. Technol.* **2017**, *95*, 311–316.
4. Ashby, M.; Cebon, D. Materials selection in mechanical design. *J. Phys. IV* **1993**, *3*, 1–9.
5. Brockmöller, T.; Gembarski, P.; Mozgova, I.; Lachmayer, R. Design Catalogue in a CAE Environment for the Illustration of Tailored Forming. In Proceedings of the 59th Ilmenau Scientific Colloquium, Technische Universität Ilmenau, Ilmenau, Germany, 11–15 September 2017.
6. Siqueira, R. Design and Optimization Method for Manufacturable Multi-material Components. Ph. D Thesis, Leibniz Universität Hannover, Hannover, Germany, 2019.
7. Cutkosky, M.; Sangbae, K. Design and fabrication of multi-material structures for bioinspired robots. *Philos. Trans. R. Soc. Lond. Ser. A* **2009**, *367*, 1799–1813.
8. Altach, J.; Bader, B.; Fröhlich, T.; Klaiber, D.; Vietor, T. Approach to the systematic categorization and qualitative evaluation of multi-material designs for use in vehicle body structures. *Procedia CIRP* **2019**, *84*, 908–915.
9. Kleemann, S.; Fröhlich, T.; Türck, E.; Vietor, T. A methodological approach towards multi-material design of automotive components. *Procedia CIRP* **2017**, *60*, 68–73.
10. Gouker, R.; Gupta, S.; Bruck, H.; Holzschuh, T. Manufacturing of multi-material compliant mechanisms using multi-material molding. *Int. J. Adv. Manuf. Technol.* **2006**, *30*, 1049–1075.

11. Behrens, B.A.; Breidenstein, B.; Duran, D.; Herbst, S.; Lachmayer, R.; Löhnert, S.; Matthias, T.; Mozgova, I.; Nürnberger, F.; Prasanthan, V.; et al. Simulation-Aided Process Chain Design for the Manufacturing of Hybrid Shafts. *J. Heat Treatm. Mater.* **2019**, *74*, 115–135.

12. Suh, N. *Complexity: Theory and Applications*; Oxford University Press: New York, NY, USA, 2005.

13. Gembarski, P.; Lachmayer, R. Complexity Management of Solution Spaces in Mass Customization. In Proceedings of the 8th International Conference on Mass Customization and Personalization-Community of Europe (MCP-CE 2018), Novi Sad, Serbien, 19–21 September 2018, pp. 123–131.

14. Gembarski, P.; Lachmayer, R. Solution space development: conceptual reflections and development of the parameter space matrix as planning tool for geometry-based solution spaces. *Int. J. Ind. Eng. Manag.* **2018**, *3*, 4.

15. Vajna, S.; Weber, C.; Zeman, K.; Hehenberger, P.; Gerhard, D.; Wartzack, S. *CAx für Ingenieure—Eine Praxisbezogenen Einführung*, 3rd ed.; Springer: Berlin, Germany, 2018.

16. Huang, G.Q. *Design for X: Concurrent Engineering Imperatives*; Springer Science+Business Media: Dordrecht, The Netherlands, 1996.

17. Boothroyd, G. Product design for manufacture and assembly. *Comput. Aided Des.* **1994**, *26*, 505–520.

18. Behrens, B.A.; Goldstein, R.; Guisbert, D.; Duran, D. Thermomechanical Processing of Friction Welded Steel-Aluminum Billets to Improve Joining Zone Properties. In Proceedings of the 4th International Conference on Heat Treatment and Surface Engineering in Automotive Applications (Thermal Processing in Motion), Spartanburg, SC, USA, 5–7 June 2018.

19. Denkena, B.; Bergmann, B.; Breidenstein, B.; Prasanthan, V.; Witt, M. Analysis of potentials to improve the machining of hybrid workpieces. *Prod. Eng.* **2018**, *13*, 11–19. [CrossRef]

20. Gembarski, P.C. Komplexitätsmanagement mittels wissensbasiertem CAD—Ein Ansatz zum unternehmenstypologischen Management konstruktiver Lösungsräume. Ph. D Thesis, Leibniz Universität Hannover, Hannover, Germany, 2018.

21. Biedermann, M.; Meboldt, M. Computational design synthesis of additive manufactured multi-flow nozzles. *Addit. Manuf.* **2020**, *35*, 1–9. [CrossRef]

22. Li, H. Generative Design Approach for Robust Solution Development. Ph. D Thesis, Leibniz Universität Hannover, Hannover, Germany, 2020.

23. Stokes, M. *Managing Engineering Knowledge - MOKA: Methodology for Knowledge Based Engineering Applications*; Professional Engineering Publishing Limited: London, UK, 2001.

24. Kulon, J.; Broomhead, P.; Mynors, D. Applying knowledge-based engineering to traditional manufacturing design. *Int. J. Adv. Manuf. Technol.* **2006**, *30*, 945–951. [CrossRef]

25. Hopgood, A.A. *Intelligent Systems for Engineers and Scientists*; CRC Press: Boca Raton, FL, USA, 2016.

26. Koch, S.; Behrens, B.A.; Hübner, S.; Scheffler, R.; Wrobel, G.; Peßow, M.; Bauer, D. 3D CAD modeling of deep drawing tools based on a new graphical language. *Comput. Aided Des. Appl.* **2018**, *15*, 619–630. [CrossRef]

27. Gembarski, P.C. On the Conception of a Multi-agent Analysis and Optimization Tool for Mechanical Engineering Parts. In *Agents and Multi-Agent Systems: Technologies and Applications 2020*; Springer: Berlin/Heidelberg, Germany, 2020; pp. 93–102.

28. Kim, H.; Altan, T.; Sevenler, K. Computer-aided part and processing-sequence design in cold forging. *J. Mater. Process. Technol.* **1992**, *33*, 57–74. [CrossRef]

29. Caporalli, Â.; Gileno, L.; Button, S. Expert system for hot forging design. *J. Mater. Process. Technol.* **1998**, *80–81*, 131–135. [CrossRef]

30. Boyle, I.; Rong, Y.; Brown, D.C. A review and analysis of current computer-aided fixture design approaches. *Rob. Comput. Integr. Manuf.* **2011**, *27*, 1–12. [CrossRef]

31. Findik, F. Recent developments in explosive welding. *Mater. Des.* **2010**, *32*, 1081–1093. [CrossRef]

32. Han, J.; Ahn, J.; Shin, M. Effect of interlayer thickness on shear deformation behavior of AA5083 aluminum alloy/SS41 steel plates manufactured by explosive welding. *J. Mater. Sci.* **2003**, *38*, 13–18. [CrossRef]

33. Hirtler, M.; Jedynak, A.; Sydow, B.; Sviridov, A.; Bambach, M. A Study on the Mechanical Properties of Hybrid Parts Manufactured by Forging and Wire Arc Additive Manufacturing. *Procedia Manuf.* **2020**, *47*, 1141–1148. [CrossRef]

34. Vaezi, M.; Chianrabutra, S.; Mellor, B.; Yang, S. Multiple material additive manufacturing-Part 1: A review. *Virtual Phys. Prototyp.* **2013**, *8*, 19–50. [CrossRef]

35. Leiber, R. Hybridschmieden bringt den Leichtbau voran. *Alum. Prax.* **2011**, *7*, 7–8.

36. Kriwall, M.; Stonis, M.; Bick, T.; Treutler, K.; Wesling, V. Dependence of the Joint Strength on Different Forming Steps and Geometry in Hybrid Compound Forging of Bulk Aluminum Parts and Steel Sheets. *Procedia Manuf.* **2020**, *47*, 356–361. [CrossRef]

37. Behrens, B.A.; Chugreev, A.; Selinski, M.; Matthias, T. Joining zone shape optimisation for hybrid components made of aluminium-steel by geometrically adapted joining surfaces in the friction welding process. *AIP Conf. Proc.* **2019**, *2113*, 040027.

38. Maalekian, M. Friction welding - critical assessment of literature. *Sci. Technol. Weld. Join.* **2007**, *12*, 738–759. [CrossRef]

39. Nothdurft, S.; Springer, A.; Kaierle, S.; Ohrdes, H.; Twiefel, J.; Wallaschek, J.; Mildebrath, M.; Maier, H.; Hassel, T.; Overmeyer, L. Laser welding of dissimilar low-alloyed steel-steel butt joints and the effects of beam position and ultrasound excitation on the microstructure. *J. Laser Appl.* **2018**, *30*, 032417. [CrossRef]

40. Chugreeva, A.; Mildebarth, M.; Diefenbach, J.; Barroi, A.; Lammers, M.; Hermsdorf, J.; Hassel, T.; Overmeyer, L.; Behrens, B.A. Manufacturing of High-Performance Bi-Metal Bevel Gears by Combined Deposition Welding and Forging. *Metals* **2018**, *8*, 898. [CrossRef]

41. Behrens, B.A.; Klose, C.; Chugreev, A.; Heimes, N.; Thürer, S.; Uhe, J. A Numerical Study on Co-Extrusion to Produce Coaxial Aluminum-Steel Compounds with Longitudinal Weld Seams. *Metals* **2018**, *8*, 717. [CrossRef]

42. Kruse, J.; Jagodinski, A.; Langner, J.; Stonis, M.; Behrens, B.A. Investigation of the joining zone displacement of cross-wedge rolled serially arranged hybrid parts. *Int. J. Mater. Form.* **2019**, 1–13, doi:10.1007/s12289-019-01494-3. [CrossRef]

43. Behrens, B.A.; Bonhage, M.; Bohr, D.; Duran, D. Simulation Assisted Process Development for Tailored Forming. *Mater. Sci. Forum* **2019**, *949*, 101–111. [CrossRef]

44. Herbst, S.; Maier, H.; Nürnberger, F. Strategies for the Heat Treatment of Steel-Aluminium Hybrid Components. *J. Heat Treatm. Mater.* **2018**, *73*, 268–282. [CrossRef]

45. Behrens, B.A.; Chugreev, A.; Matthias, T.; Nothdurft, S.; Hermsdorf, J.; Kaierle, S.; Ohrdes, H.; Twiefel, J.; Wallaschek, J.; Mildebrath, M.; et al. Investigation of the composite strength of hybrid steel-steel semi-finished products manufactured by laser beam welding and friction welding. *IOP Conf. Ser. Mater. Sci. Eng.* **2018**, *461*, 012049. [CrossRef]

46. Behrens, B.A.; Bouguecha, A.; Vucetic, M.; Peshekhodov, I.; Matthias, T.; Kolbasnikov, N.; Sokolov, S.; Ganin, S. Experimental investigations on the state of the friction-welded joint zone in steel hybrid components after process-relevant thermomechanical loadings. *AIP Conf. Proc.* **2016**, *1769*, 130013.

47. Behrens, B.A.; Chugreev, A.; Matthias, T. Characterisation of the joining zone of serially arranged hybrid semi-finished components. *AIP Conf. Proc.* **2018**, *1960*, 040002.

48. Goldstein, R.; Behrens, B.A. Role of Thermal Processing in Tailored Forming Technology for Manufacturing Multi-Material Components. In Proceedings of the 29th ASM Heat Treating Society Conference, Columbus, OH, USA, 24–26 October 2017.

49. Behrens, B.A.; Bouguecha, A.; Frischkorn, C.; Chugreeva, A.; Duran, D. Angepasste Erwärmungs- und Umformverfahren für hybride Massivbauteile 2016. In Proceedings of the 23th Sächsische Fachtagung Umformtechnik (SFU), Dresden, Germany, 7–8 December 2016; pp. 184–193.

50. Behrens, B.A.; Amiri, A.; Duran, D.; Nothdurft, S.; Hermsdorf, J.S.; Kaierle, S.; Ohrdes, H.; Wallaschek, J.; Hassel, T. Improving the mechanical properties of laser beam welded hybrid workpieces by deformation processing. *AIP Conf. Proc.* **2019**, *2113*, 040025.

51. Denkena, B.; Breidenstein, B.; Prasanthan, V. Influence of Tool Properties on the Thermomechanical Load during Turning of Hybrid Components and the Resulting Surface Properties. *J. Heat Treatm. Mat.* **2018**, *73*, 223–231. [CrossRef]

52. Breidenstein, B.; Denkena, B.; Meyer, K.; Prasanthan, V. Influence of subsurface properties on the application behavior of hybrid components. *Procedia CIRP* **2020**, *87*, 302–308. [CrossRef]

53. Lindemann, U. *Methodische Entwicklung Technischer Produkte: Methoden Flexibel und Situationsgerecht Anwenden*; Springer: Berlin/Heidelberg, Germany, 2009.

54. Pahl, G.; Beitz, W.; Feldhusen, J.; Grote, K. *Engineering Design: A Systematic Approach*; Springer: London, UK, 2007.

55. Chandrasegaran, S.; Ramani, K.; Sriram, R.; Horváth, I.; Bernard, A.; Harik, R.; Gao, W. The evolution, challenges, and future of knowledge representation in product design systems. *Comput. Aided Des.* **2013**, *45*, 204–228. [CrossRef]

56. Vajna, S.; Weber, C.; Bley, H.; Zeman, K. *Integrated Design Engineering*; Springer-Verlag: Berlin/Heidelberg, Germany, 2014.

57. Juan, Y.C.; Ou-Yang, C.; Lin, J.S. A process-oriented multi-agent system development approach to support the cooperation-activities of concurrent new product development. *Comput. Ind. Eng.* **2009**, *57*, 1363–1376. [CrossRef]

58. Stroud, I.; Nagy, H. *Solid Modelling and CAD Systems: How to Survive a CAD System*; Springer Science & Business Media: Berlin/Heidelberg, Germany, 2011.

59. Riesenfeld, R.; Haimes, R.; Cohen, E. Initiating a CAD renaissance: Multidisciplinary analysis driven design: Framework for a new generation of advanced computational design, engineering and manufacturing environments. *Comput. Methods Appl. Mech. Eng.* **2015**, *284*, 1054–1072. [CrossRef]

60. Shah, J. Designing with parametric cad: Classification and comparison of construction techniques Geometric Modelling. In *Proceedings of the International Workshop on Geometric Modelling*; Springer: Berlin/Heidelberg, Germany, 2001; pp. 53–68.

61. Weber, C. Looking at DFX and Product Maturity from the Perspective of a New Approach to Modelling Product and Product Development Processes. In *The Future of Product Development*; Krause, F.L., Ed.; Springer: Berlin/Heidelberg, Germany, 2007; pp. 85–104.

62. Verhagen, W.; Bermell-Garcia, P.; van Dijk, R.; Curran, R. A critical review of Knowledge-Based Engineering: An identification of research challenges. *Adv. Eng. Inf.* **2012**, *26*, 5–15. [CrossRef]

63. LaRocca, G. Knowledge based Engineering: Between AI and CAD. Review of a Language based Technology to Support Engineering Design. *Adv. Eng. Inf.* **2012**, *26*, 159–179.

64. Milton, N. *Knowledge Technologies*; Polimetrica S.a.s.: Monza, Italy, 2008; Volume 3.

65. Gembarski, P.C.; Li, H.; Lachmayer, R. Template-based modelling of structural components. *Int. J. Mech. Eng. Robot. Res.* **2017**, *6*, 336–342. [CrossRef]

66. Plappert, S.; Hoppe, L.; Gembarski, P.; Lachmayer, R. Application of Knowledge-Based Engineering for Teaching Design Knowledge to Design Students. In *Proceedings of the Design Society: DESIGN Conference*; Cambridge University Press: Cambridge, UK, 2020; Volume 1, pp. 1795–1804.

67. Hirz, M.; Dietrich, W.; Gfrerrer, A.; Lang, J. *Integrated Computer-Aided Design in Automotive Development*, 1st ed.; Springer: Berlin/ Heidelberg, Germany, 2013.

68. Skarka, W. Application of MOKA methodology in generative model creation using CATIA. *Eng. Appl. Artif. Intell.* **2007**, *20*, 677–690. [CrossRef]

69. Wang, H.; Rong, Y. Case based reasoning method for computer aided welding fixture design. *Comput. Aided Des.* **2008**, *40*, 1121–1132. [CrossRef]

70. Alacrón, R.; Chueco, J.; Garcia, J.; Idoipe, A. Fixture knowledge model development and implementation based on a functional design approach. *Rob. Comput. Integr. Manuf.* **2010**, *26*, 56–66.

71. LaRocca, G.; van Tooren, M. Knowledge-based engineering to support aircraft multidisciplinary design and optimization. *Proc. Inst. Mech. Eng. Part G J. Aerosp. Eng.* **2010**, *224*, 1041–1055. [CrossRef]

72. Roach, G.M.; Cox, J.J. A Case Study of the Product Design Generator. In *Product Platform and Product Family Design*; Springer: New York, NY, USA, 2006; pp. 499–512.

73. Andrae, R.; Köhler, P. Simulation-oriented Transformation of CAD Models. *Comput. Aided Des. Appl.* **2016**, *13*, 340–347. [CrossRef]

74. Hvam, L.; Mortensen, N.H.; Riis, J. *Product Customization*; Springer Science & Business Media: Berlin/Heidelberg, Germany, 2008.

75. Sauthoff, B. Generative parametrische Modellierung von Strukturkomponenten für die Technische Vererbung. Ph. D Thesis, Leibniz Universität Hannover, Hannover, Germany, 2017.

76. Cuillière, J.C.; François, V.; Nana, A. Automatic construction of structural CAD models from 3D topology optimization. *Comput. Aided Des. Appl.* **2018**, *15*, 107–121. [CrossRef]

77. Brockmöller, T.; Mozgova, I.; Lachmayer, R. An Approach to analyse the Potential of Tailored Forming by TRIZ-Reverse. In *21st International Conference on Engineering Design*; University of British Columbia: Vancouver, BC, Canada, 2017.

78. Siqueira, R.; Mozgova, I.; Lachmayer, R. Development of a Topology Optimization Method for Tailored Forming Multi-Material Design. In Proceedings of the 24th ABCM International Congress of Mechanical Engineering, Curitiba, Brasil, 3–8 December 2017.

79. Vatanabe, S.; Lippi, T.; de Lima, C.; Paulino, G.; Silva, E. Topology optimization with manufacturing constraints: A unified projection-based approach. *Adv. Eng. Softw.* **2016**, *100*, 97–112. [CrossRef]

80. Yang, X.; Xie, Y.; Steven, G.; Querin, O. Bidirectional Evolutionary Method for Stiffness Optimization. *AIAA J.* **1999**, *37*, 1483–1488. [CrossRef]

81. Siqueira, R.; Mozgova, I.; Lachmayer, R. An Interfacial Zone Evolutionary Optimization Method with Manufacturing Constraints for Hybrid Components. *J. Comput. Des. Eng.* **2018**, *6*, 387–397.

82. Sauthoff, B.; Lachmayer, R. Generative Design Approach for Modelling of Large Design Spaces. In *Proceedings of the 7th World Conference on Mass Customization, Personalization and Co-Creation (MCPC 2014)*; Brunoe, T., Nielsen, K., Joergensen, K., Taps, S., Eds.; Springer: Berlin/Heidelberg, Germany; New York, NY, USA; Dordrecht, The Netherlands; London, UK, 2014; pp. 241–251.

83. Li, H.; Brockmöller, T.; Gembarski, P.; Lachmayer, R. An Investigation of a Generative Parametric Design Approach for a Robust Solution Development. In *Proceedings of the Design Society: DESIGN Conference*; Cambridge University Press: Cambridge, UK, 2020; Volume 1, pp. 315–324.

84. Brockmöller, T.; Siqueira, R.; Mozgova, I.; Lachmayer, R. Rechnergestützte Entwicklungsumgebung zur Konstruktion von Tailored-Forming-Bauteilen. In *DS 98: Proceedings of the 30th Symposium Design for X (DFX 2019)*; Krause, D., Paetzold, K., Wartzack, S., Eds.; TuTech Verlag: Hamburg, Germany, 2019; pp. 195–206.

85. Gembarski, P.C.; Sauthoff, B.; Brockmöller, T.; Lachmayer, R. Operationalization of Manufacturing Restrictions for CAD and KBE-Systems. In *Proceedings of the DESIGN 2016, 14th International Design Conference*; Marjanović, D., Ed.; Tools, Practice & Innovation; The Design Society: Glasgow, UK, 2016; Volume 2, pp. 621–630.

86. Roth, K. Design catalogues and their usage. In *Engineering Design Synthesis*; Chakrabarti, A., Ed.; Springer: London, UK, 2002.

87. Franke, H.J.; Löffler, S.; Deimel, M. Increasing the Efficiency of Design Catalogues by Using Modern Data Processing Technologies. In Proceedings of the DS 32: Proceedings of the DESIGN 2004, 8th International Design Conference, Dubrovnik, Croatia, 18–21 May 2004.

88. DIN509:2006-12. *Technical Drawings—Relief Grooves—Types and Dimensions—Tolerances*; Beuth: Berlin, Germany, 2006.

89. Töller, F.; Löhnert, S.; Wriggers, P. Bulk material models in Cohesive Zone Elements for simulation of joining zones. *Finite Elem. Anal. Des.* **2019**, *164*, 42–54. [CrossRef]

Publisher's Note: MDPI stays neutral with regard to jurisdictional claims in published maps and institutional affiliations.

Article

Numerical Simulation and Experimental Validation of the Cladding Material Distribution of Hybrid Semi-Finished Products Produced by Deposition Welding and Cross-Wedge Rolling

Jens Kruse [1,*], Maximilian Mildebrath [2], Laura Budde [3], Timm Coors [4], Mohamad Yusuf Faqiri [2], Alexander Barroi [3], Malte Stonis [1], Thomas Hassel [2], Florian Pape [4], Marius Lammers [3], Jörg Hermsdorf [3], Stefan Kaierle [3], Ludger Overmeyer [3] and Gerhard Poll [4]

[1]   Institut für Integrierte Produktion Hannover gGmbH, Hollerithallee 6, 30419 Hannover, Germany; stonis@iph-hannover.de

[2]   Leibniz Universität Hannover, Institute of Materials Science, An der Universität 2, 30823 Garbsen, Germany; mildebrath@iw.uni-hannover.de (M.M.); faqiri@iw.uni-hannover.de (M.Y.F.); hassel@iw.uni-hannover.de (T.H.)

[3]   Laser Zentrum Hannover e.V., Hollerithallee 8, 30419 Hannover, Germany; l.budde@lzh.de (L.B.); a.barroi@lzh.de (A.B.); m.lammers@lzh.de (M.L.); j.hermsdorf@lzh.de (J.H.); s.kaierle@lzh.de (S.K.); l.overmeyer@lzh.de (L.O.)

[4]   Leibniz University Hannover, Institute of Machine Design and Tribology, An der Universität 1, 30823 Garbsen, Germany; coors@imkt.uni-hannover.de (T.C.); pape@imkt.uni-hannover.de (F.P.); poll@imkt.uni-hannover.de (G.P.)

*   Correspondence: kruse@iph-hannover.de; Tel.: +49-511-27976-0

Received: 25 August 2020; Accepted: 2 October 2020; Published: 6 October 2020

Abstract: The service life of rolling contacts is dependent on many factors. The choice of materials in particular has a major influence on when, for example, a ball bearing may fail. Within an exemplary process chain for the production of hybrid high-performance components through tailored forming, hybrid solid components made of at least two different steel alloys are investigated. The aim is to create parts that have improved properties compared to monolithic parts of the same geometry. In order to achieve this, several materials are joined prior to a forming operation. In this work, hybrid shafts created by either plasma (PTA) or laser metal deposition (LMD-W) welding are formed via cross-wedge rolling (CWR) to investigate the resulting thickness of the material deposited in the area of the bearing seat. Additionally, finite element analysis (FEA) simulations of the CWR process are compared with experimental CWR results to validate the coating thickness estimation done via simulation. This allows for more accurate predictions of the cladding material geometry after CWR, and the desired welding seam geometry can be selected by calculating the cladding thickness via CWR simulation.

Keywords: tailored forming; cross-wedge rolling; welding; PTA; LMD-W; forming; rolling; coating

---

1. Introduction

Many technical applications place partly contradictory demands on the construction material used. For example, the parts may need to be lightweight, low-cost, and at the same time, increase performance. Material selection then becomes a multi-criteria decision problem in which factors such as production, costs, or environmental effects are considered [1–3]. The goals of the engineer may be conflicting [4]. The combination of multiple material types in a single component has the potential to achieve several goals at once, for example load bearing capacity can be varied through local adaptation of material properties to reduce the overall amount of material required. Additive manufacturing, casting, and welding are currently used for this purpose [5–7].

In sheet metal forming, the use of welded sheet blanks, also known as tailored welded blanks, has been industrially established for over 20 years [8,9]. They are used in large quantities in the automotive industry for the production of body parts such as doors and side members [7]. However, the technology for this type of processing cannot directly be applied to a multi-material component with a three-dimensional geometry.

1.1. Tailored Forming Approach

Tailored forming involves deposition welding (see Section 2.1 and Section 2.2) and subsequent forming (see Section 2.3) to produce multi-material solid components [10]. An exemplary component of this process chain is shown in Figure 1 (left). Here, a multi-material shaft consisting of a cladding material (red) and a base material (blue) is shown. The deposition-welded part of the shaft (red) serves as a bearing race for a cylindrical roller bearing (CRB, yellow). Due to an external radial load on the CRB, high-contact normal stress exceeding 2 GPa [11,12] occurs in the contact between the rolling element and the shaft. According to the Hertz–Belyaev theory [13,14], the maximum equivalent stress occurs in a material volume below the stressed surface. When a loaded rolling element rolls over a point on the raceway surface, the maximum shear stress in the subsurface varies between 0 and τ_{max}, see Figure 1, right. This cyclic loading of the material volume initiates and propagates fatigue cracks in the high-cycle regime ($>>10^6$ cycles), which is referred to as rolling contact fatigue (RCF) [15]. RCF eventually leads to material removal and, if a crack propagates to the surface and forms a chip/pitting, failure of the component. Lundberg and Palmgren [16] assumed the maximum orthogonal shear stress τ_O to be significant in causing fatigue failure. Other authors consider the von Mises–Hencky distortion energy theory [17] and the scalar von Mises stress to be better for predicting RCF failure, with the latter being directly proportional to the octahedral shear stress τ_{oct}. Figure 1, right, shows that the maximum shear stress occurs at depth z of approximately 0.5b to 0.8b. Here, RCF occurs in a highly localized volume of stressed material, so a high-strength material is required there. The remaining part of the component can be made of a less solid material with higher ductility and lower price.

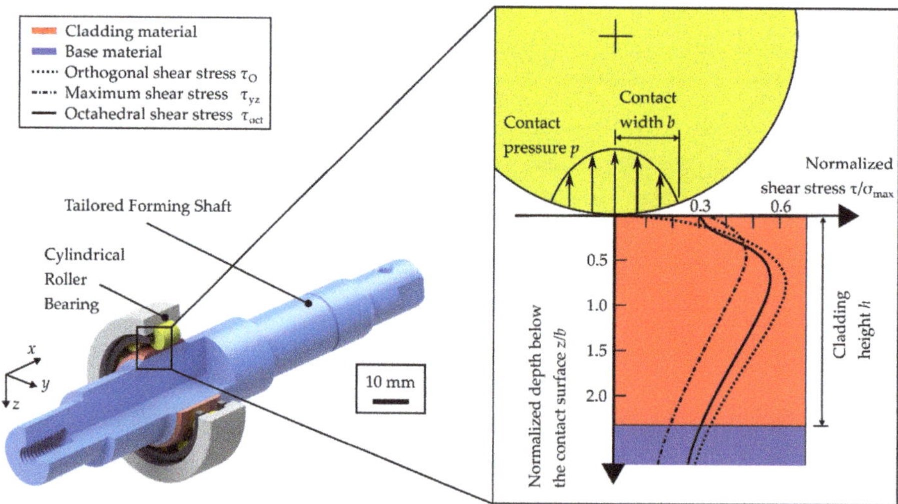

Figure 1. Partial section of tailored forming shaft with mounted cylindrical roller bearing (**left**) and loaded material volume (**right**).

1.2. Welding and Forming

Deposition welding is particularly suitable as a joining step, as it creates a substance-to-substance bond between the cladding material and base material. As already mentioned, the maximum stress zone of a roller bearing is under the contact surface. Surface hardening processes such as nitriding [18] do not reach the necessary layer thicknesses to withstand the expected loads. Through deposition welding, it is possible to produce a cladding layer, which can be adjusted to the applied forces and stresses by alloy design and thickness selection. It is also possible to deposit many kinds of metallic alloys, sometimes even those considered non-weldable because of their high carbon equivalent values (CEV). Further advantages of deposition welding are a low dilution rate and a high degree of automation [19].

The chosen deposition welding procedures are laser metal deposition welding with wire (laser metal deposition welding (LMD-W)) and plasma powder transferred arc welding (PTA). The welding processes were chosen for their different advantages. Laser metal deposition welding allows precise control of the seam geometry and produces low material dilution. PTA welding, on the other hand, allows significantly higher deposition rates but also introduces more heat into the components.

In laser metal-wire deposition welding, a wire is preheated using conductive heating and the laser radiation provides the energy necessary for melting the wire and the substrate material. The preheating of the wire reduces the required laser power, which leads to less heating of the substrate and thus to lower degrees of dilution between base material and applied material. Additional advantages of laser hot-wire processes are a better process stability and less sensitivity to tolerances in wire alignment. The low degrees of dilution ensure high quality of layers, so in most cases, the desired chemical or physical properties, such as corrosion resistance, wear resistance, and hardness, are achieved from the very first layer [20–22].

Plasma powder transferred arc welding is a thermal process for applying wear- and corrosion-resistant layers on surfaces of metallic materials. During the PTA welding process, a tungsten electrode creates a plasma arc with high energy density, which melts the surface of the base material. At the same time, the cladding material in powder form is streamed into the arc and molten. During solidification, a substance-to-substance bond between the cladding material and the base material is created. The whole welding process is performed with argon as the shielding gas.

In order to further enhance material- and process-related advantages, a subsequent forming of the joined semi-finished product is necessary.

It is possible to simulate forming processes in addition to welding processes with FEM. The prediction of the resulting heat flows is of particular importance. The mapping of heat flows allows an estimation of the heat-affected zones and the resulting residual stresses in the material. This is important, because it allows for prediction of component distortion. An example of how this can be achieved is given by Lostado Lorza et al. [23]. Due to the production process of the components used in this work, a simulation of the welding influence is not necessary. As Blohm et al. [24] showed, a hot forming process after a welding process could completely eliminate the heat effects of welding on the steel structure. The reason for this is the complete re-austenitization of the microstructure with subsequent microstructure formation, which neutralizes previous residual stresses. As shown in Figure 2, this microstructure transformation can also be observed in the samples of this work. Figure 2a shows the microstructure of the base material C22.8 after welding. A classical Widmanstätten microstructure of long ferrite needles can be seen, and the impact of the heat from welding is obvious. The 100Cr6 is solidified in a pearlitic structure and is located in the darker region in the upper part of the picture. After cross-wedge rolling (CWR), the microstructure in the base material and in the coating material is much finer-grained than after welding and corresponds to a typical microstructure from forming (Figure 2b). In both microstructure states, a defined orientation of the grains or inhomogeneity cannot be identified, so for the simulation of the forming process, isotropic forming properties are assumed.

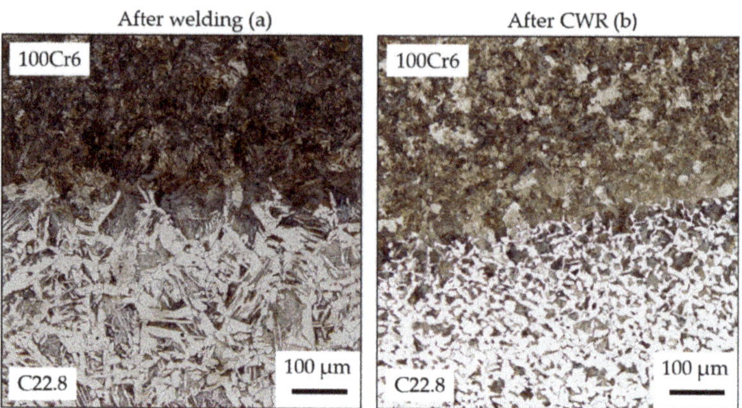

Figure 2. Microstructure of the joining zone after welding (**a**) and after cross-wedge rolling (CWR) (**b**).

1.3. Bearing Fatigue Life of Tailored Forming Machine Elements

Previous investigations [25,26] have shown that the performance of tailored forming machine elements, in particular their fatigue life, is dependent on the layer height of the cladding material. The cladding height h is defined as the radial distance between the rolling element/shaft contact point at one end, and the cladding layer/base material interface at the other, see Figure 1 (right).

Figure 3 shows the calculated bearing fatigue life for different radial loadings and cladding heights with test data for a monolithic specimen as reference [26]. Here, fatigue-life simulations were carried out for a radial loading on the tested CRB of $F_{rad} = 2$ kN with a resulting Hertzian contact pressure of $p_{max} = 1.8$ GPa. The calculated bearing life L_{50}, where 50% of the specimens are expected to have failed, is $L_{50} = 23.5 \times 10^6$ revolutions. This is within an error margin of 16% of the bearing fatigue life from the experimental studies. The basic trends of the calculated probability of survival are represented by the experimental values L_{10} and L_{63}. A too-thin cladding layer reduces fatigue life of multi-material machine elements by a factor of 3. With a cladding height of $h > 0.5$ mm the difference in fatigue life compared to monolithic parts is within a 15% margin. These preliminary results show that a minimum cladding height in dependence of the load is necessary to achieve the same fatigue life as a monolithic component.

The approach for hybrid roller bearings is particularly interesting for large scales. Large-diameter bearings are currently manufactured from classic forging materials such as 42CrMo4, which are not specifically designed for roller bearing applications. By coating the running surfaces with a small amount of a high-performance material, it would be possible to increase the service life enormously or to use the bearings in much more corrosive environments.

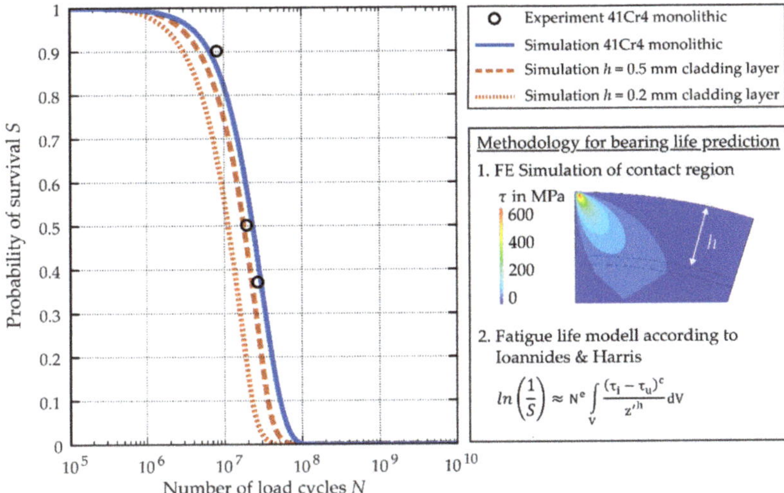

Figure 3. Bearing fatigue-life simulation for different cladding layer thicknesses.

2. Materials and Methods

For the reasons given in the previous chapter, the production process for manufacturing tailored forming machine elements must meet the following conditions to be implementable in industrial applications:

- Depending on the application, even a small amount of high-performance material may be sufficient to produce a multi-material component with a performance comparable to that of conventional manufacturing processes. However, the cladding layer in the region subject to rolling contact loading must not be too thin.
- The process must have an economically viable application rate and quality. This requires precise knowledge of the application and its loads.
- It must be possible to set the layer height as the decisive target value for production within narrow limits.

The aim of this paper therefore is to achieve a precise adjustment of the cladding thickness h by controlling the production process, see Figure 4. For this purpose, first, two different joining processes—laser metal deposition welding and plasma-transferred arc welding—are presented in Sections 2.1 and 2.2, and the characteristics that particularly influence cladding thickness are empirically examined using welding test rigs. Subsequently, cross-wedge rolling as a forming process is described in Section 2.3, and the decisive variables that influence layer height are worked out. Extensive numerical analyses are performed, which are validated by experimental investigations on a test rig. The purpose of these cross-wedge rolling experiments is to determine the input geometry with regard to width and height of the cladding layer in advance. In this way, the fatigue-life requirements for the application can be reconciled with the target parameters for deposition welding.

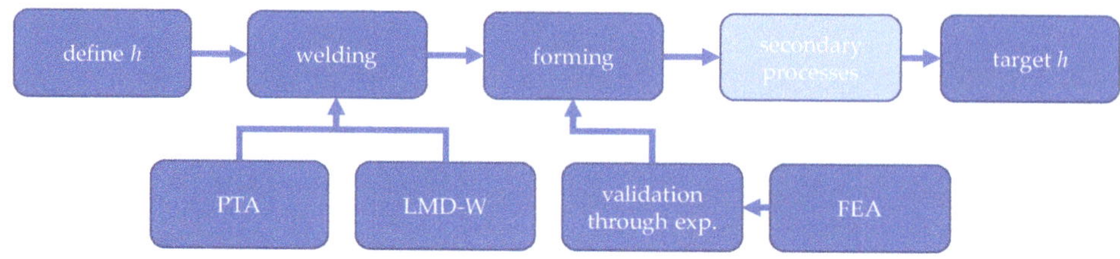

PTA: Plasma Powder Transferred Arc Welding
LMD-W: Laser Metal Deposition Welding with Wire
FEA: Finite Element Analysis
h: cladding thickness of welded material
exp.: Experimental investigations

Figure 4. Methods for defining cladding material distribution.

Two different cladding materials are used for the investigation. The martensitic chrome silica steel X45CrSi9-3 is used as cladding material for the laser metal deposition by wire process. The cladding material for the PTA process consists of the roller bearing steel 100Cr6, which has a CEV > 1 and is therefore difficult to weld [27]. The material has a high resistance against cyclic loading. The 100Cr6 powder is filtered with a sieving unit. Only powder consisting of metal grains with a diameter between 63 µm and 200 µm is used for the welding [28]. This corresponds to the current industrial standard for additional materials in powder form that are used for welding. The basic material of the shafts consists of the unalloyed and heat-resistant steel C22.8, which is mainly used in valve construction and is considered to be easily to weld. The chemical composition of base material and cladding material are shown in Table 1.

Table 1. Chemical composition in wt.% of C22.8, X45CrSi9-3, and 100Cr6.

Material / Element	C	Si	Mn	P	S	Cr	Ni
C22.8	0.17–0.24	<0.40	0.40–0.70	<0.045	<0.045	<0.40	-
X45CrSi9-3	0.4–0.5	2.7–3.3	≤0.6	≤0.004	≤0.03	8.0–10.0	≤0.5
100Cr6	0.93–1.05	0.15–0.35	0.25–0.45	≤0.025	≤0.015	1.35–1.60	-

2.1. Laser Metal Deposition Welding with Wire

Hybrid semi-finished products are manufactured by deposition welding of the martensitic chrome silica steel X45CrSi9-3 onto a base cylinder made of the unalloyed steel C22.8. A coaxial laser hot-wire welding head is used to produce the hybrid components. Compared to lateral wire feeding, coaxial wire feeding has the advantage that the process is independent of the welding direction and is therefore suitable for the manufacturing of complex components [29,30].

Base cylinders with a diameter of 27 mm and 29 mm are used as substrate. The substrate is sandblasted and then cleaned with ethanol and kept at room temperature without any preheating.

For manufacturing of the hybrid semi-finished products, the base cylinder is placed in a rotary axis, which can be moved in the X–Y plane. By superimposing the rotational movement of the base cylinder and the movement of the rotational axis in the x-direction, spiral seams are applied to the base cylinder. The processing head does not move during welding. The welded layer is positioned in the middle of the base cylinder (see Figure 5).

Figure 5. Shaft with cladding applied by laser metal deposition with wire.

Hybrid semi-finished products with various geometries are produced to validate the simulation of the layer distribution during CWR. Cladding layers consisting of 6 or 11 adjacent seams with a width of 8 or 15 mm, respectively, are applied to these using the before-mentioned setup. The claddings are applied in one or two layers, whereby the two layers where welded unidirectionally, with a short break of 90–120 s between layers.

An overview of the welding parameters is shown in Table 2.

Table 2. Welding parameters for laser metal deposition with wire.

Parameter	LMD-W [1]
Welding speed	1200 mm/min
Current	110 A
Wire feed rate	2.8 m/min
Laser power	2.3 kW
Shielding gas flow	8 L/min
Wire diameter	1.0 mm

[1] Laser Metal Deposition Welding with Wire.

Examples of the seam geometry after deposition welding are shown in Figure 6. The seam height is approx. 1.2 mm for single-layer deposition and approx. 2.4 mm for double-layer deposition.

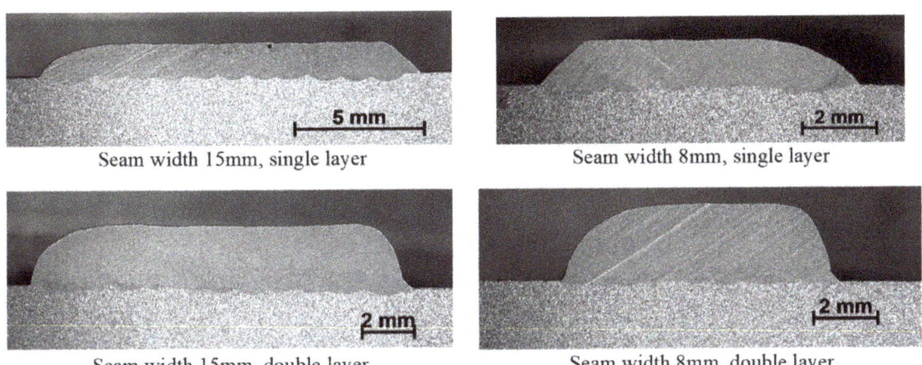

Figure 6. Examples of the seam geometry after deposition welding.

For each set of parameters, one shaft after welding and one shaft after CWR is cut, and the layer distributions are evaluated. The values for characteristics of the welded layer are used as input variable data for the simulation.

2.2. Plasma Powder Transferred Arc Welding

The PTA welding process is carried out by a six-axis REIS RV industrial robot (KUKA AG, Augsburg, Germany), where two additional axes are realized by a turn and tilt table. The welding torch is the Kennametal Stellite HPM 302 (Kennametal GmbH, Rosbach, Germany), which is water-cooled.

First, the bar blanks are turned and cut to size. The finished components have a length of 150 mm and a diameter of 27 mm. Prior to the welding process, the components are cleaned with acetone to get clean surfaces. The components are welded at room temperature. For the welding process, the shaft is fixed in the additional axis and aligned parallel to the ground. The welding head is in a horizontal welding position aligned with the shaft. During the welding process, the welding head only moves with a defined pendulum movement in the y-direction, while the shaft rotates 6 times around its own axis. The result is a spiral weld seam, which is 30 mm long in total (see Figure 7). The weld seam is welded a little longer than required, because the beginning and the end of the seam cannot be welded in an optimum way for technical reasons.

Figure 7. Welding seam after cladding by PTA.

To achieve an even application layer, it is important that the weld pool has enough time to solidify. Because of the small diameter and the rotation of the shaft, there is a risk that the weld pool does not cool down in time and drops to the ground due to gravity. Therefore, the torch is moved slightly off the center-point, which increases the arc length and gives the weld pool more time to cool down. In order to apply the weld seam as homogeneously as possible and to avoid pores, the welding torch oscillates over a short distance of 4.5 mm in the welding direction with a frequency of 1 Hz. This slightly increases the dynamics of the weld pool and allows degassing of the melt, which reduces the chance of pore formation. The whole welding process takes about 5 min, and the shaft reaches 700 °C. In order to counteract a strong heating of the cladding and base material, the temperatures are dynamically adjusted during the welding process. At the beginning, a current of 120 A is selected to generate a fast heating of the shaft. Further in the process, the current is gradually reduced in order to keep the dilution low and as constant as possible (see Figure 8). An overview of the general welding parameters is shown in Table 3.

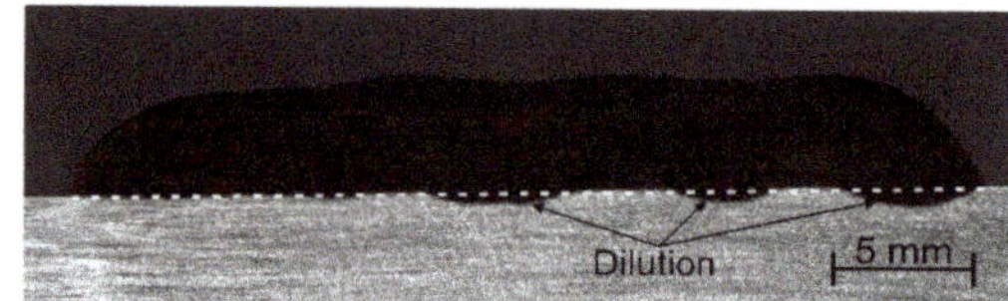

Figure 8. Dilution during the welding process.

Table 3. Welding parameters for plasma powder transferred arc welding.

Parameter	Value
Shielding gas flow	10 L/min
Plasma gas flow	1.5 L/min
Transport gas flow	6 L/min
Welding speed	0.06 m /min
Length of the welding seam	30 mm
Current	120–100 A
Voltage	27–25 V (depends on current)
Powder material	100Cr6
Grid size of powder particles	0.06–0.2 mm
Deposition rate	0.9 kg/h

2.3. Cross-Wedge Rolling

Cross-wedge rolling is an efficient way to distribute masses on work pieces as a preform operation prior to forging or milling [31]. During the CWR process, material is axially shifted by wedge geometries on the tool surface (see Figure 9). Depending on the wedge geometry, reductions in diameters up to 90% can be achieved with high reproducibility [31]. CWR is one of the forming processes investigated within the tailored forming process chain, next to die forging and impact extrusion. The forming parameters of the CWR process have impact on the quality of the hybrid parts produced within the process chain, especially with regard to machining. Depending on the forming temperature and heating strategy, the forming velocities and tool spacing can have an influence on the quality of the rolled part [32,33]. Figure 9 shows the typical process of cross-wedge rolling, which is being investigated experimentally and via simulation.

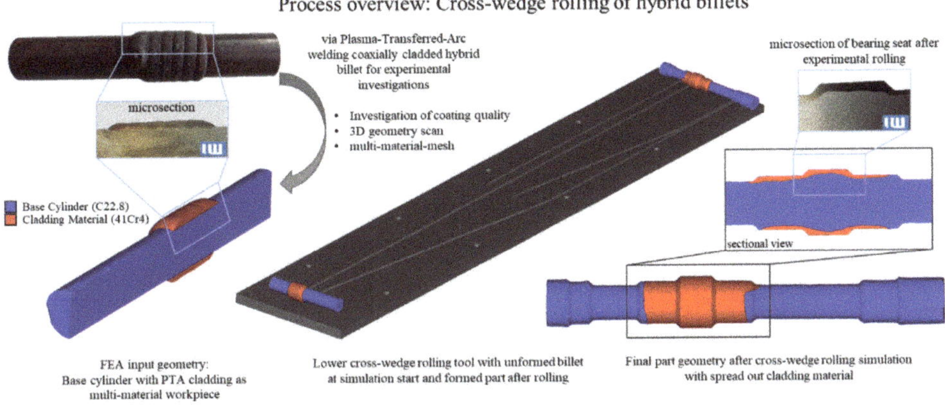

Figure 9. Process overview of cross-wedge rolling of hybrid work pieces.

In this paper, the influence of the CWR process on the forming behavior of differently welded semi-finished workpieces is investigated with regard to its predictability via simulations and its potential influence on the service lifetime of hybrid parts. For this, differently shaped work pieces were welded, as shown in Section 2.1 and Figure 6. Out of the various types of workpieces, several were formed by CWR. At least three of the work pieces were cut in half after CWR, so that the material distribution of the cladding material could be examined. The CWR process was the same as that shown in Figure 9. The other halves of the formed work pieces then continued along the process chain until the final step of the research was reached: service life investigations.

2.3.1. Cross-Wedge Rolling Simulation

Before any experimental investigations of cross-wedge rolling were conducted, the process was first simulated, to save resources and time. As mentioned before, finite element analysis (FEA) can accurately depict the CWR process [33,34]. The simulations were calculated within the commercial FEA software Forge NxT 3.0, developed by TRANSVALOR S.A. (CS 40237 Biot, 06904 Sophia Antipolis cedex, France). No custom subroutines were used. The simulations were set up within the graphical user interface of the software. No modifications to the software were made.

The tool geometries within the simulation are based on the Computer Aided Design (CAD) files of the tools that are used for the experimental trials. The advantage of the symmetry of parts in the area of the bearing seat was used to reduce calculation time (Figure 10).

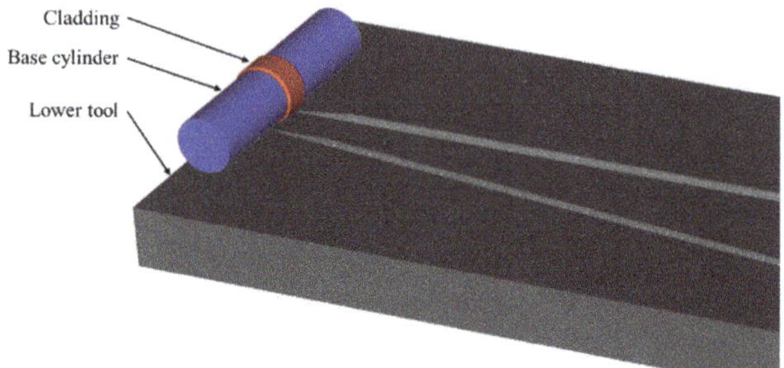

Figure 10. CAD model of hybrid work piece and lower cross-wedge rolling tool as FEA input.

Only the first 1.5 s of the 11-s forming process were investigated, because that is when the bearing seat geometry is formed. After that, the rest of the part is formed. Figure 11 shows the final simulation step used to investigate the cladding distribution after forming the bearing seat region. Figure 12 shows the process at different stages during simulation. The kinematics are analogous to the forming of the area in the experimental trials.

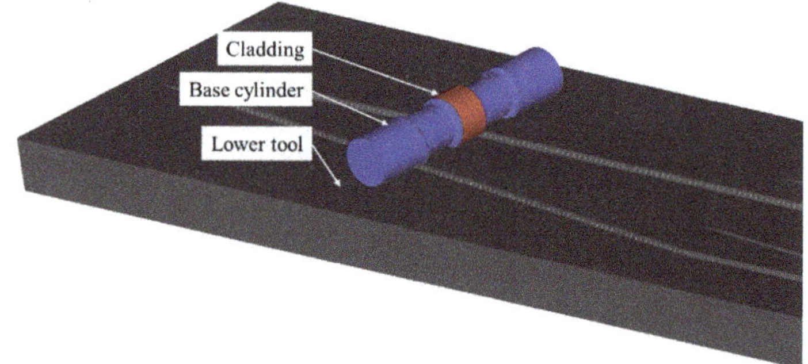

Figure 11. Simulation result of hybrid work piece cross-wedge rolling tool after 1.5 s process time.

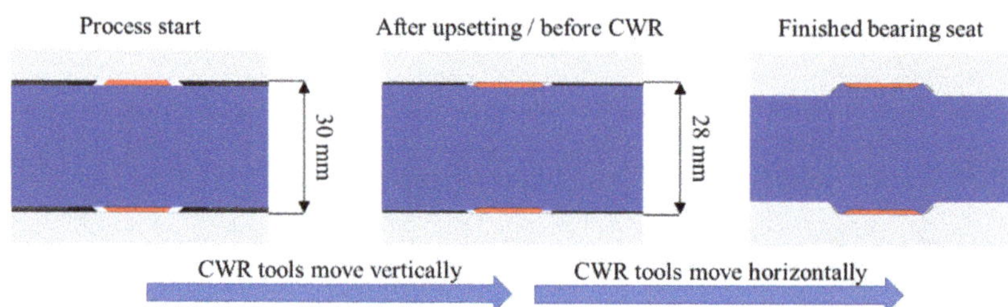

Figure 12. Cross-wedge rolling process at different time steps, where the base cylinder is in blue and the cladding in red.

To further improve calculation duration, different mesh sizes of tetrahedral elements were used within each billet. The mesh sizes were analytically investigated within a mesh sensitivity analysis in previous work. Nonlinear shape with coupled thermal-mechanical functions were used. Especially for hybrid parts, choosing the optimal mesh sizes improves calculation times drastically while still giving significant results. In Forge NxT, contact is considered using a velocity field penalization method, which allows for a slight penetration of the part into the die. This is the only method/algorithm available by default [35].

The quality of a volume mesh generated in Forge NxT is measured using the surface and the volume shape factor. "The surface shape factor is the simplest criteria and should be investigated first. It indicates whether the element is close to (=1) or far away from (<< 1) the ideal equiangular triangle shape. It can also detect degenerate (<0), flat (=0), or quasi-flat ($\rightarrow$0) elements." [35]

The volume shape factor "determines whether the elements are close to (=1) or far away from (<< 1) the ideal equiangular tetrahedron shape. It can also detect degenerate (<0), zero volume (=0), or quasi-flat ($\rightarrow$0) elements." [35]

The remeshing algorithm is triggered if the quality is below 0.4. The quality is computed as a ratio between the surface and the square perimeter of an element [35]. The volume and surface shape factor were above 0.75 (0.4 is considered the recommended minimum [35]).

The distortion value was calculated as approximately 0.82. In order to measure element distortion while eliminating the scale factor between parent and actual element referential, a "normalized" value of the Jacobian, defined as the product of the original Jacobian and a correction term K, was computed. This is automatically done by Forge and can be displayed to the user [35].

The bearing seat area was meshed with a mesh size of 1 mm, and the rest of the work piece was discretized with a mesh size of 3 mm. For the initial material distribution, the multi-material-set feature of Forge NxT 3.0 was used. Early simulations within subproject B1 (before Forge NxT 3.0 was released) were realized by modeling the base cylinder and the cladding material as separate objects within the simulation, which were then linked by bilateral sticking. The heat transfer between cladding and base cylinder was not satisfactory [32]. As a result of close communication with Transvalor with regard to the needs of simulation for hybrid work pieces, the multi-material-set feature was improved in Forge NxT 3.0. Therefore, it is now possible to load one geometry as work piece and define the segments of the work piece that are supposed to be made of a different material. This results in one mesh in which certain elements can be of a different material from the others. This approach also improves the accuracy of remeshing, which is indispensable, due to the high degrees of deformation. The flow curves for the material C22.8 were created using a Gleeble 3800-GTC (Dynamic Systems Inc., Poestenkill, New York, USA) and a DIL 805A/D+T dilatometer (TA Instruments, New Castle, USA). The cladding material was taken from the Forge NxT Database and was fitted for the Hensel and Spittel approach. This model describes the forming behavior of the material in dependence of the temperature T, the effective strain, the flow behavior by parameter ε, and the strain rate dε/dt. Within Forge NxT, the Hensel and Spittel equation is simplified [35,36], resulting in a viscoplastic material model. The flow curves and parameters of several materials were input in Forge NxT. The initial work piece temperature was set to 1250 °C and the tool surface temperature to 250 °C. The tool velocity for rolling was set to 250 mm/s for each tool. The translational movement of the tools was generated by the hydraulic press preset. Before rolling, the work piece is slightly upset with vertical force. This is realized by the upper tool moving towards the lower tool for 1 s, reducing the spacing of the tools from 32 to 28 mm. The heat transfer between tools and work piece was set to the preset "steel hot medium", resulting in an alpha value of 10^4 W/m^{2}K. The thermal effusivity of the tools was set to 11.76 kJ/m^2·K·s$^{0.5}$. The ambient air temperature was set to a constant temperature of 50 °C. The chosen preset "steel hot medium" and the heat-exchange algorithm within Forge NxT take conduction, convection and radiation between the tools, work piece, and ambient air into account. The friction between work piece and tools were set to the preset "very high Tresca", resulting in a value m = 0.8, which proved to be a good approximation of friction for hybrid CWR processes [32]. As default, thermal expansion calculation is not enabled for material flow simulations in Forge NxT. Additionally, as default, rigid dies are used to simulate a CWR process. To assure the accuracy of the simulation model used, simulations comparing these influences were conducted and the cladding thickness for each set of parameters was measured within Forge NxT (see Figure 13).

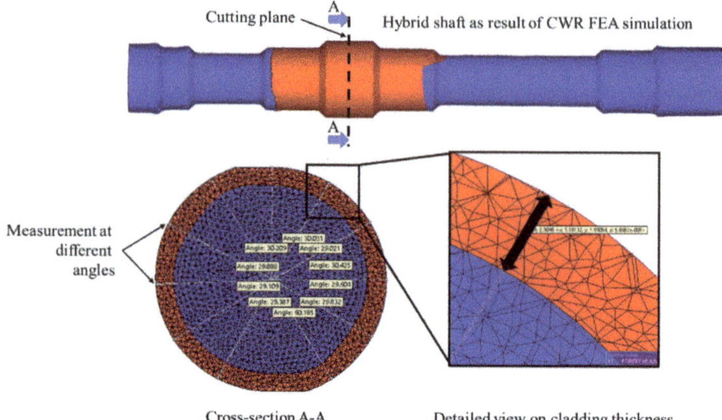

Figure 13. Measurement of the cladding thickness in the center of the bearing seat.

For this purpose, three different simulations setups (Table 4) were calculated up to the point where the cross-section of the bearing seat is nearly completely shaped. The results were compared. As Table 5 shows, the deviations between the results of the different FEA setups is small (deviation of < 2.5 % = 0.058 mm for the avg. cladding thickness), keeping in mind that the average mesh size in this cross-section area is approximately 1 mm. Therefore, the accuracy with the default settings is sufficient to predict the material flow behavior under justifiable expenditure of calculation time (Table 4). For setup #3, the calculation took more than 160 h, which is too long when simulating an array of variations.

Table 4. Comparison matrix for FEA calculation time of cross-wedge rolling of work piece.

Case ID	Thermal Expansion Calculation [1]	Tool Behavior	Thermal Expansion due to Initial Heating [1]	Estimated Normalized Calculation Time
#1	No	rigid	No	100%
#2	Yes	rigid	Yes	182%
#3	Yes	deformable	Yes	449%

[1] of work piece.

Table 5. Deviation comparison of different FEA setup results of the cross-wedge rolling process.

Simulation Result	Setup	Cross Section Surface of:		Max. Coating Thickness	Min. Coating Thickness	Avg. Coating Thickness
		Cladding	Base Cylinder			
	#1: No thermal expansion; Rigid tools	644.082 mm^2	452.646 mm^2	2.263 mm	2.854 mm	2.439 mm
	#2: Thermal expansion; Rigid tools	644.912 mm^2	455.057 mm^2	2.101 mm	2.830 mm	2.426 mm
	Δ to Setup #1	0.10%	0.50%	7.20%	0.80%	0.50%
	#3: Thermal expansion; Deformable tools	643.981 mm^2	430.330 mm^2	2.131 mm	2.872 mm	2.381 mm
	Δ to Setup #1	0.00%	5.20%	6.20%	0.60%	2.40%

Due to the little impact of the thermal expansion and tool elasticity, the simulations were set up with rigid dies and no thermal expansion calculation. The parameters used within the simulation are shown in Table 6.

Table 6. Simulation parameters used within the model of the hybrid cross-wedge rolling process.

Material	Work Piece Geometry	Tool Velocity	Work Piece Temperature
Base cylinder: C22.8 Cladding: X45CrSi9-3 or 100Cr6	Ø 27 or 29 mm 8 or 15 mm width; 1.2; 2.4; or 2.5 mm height	240 mm/s	1250 °C

Besides the work piece geometry, all simulation parameters remained the same for all investigations. Four types of cladding geometries were used for the material X45CrSi9-3 and combined with two different base cylinder diameters (27 and 29 mm), which results in eight work piece variants. For 100Cr6, three cladding geometries were used and a 27 mm diameter due to limited availability of base cylinders in the chosen dimensions. After the hybrid semi-finished work pieces were welded, they were cooled down at ambient air conditions. The cooled-down work pieces were cut in half, and the different cladding thicknesses were measured (Figure 6). The geometry of all 100Cr6 parts was created by milling similar welded work pieces (Figures 7 and 14) to three different geometries: 10, 15, and 20 mm seam width and 2.5 mm seam height (Figure 14). This was done due to the less accurate layer application of PTA welding compared to the hot-wire deposition welding. This was the most reproducible way to immediately get similar cladding geometries.

Figure 14. Work piece 100Cr6/C22.8 milled to 27 mm diameter and 10 mm cladding material seam.

The average thickness of each cladding layer and its width were measured and input for the simplified geometry of the simulation. The various adapted geometries are shown in Figures 15 and 16.

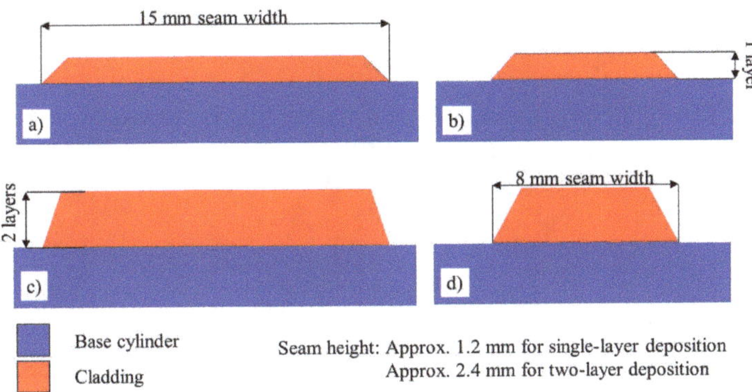

Figure 15. CAD model of different cladding (red) distributions on base cylinder (blue) forX45CrSi9-3, (**a**) 1 layer at 15 mm seam width (**b**) 1 layer at 8 mm seam width (**c**) 2 layers at 15 mm seam width (**d**) 2 layers at 8 mm seam width.

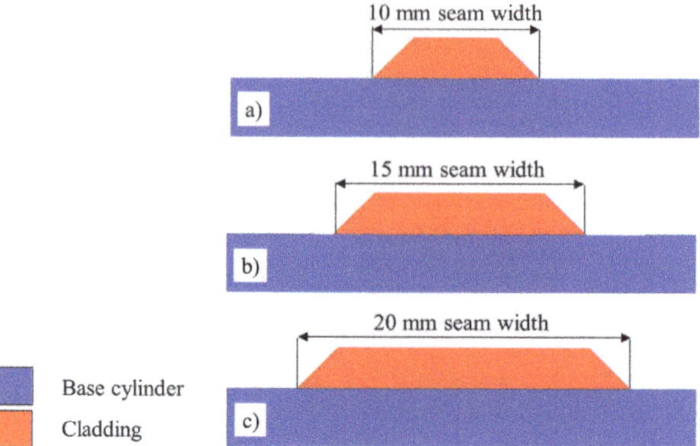

Figure 16. CAD model of different cladding (red) distributions on base cylinder (blue) for 100Cr6, (**a**) 10 mm seam width (**b**) 15 mm seam width (**c**) 20 mm seam width.

After the simulations were calculated, the material distribution was analyzed. The results of the simulation will be discussed in Section 3.

2.3.2. Cross-Wedge Rolling Experiment

For experimental investigations, the CWR module (a self-built test stand) of the Institut für Integrierte Produktion Hannover gGmbH (IPH) was used (see Figure 17). The module consists of a machine frame in which two sleds can glide by translatory motion. The sleds have mounting holes for the CWR tools. The translational (horizontal) movement of the sleds is created by hydraulic cylinders, providing each sled with up to 125 kN of force. The vertical force, which is necessary to ensure constant spacing between the tools, regardless of the forming forces, is generated by the hydraulic press into which the module is mounted. The 6,300 kN press (manufactured by NEFF) is set to 50 kN of closing force. The minimum spacing is ensured by spacing discs with standardized heights that function as a vertical end-stop.

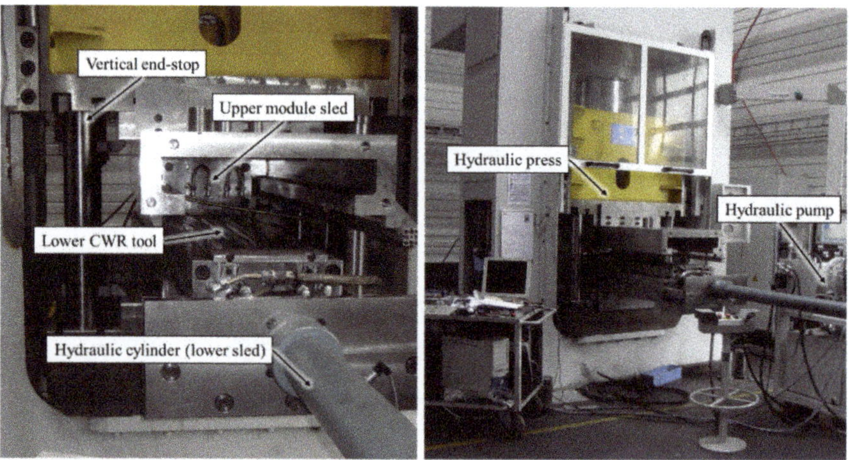

Figure 17. Cross-wedge rolling test stand at the Institut für Integrierte Produktion Hannover.

Prior to CWR, the parts need to be heated. This can be done using either the furnace or the IPH induction heating unit (manufactured by EMA-TEC GmbH). The advantage of the furnace is the close to perfectly homogenous heating of the part, whereas induction heating creates slight gradients of temperature within the work piece. Nevertheless, the short amount of heating time (60 s) used compared to the furnace (at least 40 min) results in less scale on, and less surface decarburization in the surface of, the work piece. With extensive heating, too little carbon may remain to harden the surface of the bearing seat after rolling and milling. Therefore, the work pieces were heated by induction heating for this research. The work pieces were heated for 60 s, resulting in approx. 0.4 kWh of energy induction into the work piece, which led to a peak temperature of 1300 °C, before the 20 s for transport into the CWR module. After transport, the work piece retained a temperature around 1250 °C. The tools were heated by heating cartridges to a temperature of 200 °C. The bottom of each tool was isolated with a polymer plate made of AS600M to reduce temperature leakage. Figure 18a shows a variety of hybrid work pieces with different amounts of the cladding material shown in Figure 6.

(a)   (b)

Figure 18. (**a**) Work pieces with different amounts of cladding; (**b**) (cold) work piece in starting position.

The work pieces were placed with their center aligned to the center of the bearing seat forming wedges. A mechanical end-stop ensured correct positioning (Figure 18b). After positioning, the upper tool was lowered onto the end-stops with 50 kN of force, resulting in a defined rolling gap of 28 mm. As soon as the desired gap width was reached, the hydraulic cylinders of the tool sleds were moved to start the rolling process, which took about 9 s. After rolling, the work pieces were taken out of the CWR module and placed onto a steel tray to slowly cool in air with free convection to prevent hardening and distortion. When cooled down, the work pieces were cut in half to examine the material distribution and to check for defects, as shown in Figure 9.

3. Results

The cladding distribution and thickness in the simulation can be compared to experimental results for every parameter combination. From now on, the parameter combinations will be given as a string based on the diameter of the base cylinder, the cladding width and number of layers (e.g., 27_15_1). The simulation pictures for X45CrSi9-3 were taken at 1.25 s of process time. The pictures of the experiment were taken after completion of the CWR process and subsequent splitting in half.

As shown in Figures 19–26, the results of the simulations are similar to the experimental findings. Especially for the parameter combinations with 27 mm base cylinder diameter and one cladding layer, the shape and the measurements are very similar (Figures 19–21; Figure 25). The deviations are within the margin of error. The parameter combination 27_15_1 (Figure 19) shows a deviation of 0.45% in cladding width and 3.47% in cladding height, resulting in only 0.04 mm difference in cladding height between simulation and experiment, which is almost indistinguishable.

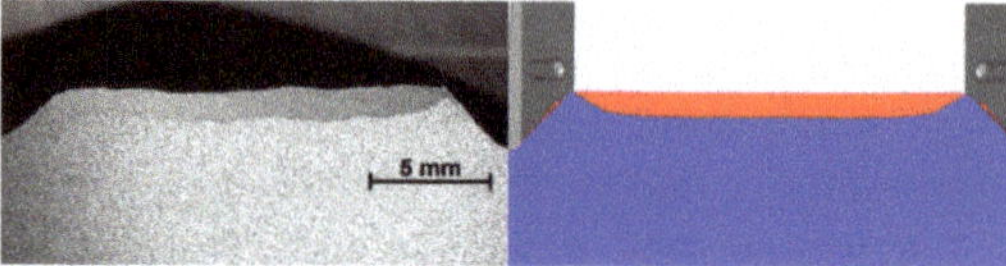

X45CrSi9-3	Experiment	Simulation	Δ
Cladding width in mm	17.6	17.52	0.45%
Cladding height in mm	1.11	1.15	3.47%

Figure 19. Comparison experiment/simulation; 27 mm base cylinder, 15 mm cladding width, 1 layer.

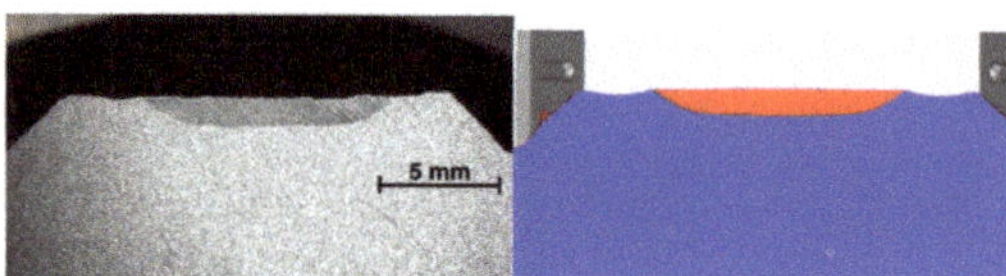

X45CrSi9-3	Experiment	Simulation	Δ
Cladding width in mm	10.59	11.44	7.43%
Cladding height in mm	1.14	1.21	5.79%

Figure 20. Comparison experiment/simulation; 27 mm base cylinder, 8 mm cladding width, 1 layer.

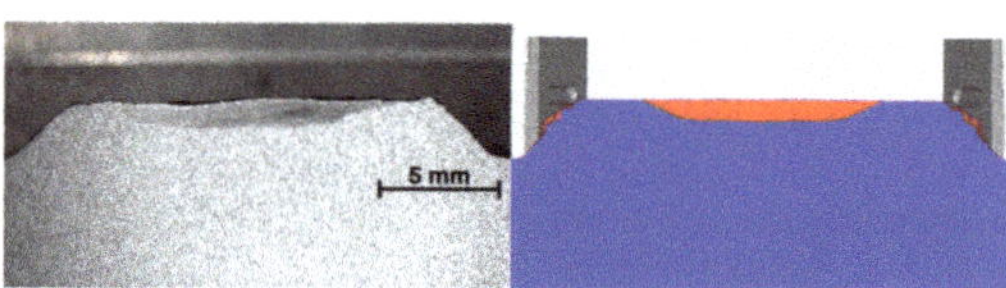

X45CrSi9-3	Experiment	Simulation	Δ
Cladding width in mm	11.92	12.02	0.83%
Cladding height in mm	1.17	1.07	9.35%

Figure 21. Comparison experiment/simulation; 29 mm base cylinder, 8 mm cladding width, 1 layer.

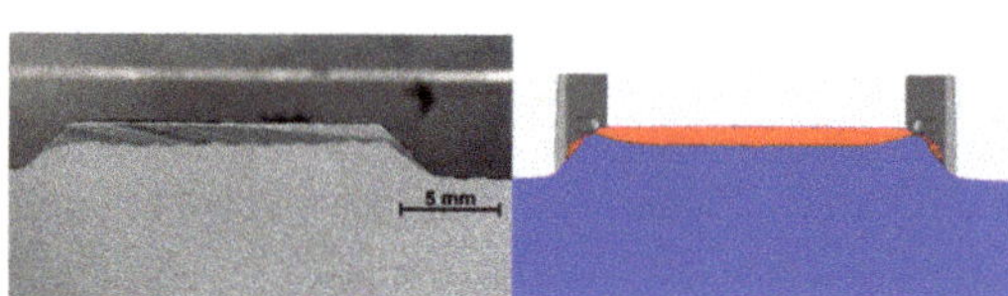

X45CrSi9-3	Experiment	Simulation	Δ
Cladding width in mm	20.16	19.1	5.55%
Cladding height in mm	1.25	1.13	10.62%

Figure 22. Comparison experiment/simulation; 29 mm base cylinder, 15 mm cladding width, 1 layer.

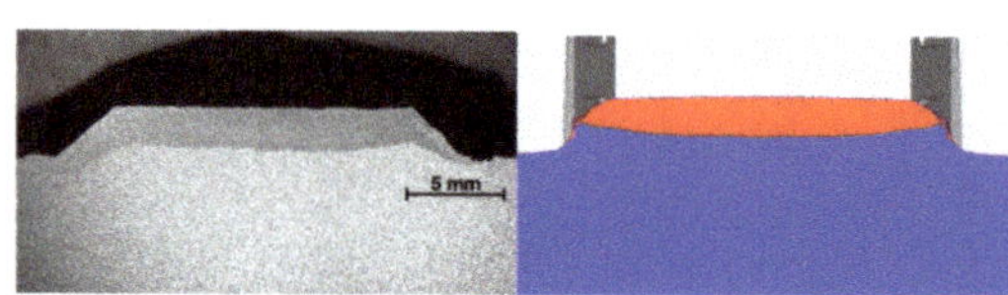

X45CrSi9-3	Experiment	Simulation	Δ
Cladding width in mm	23.89	21.7	10.09%
Cladding height in mm	2.06	2.31	10.82%

Figure 23. Comparison: experiment/simulation; 29 mm base cylinder, 15 mm cladding width, 2 layers.

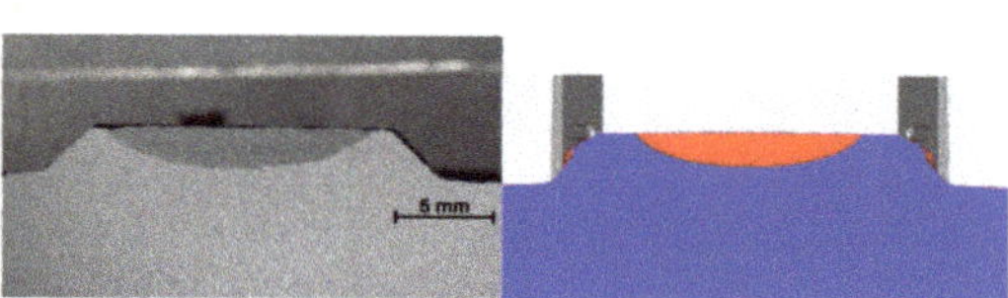

X45CrSi9-3	Experiment	Simulation	Δ
Cladding width in mm	13.72	13.9	1.29%
Cladding height in mm	2.06	1.98	4.04%

Figure 24. Comparison experiment/simulation; 29 mm base cylinder, 8 mm cladding width, 2 layers.

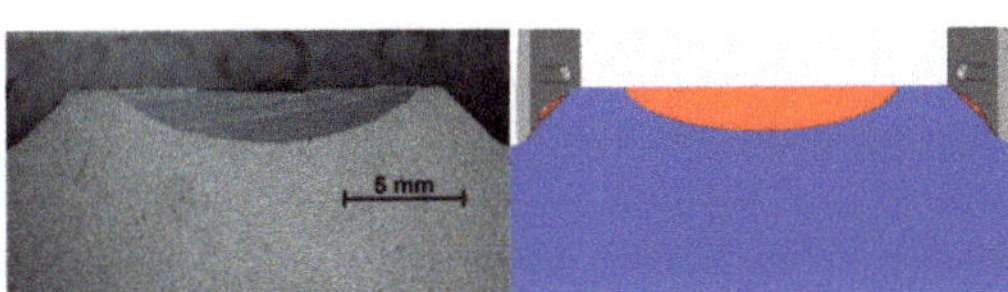

X45CrSi9-3	Experiment	Simulation	Δ
Cladding width in mm	12.58	13.06	3.68%
Cladding height in mm	2.11	2.09	0.96%

Figure 25. Comparison experiment/simulation; 27 mm base cylinder, 8 mm cladding width, 2 layers.

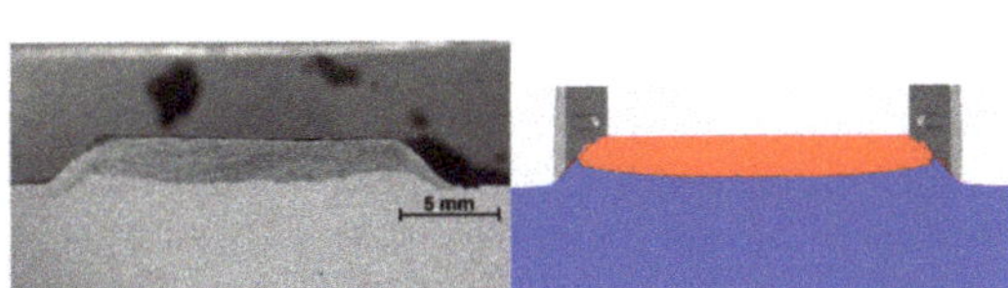

X45CrSi9-3	Experiment	Simulation	Δ
Cladding width in mm	22.64	21.06	7.50%
Cladding height in mm	2.28	2.45	6.94%

Figure 26. Comparison experiment/simulation; 27 mm base cylinder, 15 mm cladding width, 2 layers.

For the 100Cr6 work pieces (Figures 27–29, simulation picture taken at 2.0 s process time), the deviations between simulation and experiment decrease with increasing amount of cladding material. The overall shape of the cladding after rolling is comparable to the results of the simulation and, in the best case (Figure 29), below 3% deviation. Comparing the 15 and 20 mm cladding widths, it can be shown that the increase in cladding height is marginal. The additional cladding material just flows out of the bearing seat area without any use in future application. Which height will be sufficient for 100Cr6 hybrid shafts will be investigated in future research.

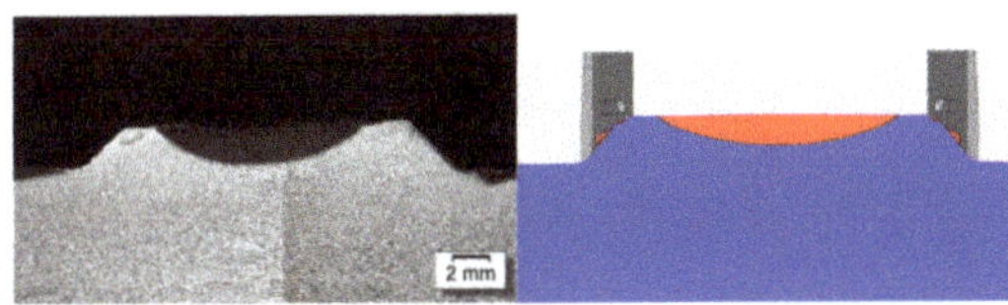

100Cr6	Experiment	Simulation	Δ
Cladding width in mm	14.74	13.21	11.58%
Cladding height in mm	1.88	2.50	32.98%

Figure 27. Comparison experiment/simulation 100Cr6, 27 mm base cylinder, 10 mm cladding width.

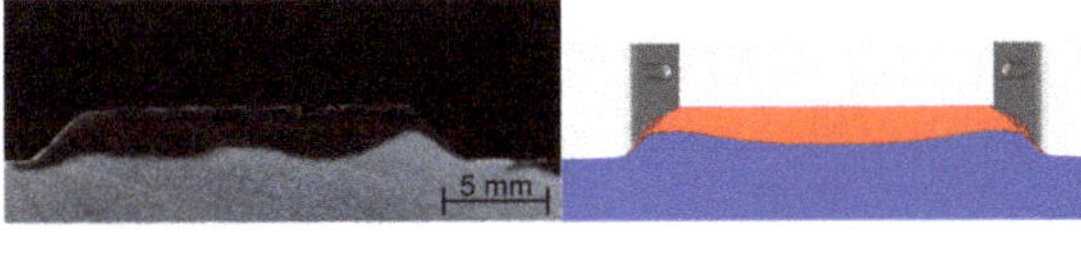

100Cr6	Experiment	Simulation	Δ
Cladding width in mm	21.29	22.21	4.32%
Cladding height in mm	2.16	2.40	10.00%

Figure 28. Comparison experiment/simulation 100Cr6, 27 mm base cylinder, 15 mm cladding width.

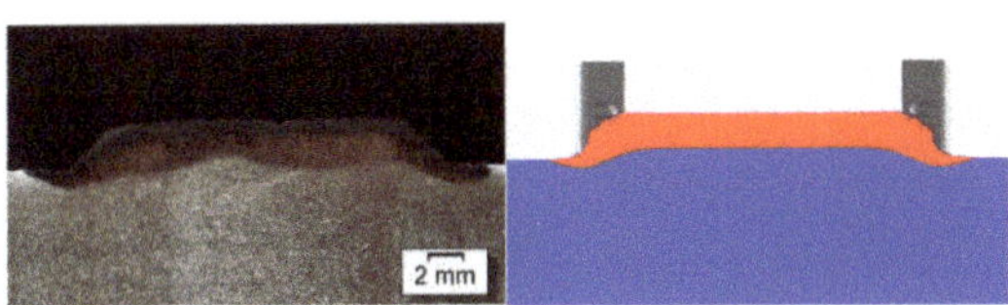

100Cr6	Experiment	Simulation	Δ
Cladding width in mm	27.41	26.61	2.92%
Cladding height in mm	2.25	2.20	2.22%

Figure 29. Comparison experiment/simulation 100Cr6, 27 mm base cylinder, 20 mm cladding width.

The parameter combinations with 29 mm base cylinder bear more cladding after welding, since the diameter is increased, while the same initial layer height is kept. Therefore, more cladding material is spread, resulting in too much cladding material flowing over the edges of the bearing seat, when two layers of cladding combined with 15 mm cladding width are used (Figure 26). When comparing simulation and experimental results for this parameter combination, the deviation in geometry is visually apparent. As for the experimental results, the cladding material gushed over the edge of the bearing seat, forming a curvilinear tail (Figure 26). In the corresponding simulation, the cladding material remains completely within the borders of the bearing seat. No significant amount of cladding material is distributed over to the next shaft segment.

Even though this amount of cladding material would not be considered for future investigations due to its wastefulness, the accuracy of the simulation models needed to be confirmed. Therefore, simulations with even more cladding material were calculated to see at which point the simulation diverges from the experimental findings. For this, additional hybrid work pieces were welded. This time, the diameter of the base cylinder was set to 29 mm and the cladding width to 20 mm. The simulation was again analyzed at 1.25 s of process time.

The cladding material distribution is similar between simulation and experiment (Figure 30) for 29_20_3S work pieces at 1.25 s process time. To investigate this, more simulations were calculated, varying only the cladding material width in 1 mm steps between 15 and 20 mm. The base cylinder diameter was set to 27 mm (Figure 31).

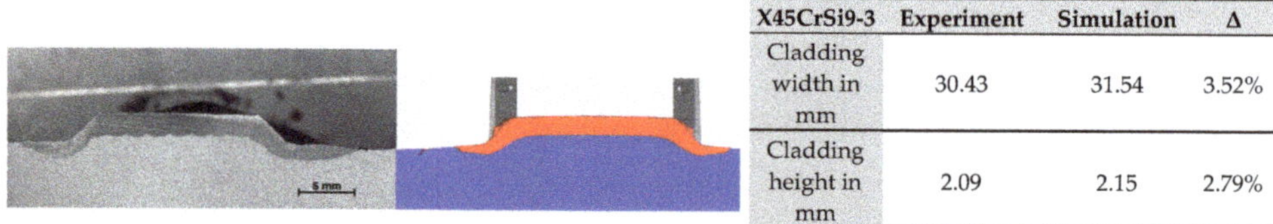

X45CrSi9-3	Experiment	Simulation	Δ
Cladding width in mm	30.43	31.54	3.52%
Cladding height in mm	2.09	2.15	2.79%

Figure 30. Comparison experiment/simulation; 29 mm base cylinder, 20 mm cladding width, 3 layers.

Figure 31. Different amount of cladding material resulting in differently shaped bearing seat material distribution at 1.25 s process time.

To further investigate possible reasons for the simulation model not fitting experimental results for certain amounts of cladding material, the cladding distribution was analyzed for different time steps of the simulation to ensure the simulation had been calculated to a sufficient process duration. Figure 32 shows the cladding material distribution for a 29_15_2S work piece over time as simulation output. It can be shown that the cladding material changes its shape over time during the CWR process. Whereas the cladding material of work pieces with small amounts of cladding material, e.g., 27_8_1S, remains within the bearing seat and is fully formed after 1.25 s of process time, the large cladding material amounts of the 29_15_2S work piece continue to change shape with further work piece rotations (Figure 32).

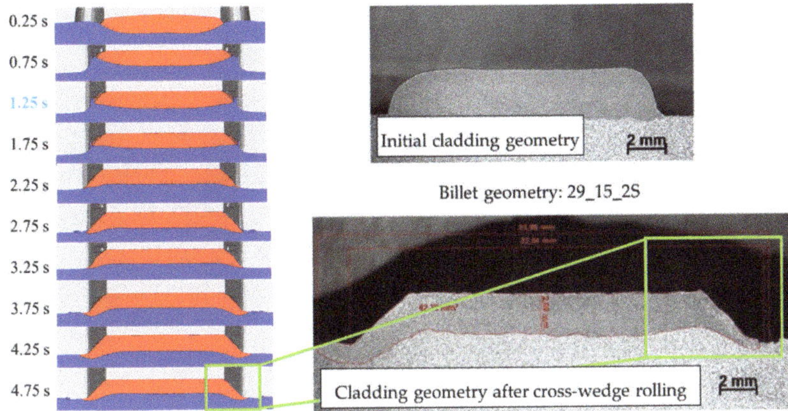

Figure 32. Analysis of cladding material distribution during cross-wedge rolling over time—cladding material X45CrSi9-3 (red) on C22.8 (blue).

The reason for this behavior becomes clear when inspecting the cross-section of the bearing seat segment of the shaft during forming at different process time steps within the simulation (Figure 33).

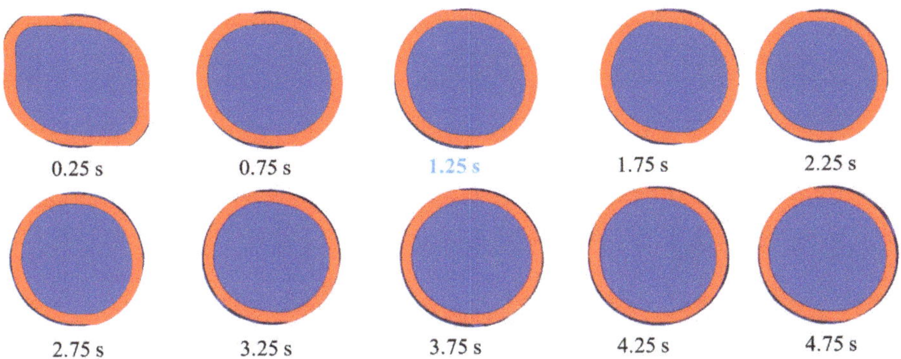

Figure 33. Over time analysis of the cladding material distribution within the cross-section of the work piece during cross-wedge rolling—cladding material X45CrSi9-3 (red) on C22.8 (blue).

At the beginning of the forming process, right after the closing of the tools, the work piece is deformed from a cylindrical shape to an ellipsoid shape (~0.25 s, Figure 33). This is intended to increase the degree of deformation to improve the material properties of the cladding material after the forming process. For small amounts of cladding material, mainly the cladding itself is deformed and the base cylinder remains a cylindrical shape. Due to the large amount (≥15 mm width) of cladding material, the whole bearing seat segment becomes deformed. For work pieces with 8 mm cladding width, 1.25 s process time was sufficient to return the work piece to a cylindrical shape in the area of the bearing seat after the initial upsetting. This results in more work piece rotations necessary to completely shape the bearing seat cladding. Therefore, when calculated to a further process time, the simulation model is sufficient again (Figure 32). Additionally, incorrect insertion of work pieces in the CWR module can result in unsymmetrical cladding distribution. If the influence of the work piece positioning is to be investigated, a non-symmetric simulation setup will be required. Then, even incorrectly inserted work pieces could be simulated with high accuracy, as Figure 34 qualitatively shows.

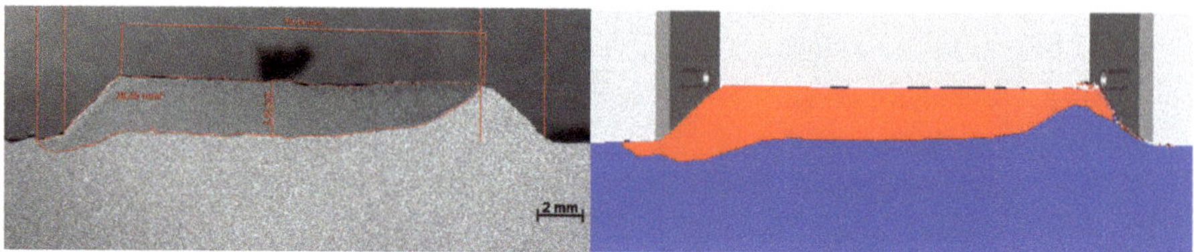

Figure 34. Unsymmetrical cladding distribution (experiment left, simulation right) due to positioning error.

When comparing the deviations between the different cladding geometries, especially the height, several influences can be identified. As shown in Figure 19 to Figure 29, the overall shape of the cladding material distribution is comparable, although the cladding height deviates more than 30% in some cases (e.g., 100Cr6). To analyze the effects of the initial work piece geometry on the cladding height deviation, the standardized and main effects were determined (Figure 35). It can be seen that the diameter of the work piece's base cylinder has a significant effect on the deviation between simulation and experiment with regard to the cladding height (Figure 35a). The combination of seam height and seam width (total cladding volume) has less but still significant effect. The seam width alone has barely any effect, and the seam height by itself is not significant at all. Figure 35b shows that with increasing base cylinder diameter, the deviation also increases. The deviation also increases for the higher seam widths. This can be explained by more material spreading out from the bearing seat area of the work piece. The more that material is kneaded and spread out from the center part of the work piece, the larger the deviation between simulation and experiment becomes. This is analog to the previous findings, where a larger amount of cladding material resulted in larger geometry deviations (Figure 32).

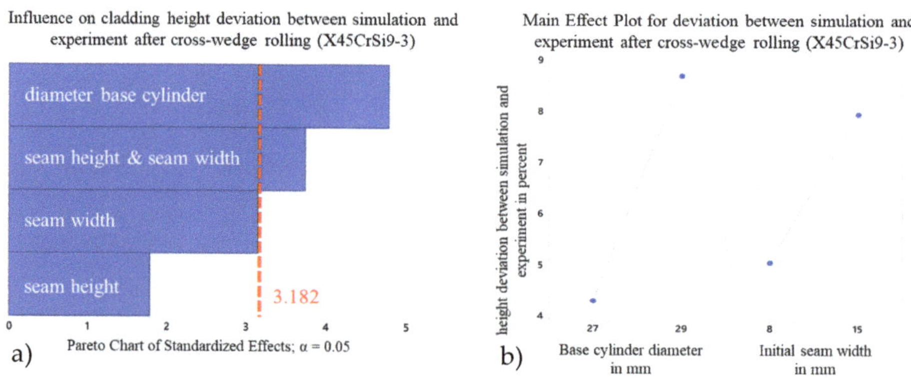

Figure 35. Influence of the cladding material amount on the accuracy of the simulation (X45CrSi9-3); (**a**) Pareto chart of standardized effects, (**b**) Main effect plot for simulation and experiment deviation

Since only the width of the seam was varied for the 100Cr6 work pieces, the approach to deviation comparison is different. Figure 27 to Figure 29 show that for 10 mm of cladding material width, the cladding height deviates more between simulation and experiment than it does in the case of larger cladding widths and therefore cladding volume.

Figure 36 shows the influence of the cladding material volume on the deviation between simulation and experiment. When the cladding material geometries of the 100Cr6 and the X45CrSi9-three work pieces are compared, it is obvious that the 100Cr6 has experienced less forming, and the cladding is not spread as wide as on the X45CrSi9-3. This can be explained by the higher flow curves of the 100Cr6 welding material compared to the database material stored in Forge NxT. Due to this, simulations with flow curves of the actual welding material should be considered in the future to improve FEA prediction accuracy.

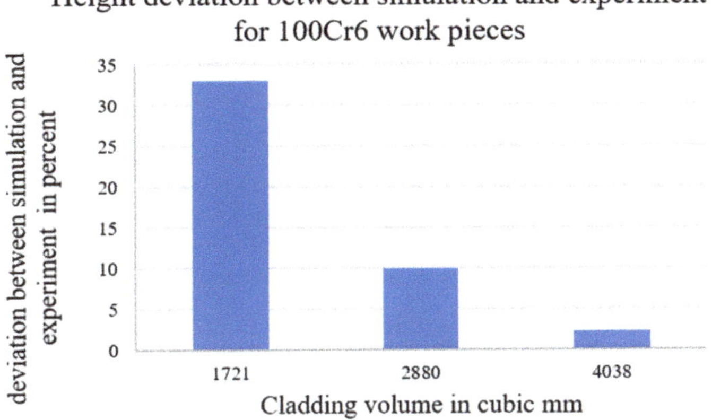

Figure 36. Influence of the cladding material amount on the accuracy of the simulation (100Cr6).

4. Discussion and Conclusions

When comparing experiments with simulation results of the cross-wedge rolling process of hybrid material work pieces, it is not sufficient to analyze only the first seconds of the forming process. Even though the axial material shifting by the wedges is technically over after the first rotation, there is still axial material flow due to excess volume remaining within the bearing seat area when using work pieces with too much cladding material. The takeaway from this would be that the time of the process simulated should be adapted to the cylindrical shape of the work piece. Only when the bearing seat is cylindrical within the simulation have the calculations progressed far enough to make a statement about the cladding distribution. When the whole process was computed, which takes almost 10 times longer than the first 1.25 s of the process, an accurate prediction of the cladding material distribution could be made. Since such a large amount cladding material would not be used for this application, such long calculation times are not something to consider. For smaller amounts of X45CrSi9-3 cladding material, the simulation model is sufficient and the method of cladding material thickness prediction is valid. For 100Cr6, more detailed simulations with more variations should be carried out. Considering that these were the first investigations with this not conventionally weldable material, the results are reasonably good.

It could be shown that the simulation results of cross-wedge rolling with coaxially arranged hybrid work pieces are in good agreement with the experimental results. In previous work, the basic suitability of the simulations for serially arranged work pieces [32] or work pieces with constant cladding thickness [37] but varied process parameters were investigated. Building on these findings, the research in this work concludes the investigations with regard to "if" and "how accurately" these multi-material simulations can predict the material flow during cross-wedge rolling. These numerical predictions of material flow can be used to weld work pieces with exact amounts of cladding materials to save material and ensure optimal layer thickness within the functional area of the part.

Too much cladding results in wasted material after machining. The goal would be to use as little cladding material as possible but enough to ensure optimal service life of the part. Which amount of cladding material will result in optimal service life of the part will be investigated in further research.

Further simulations will be carried out with flow curves taken from actual welded cladding material. Currently, specimens are being prepared in order to carry out material characterization for every material used within the research project. This will help to improve the simulation model's accuracy even more.

Author Contributions: Conceptualization, J.K. and A.B.; formal analysis, J.K., L.B., M.Y.F., and A.B.; investigation, J.K., L.B., and M.Y.F.; methodology, J.K., M.M., L.B., T.C., and M.Y.F.; project administration, M.S., T.H., and F.P.; software, J.K.; supervision, M.S., T.H., L.O., and G.P.; validation, J.K., M.M., L.B., and M.Y.F.; visualization, J.K., L.B., T.C., and M.Y.F.; writing—original draft, J.K., L.B., T.C., and M.Y.F.; writing—review and editing, M.M., M.S., T.H., F.P., M.L., J.H., S.K., L.O., and G.P. All authors have read and agreed to the published version of the manuscript.

Funding: This research was funded by the Deutsche Forschungsgemeinschaft (DFG, German Research Foundation)—CRC 1153, subproject B1, A4, C3, T1—252662854.

Conflicts of Interest: The authors declare no conflict of interest.

References

1. Mohamed, A.; Celik, T. An integrated knowledge-based system for alternative design and materials selection and cost estimating. *Expert Syst. Appl.* **1998**, *14*, 329–339. [CrossRef]

2. Ljungberg, L.Y.; Edwards, K.L. Design, materials selection and marketing of successful products. *Mater. Des.* **2003**, *24*, 519–529. [CrossRef]

3. Zhou, C.-C.; Yin, G.-F.; Hu, X.-B. Multi-objective optimization of material selection for sustainable products: Artificial neural networks and genetic algorithm approach. *Mater. Des.* **2009**, *30*, 1209–1215. [CrossRef]

4. Sirisalee, P.; Ashby, M.F.; Parks, G.T.; Clarkson, P.J. Multi-criteria material selection in engineering design. *Adv. Eng. Mater.* **2004**, *6*, 84–92. [CrossRef]

5. Bandyopadhyay, A.; Heer, B. Additive manufacturing of multi-material structures. *Mater. Sci. Eng. R Rep.* **2018**, *129*, 1–16. [CrossRef]

6. Gouker, R.M.; Gupta, S.K.; Bruck, H.A.; Holzschuh, T. Manufacturing of multi-material compliant mechanisms using multi-material molding. *Int. J. Adv. Manuf. Technol.* **2006**, *30*, 1049–1075. [CrossRef]

7. Merklein, M.; Johannes, M.; Lechner, M.; Kuppert, A. A review on tailored blanks—Production, applications and evaluation. *J. Mater. Process. Technol.* **2014**, *214*, 151–164. [CrossRef]

8. Irving, B. Blank welding forces automakers to sit up and take notice. *Weld. J.* **1991**, *70*, 39–45.

9. Saunders, F.I.; Wagoner, R.H. Forming of tailor-welded blanks. *Metall. Mater. Trans. A* **1996**, *27*, 2605–2616. [CrossRef]

10. Behrens, B.-A.; Overmeyer, L.; Barroi, A.; Frischkorn, C.; Hermsdorf, J.; Kaierle, S.; Stonis, M.; Huskic, A. Basic study on the process combination of deposition welding and subsequent hot bulk forming. *Prod. Eng.* **2013**, *7*, 585–591. [CrossRef]

11. Nguyen-Schäfer, H. Contact stresses in rolling bearings. In *Computational Design of Rolling Bearings*; Springer International Publishing: Cham, Switzerland, 2016. [CrossRef]

12. Harris, T.A.; Barnsby, R.M. Life ratings for ball and roller bearings. *Proc. Inst. Mech. Eng. Part J J. Eng. Tribol.* **2001**, *215*, 577–595. [CrossRef]

13. Hertz, H. Ueber die Berührung fester elastischer Körper. *Journal für die Reine und Angewandte Mathematik* **1882**, *92*, 156–171. [CrossRef]

14. Belyaev, N.M. Local stresses in compression of elastic bodies. *Inzhenernye Sooruzheniya i Stroitelnaya Mekhanika* **1924**, 27–108.

15. Sadeghi, F.; Jalalahmadi, B.; Slack, T.S.; Raje, N.; Arakere, N.K. A Review of Rolling Contact Fatigue. *ASME J. Tribol.* **2009**, *131*, 041403. [CrossRef]

16. Palmgren, A.; Lundberg, G. Dynamic Capacity of Rolling Bearings. *Acta Polytech. Mech. Eng. Ser.* **1947**, *1*, 7.

17. Zwirlein, O.; Schlicht, H. Werkstoffanstrengung bei Walzbeanspruchung—Einfluss von Reibung und Eigenspannungen. *Materialwissenschaft und Werkstofftechnik* **1980**, *11*, 1–14. [CrossRef]

18. Förster, R.; Förster, A. *Einführung in die Fertigungstechnik*; Springer Vieweg: Berlin, Germany, 2018; p. 130. [CrossRef]

19. Schuler, V.; Twrdek, J. *Praxiswissen Schweißtechnik: Werkstoffe, Prozesse, Fertigung*, 6th ed.; Springer Vieweg: Wiesbaden, Germany, 2019; p. 268. ISBN 978-3-658-24266-4.

20. Kottman, M.; Zhang, S.; McGuffin-Cawley, J.; Denney, P.; Narayanan, B.K. Laser Hot Wire Process: A Novel Process for Near-Net Shape Fabrication for High-Throughput. *JOM J. Miner. Met. Mater. Soc.* **2015**, *67*, 622–628. [CrossRef]

21. Liu, S.; Liu, W.; Kovacevic, R. Experimental investigation of laser hot-wire cladding. *Proc. Inst. Mech. Eng. B J. Eng.* **2017**, *231*, 1007–1020. [CrossRef]

22. Nowotny, S.; Brueckner, F.; Thieme, S.; Leyens, C.; Beyer, E. High-performance laser cladding with combined energy sources. *J. Laser Appl.* **2015**, *27*, 1–7. [CrossRef]
23. Lostado Lorza, R.; Escribano García, R.; Fernandez Martinez, R.; Martínez Calvo, M.Á. Using Genetic Algorithms with Multi-Objective Optimization to Adjust Finite Element Models of Welded Joints. *Metals* **2018**, *8*, 230. [CrossRef]
24. Blohm, T.; Nothdurft, S.; Mildebrath, M.; Ohrdes, H.; Richter, J.; Stonis, M.; Langner, J.; Springer, A.; Kaierle, S.; Hassel, T.; et al. Investigation of the joining zone of laser welded and cross wedge rolled hybrid parts. *Int. J. Mater. Form.* **2018**, *11*, 829–837. [CrossRef]
25. Coors, T.; Hwang, J.-I.; Pape, F.; Poll, G. Theoretical investigations on the fatigue behavior of a tailored forming steel-aluminium bearing component. *AIP Conf. Proc.* **2019**, *2113*, 040020. [CrossRef]
26. Coors, T.; Pape, F.; Poll, G. Bearing Fatigue Life of a Multi-Material Shaft with an Integrated Raceway. *Bear. World J.* **2018**, *3*, 23–30.
27. Matthes, K.; Schneider, W. *Schweißtechnik—Schweißen von Metallischen Konstruktionswerkstoffen*, 6th ed.; Hanser Verlag: München, Germany, 2016; pp. 33–34. ISBN 978-3446405684.
28. Dilthey, U. *Schweißtechnische Fertigungsverfahren 2—Verhalten der Werkstoffe beim Schweißen*, 3rd ed.; Springer: Berlin, Germany, 2005; p. 130. ISBN 978-3-540-27402-5.
29. Lammers, M.; Budde, L.; Barroi, A.; Hermsdorf, J.; Kaierle, S. Entwicklung von Laser-Systemkomponenten optimiert für die additive Serienfertigung mittels SLM. In *Konstruktion für die Additive Fertigung*, 1st ed.; Lachmayer, R., Lippert, R.B., Kaierle, S., Eds.; Springer Vieweg: Berlin/Heidelberg, Germany, 2020; pp. 21–35. [CrossRef]
30. Pajukoski, H.; Näkki, J.; Thieme, S.; Tuominen, J.; Nowotny, S.; Vuoristo, P. High performance corrosion resistant coatings by novel coaxial cold- and hot-wire laser cladding methods. *J. Laser Appl.* **2016**, *28*, 012011. [CrossRef]
31. Pater, Z. Cross-Wedge Rolling. In *Comprehensive Materials Processing 13 Volume Set*, 1st ed.; Hashmi, S., Batalha, G.F., van Tyne, C.J., Yilbas, B., Eds.; Elsevier: Amsterdam, The Netherlands, 2014; Volume 3. [CrossRef]
32. Kruse, J.; Jagodzinski, A.; Langner, J.; Stonis, M.; Behrens, B.-A. Investigation of the joining zone displacement of cross-wedge rolled serially arranged hybrid parts. *Int. J. Mater. Form.* **2019**, *13*, 577–589. [CrossRef]
33. Pater, Z.; Tomczak, J.; Bulzak, T. New forming possibilities in cross wedge rolling processes. *Arch. Civ. Mech. Eng.* **2018**, *18*, 149–161. [CrossRef]
34. Pater, Z.; Tomczak, J.; Bulzak, T.; Wójcik, Ł.; Lis, K. Rotary compression in tool cavity—A new ductile fracture calibration test. *Int. J. Adv. Manuf. Technol.* **2020**, *106*, 4437–4449. [CrossRef]
35. Transvalor, S. *A Reference Documentation*; FORGE®NxT 3.0; Transvalor: Biot, France, 2019.
36. Hensel, A.; Spittel, T. *Kraft- und Arbeitsbedarf Bildsamer Formgebungsverfahren*, 1st ed.; Deutscher Verlag für Grundstoffindustrie: Leipzig, Germany, 1978.
37. Jagodzinski, A.; Kruse, J.; Barroi, A.; Mildebrath, M.; Langner, J.; Stonis, M.; Lammers, M.; Hermsdorf, J.; Hassel, T.; Behrens, B.-A.; et al. Investigation of the Prediction Accuracy of a Finite Element Analysis Model for the Coating Thickness in Cross-Wedge Rolled Coaxial Hybrid Parts. *Materials* **2019**, *12*, 2969. [CrossRef]

Article

Design, Setup, and Evaluation of a Compensation System for the Light Deflection Effect Occurring When Measuring Wrought-Hot Objects Using Optical Triangulation Methods

Lorenz Quentin *, Carl Reinke, Rüdiger Beermann and Markus Kästner and Eduard Reithmeier

Institute for Measurement and Automatic Control, Leibniz University Hannover, Nienburger Strasse 17, 30167 Hannover, Germany; reinke@imr.uni-hannover.de (C.R.); beermann@imr.uni-hannover.de (R.B.); kaestner@imr.uni-hannover.de (M.K.); sekretariat@imr.uni-hannover.de (E.R.)

* Correspondence: lorenz.quentin@imr.uni-hannover.de or quentin@imr.uni-hannover.de

Received: 10 June 2020; Accepted: 3 July 2020; Published: 7 July 2020

Abstract: In this paper, we present a system to compensate for the light deflection effect during the optical geometry measurement of a wrought-hot object. The acquired 3D data can be used to analyze the formed geometry of a component directly after a hot forging process without waiting for the needed cooling time to room temperature. This may be used to parameterize the process and to detect defect components early in the production process, among others. The light deflection as the deviation from the linear path of the light is caused by an inhomogeneous refractive index field surrounding the hot object. We present the design and setup for a nozzle-based forced air flow actuator, which suppresses the light deflection effect. The design process includes a simulation of the developing field, as well as of the interaction of the field with an external forced air flow. The cooling effect of the air flow is evaluated, and conclusions are drawn from the conflicting interests of good measurement conditions against the forced cooling of the hot object. The findings are then implemented in the physical setup of the suppression system. The system is evaluated using a previously established method based on optical triangulation and fringe projection. Other occurring effects and their influence on the evaluation are considered and discussed.

Keywords: tailored forming; bulk metal forming; geometry measurement; wrought-hot objects

---

1. Introduction

The increasing technical and economical requirements for components have led to mono material bulk metal components reaching the limits of their performance due to the material used. To extend those limits, the use of hybrid components is the subject of current research [1,2]. A high technological potential lays in hybrid bulk metal components that are first joined and then further processed. In the collaborative research center 1153 "Tailored Forming", such a process chain has been developed [3], following the concept "Put the right material in the right place". Due to the use of novel material combinations in hot bulk metal forming processes, the development of new evaluation techniques also needs to be considered, e.g., to validate new simulation models [4] for cross-wedge rolling [5]. The distortion induced by the combination of cooling and materials with different thermal expansion coefficients needs to be considered for the analysis of such processes. Therefore, the evaluation of the forming process by analyzing the component after cooling down to room temperature is not sufficient. Instead, the component and especially its geometry need to be measured while in the hot state directly after leaving the forming process. To this end, an optical 3D geometry measurement system based on triangulation can be used [6–8]. However, the hot object is surrounded by an inhomogeneous refractive index field, deflecting the measurement light from its linear path [9]. There have been attempts to estimate the effect [10] and methods to compensate its impact on triangulation measurements [11]. However, there are currently no in-line or close-to-line measurement systems that take the light deflection into account or compensate for it. The system presented in this paper aims to close that gap by adding a suppression device for the light deflection effect based on formed and forced air flow to an optical 3D geometry measurement system. The design, setup, and analysis of such a system is described in this paper.

2. Background

In this section, the literature for the methods to estimate the light deflection effect is revisited. The content includes the concept of the background-oriented Schlieren (BOS) method to reconstruct a refractive index field, as well as a presentation of a multi-camera fringe projection system (FPS) to quantify the influence of refractive gradients on triangulation measurements.

2.1. Background-Oriented Schlieren Method

The background-oriented Schlieren method was introduced to estimate density differences in a transparent medium from a set of images [12]. To this end, a contrast-rich and unique background was set up in front of a camera. An image was taken at zero-state as the reference, and additional images were captured during the measurement sequence. The displacement of image points due to the light deflection induced by the density gradient was quantified using suitable algorithms, e.g., Farnbäck et al. [13]. The estimation of the density gradient through the refractive index gradient required extensive knowledge about the measurement setup [14], e.g., the size and position of the refractive index field. This knowledge might be difficult to obtain in changing or unknown measurement conditions.

2.2. Using a Multi-Camera Fringe Projection System to Estimate the Direct Influence of the Light Deflection Effect on Optical Triangulation Measurements

In a previous paper [15], we used redundantly reconstructed 3D points of a multi-camera FPS to estimate the magnitude of the light deflection that was occurring when measuring a red-hot object. There, we concluded that the deviations of redundantly reconstructed points were a valid metric to quantify said effect. The validity of the method was shown using a discrete refractive index influence in the form of a glass plate. While we were not able to use the comparison of single object points, an analysis of all available reconstructed points using a histogram showed resounding results.

3. Proposed Method and Necessary Constraints

It is proposed to use a forced gas flow in order to reduce the light deflection effect of an inhomogeneous and dynamic refractive index field in an industrial environment. Beermann et al. [9] successfully used a forced laminar gas flow to lower the geometry deviations when measuring a hot rod using a laser light section method. From said article, it was concluded that the forced gas flow needed to be powerful enough to suppress the developed refractive index field while not cooling the hot component too much. To solve this conflicting interest, the velocities and flow parameters were simulated, as well as the cooling effect of such a forced gas flow to a hot component. To this end, the free convection of hot air above a hot cylindrical component needs to be simulated to gain insight into its characteristics. Additionally, the interaction of a forced gas flow with said free convection also needs to be analyzed. The result of that design process yields the necessary construction parameters for the nozzles, as well as the optimal solution for the necessary flow velocities in relation to both the suppression of the refractive index field and the cooling of the component.

The apparatus used for the forming of the forced convection should include an electronically triggered magnetic valve to open and close it conveniently from the same computer that controls the measurement system. To enable operational space for a handling system, as well as for the measurement light, the setup should use nozzle pipes to form the desired air flow at a set distance away from the specimen. The minimum distance was postulated to be approximately 500 mm.

4. Simulation

To design a forced flow apparatus to suppress the light deflection effect successfully, the described conflicting interest between the reduction of the light deflection and the forced cooling of the component needs to be addressed. To this end, the interaction of a forced air flow with the thermal field around an object was simulated to find a compromise. The simulations were conducted using ANSYS Discovery AIM. While the overall aim was to measure the geometry of crank shafts, the simulated object was simplified to a cylinder. Its main axis was concentric to the y-axis of world coordinate system (see Figure 1). It was placed in a rectangular, friction-less flow channel of dimensions $d_x = 200$ mm $\times d_z = 400$ mm. The gravitational acceleration of $g = 9.81$ m s^{-1} was assumed to point in negative z-direction. The simulation mode itself was transient and halted after $t_s = 5$ s. At that stage, a stationary state was reached.

To estimate the flow velocities induced by the density gradient around the hot object, the temperature of the cylinder was assumed to be $\vartheta_c = 1000$ °C, while the temperature of the ambient air was set to $\vartheta_\infty = 20$ °C. To simulate a simplified interaction between the nozzle flow and the heat-induced convection, a forced uniform flow was introduced into the simulation. According to Section 3, the direction of the forced flow was in the negative z-direction. The forced flow itself was modeled to be homogeneously distributed in the entire area of the flow channel used. Different forced flow velocities v_{ff} were combined with different cylinder diameters d_c to investigate its influence.

The light deflection effect was evaluated in relation to the thickness of air d_l with a temperature $\vartheta_{air} > 110$ °C. The thickness of the air layer was extracted from the simulation by measuring in a straight line up from the top center point of the cylinder (see Figure 2b). The cooling of the component was also simulated through the temperature of the topmost point on the cylinder, i.e., $\vartheta_c(z_c = max)$.

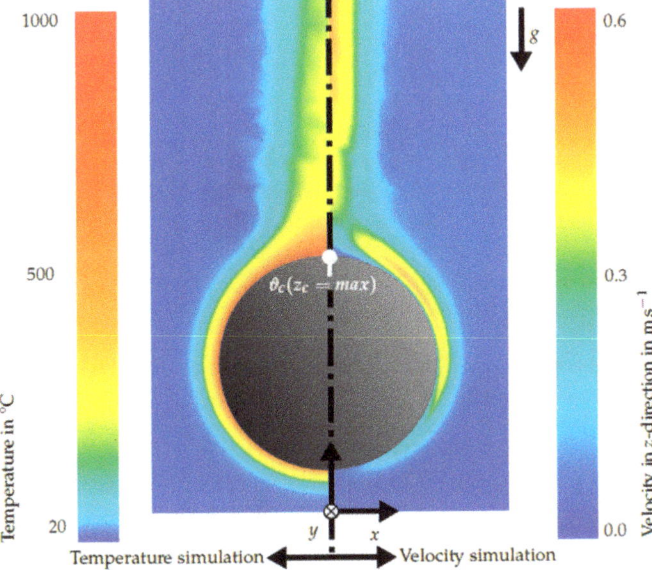

Figure 1. Comparison of the simulated convective flow over a cylinder with a diameter $d_c = 30$ mm and a temperature $\vartheta_c = 1000$ °C. The ambient temperature is $\vartheta_\infty = 20$ °C Left side: Results for the simulated temperature. Right side: Resulting air velocity in the z-direction.

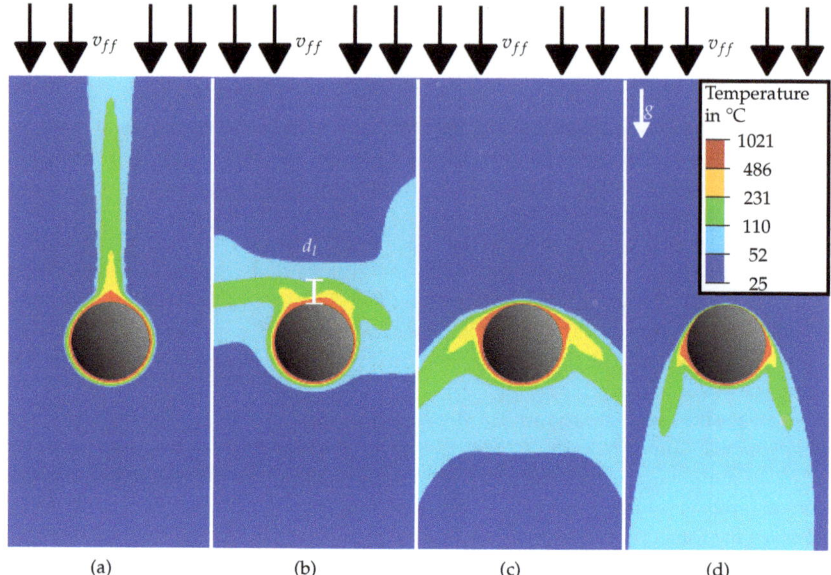

Figure 2. Simulation of the effect of different forced air flow velocities v_{ff} on the development of a heat-induced temperature field around a hot cylinder: (**a**) $v_{ff} = 0.0\,\mathrm{m\,s^{-1}}$; (**b**) $v_{ff} = 0.2\,\mathrm{m\,s^{-1}}$; (**c**) $v_{ff} = 0.55\,\mathrm{m\,s^{-1}}$; (**d**) $v_{ff} = 1.0\,\mathrm{m\,s^{-1}}$. The fixed boundary conditions are the following: diameter of cylinder $d_c = 37.5\,\mathrm{mm}$; initial temperature of cylinder $\vartheta_c = 1000\,°\mathrm{C}$, temperature of air $\vartheta_\infty = 25\,°\mathrm{C}$. The procedure for the calculation of the air layer thickness d_l is sketched in (**b**).

4.1. Simulation Results and Discussion

The developed temperature and velocity fields in the simulation without an additional forced air flow are shown in Figure 1. The size of both fields in y was not investigated here, since only a 2D cylindrical object was simulated. Overall, the shape of the convective flow over the cylindrical object was qualitatively similar to the one measured by, e.g., Beermann [14]. The simulation of the free convection field was therefore considered to be valid.

The effect of different forced air flow velocities on the development of the heat-induced temperature field is shown in Figure 2 in combination with Figure 3. Figure 2 shows the sectional view through the simulated temperature field at different forced flow velocities v_{ff}. There was an equilibrium-like state between v_{ff} and v_{cf} for $v_{ff} = 0.2\,\mathrm{m\,s^{-1}}$ (see Figure 2b). The simulation results for $v_{ff} > 1.0\,\mathrm{m\,s^{-1}}$ were conducted (see Figure 2), but are not shown here, since the differences with respect to Figure 2d were considered to be marginal.

Figure 3 shows the relation between the thickness of the air layer d_l with $\vartheta_{air} > 110\,°\mathrm{C}$ and the cooling rate ΔT_{ff} as a function of the forced air flow velocity v_{ff}. We estimated a function to fit into the air layer data (red dashed line), corresponding to the general style of:

$$d_l = \frac{a}{bv_{ff}^3 + c}. \tag{1}$$

In this equation, a, b, and c are arbitrary values not corresponding to any physical attributes. The equation itself is similar to an inverse relation between d_l and v_{ff}, as the thickness can be considered to be infinite for $d_l(v_{ff} \to 0) \to \inf$ and zero for larger values $d_l(v_{ff} \to \inf) \to 0$, which corresponded to the development of the field in Figure 2. The displayed line was considered a helpful tool for the visualization of the development of the thickness of the hot air layer; therefore, a description of the optimized parameters was omitted.

Prior to the simulation, the thickness of the air layer was expected to also be a function of the diameter of the cylinder $d_l = f(d_c)$. This seemed to be the case up to $v_{ff} \leq 0.3\,\mathrm{m\,s^{-1}}$, while being less distinct for $v_{ff} < 0.3\,\mathrm{m\,s^{-1}}$. The only small differences between the free convection velocities $v_{fc}(d_c)$ hinted at a marginal influence of the cylinder diameter on the free convection.

Considering the cooling effect of the forced flow onto the cylinder (blue lines in Figure 3), the results showed an increasing cooling effect with increasing flow velocities. The cooling effect was assumed to be inversely proportional to the radius of the cylinder, which was considered to be due to the decreasing volume-surface quotient.

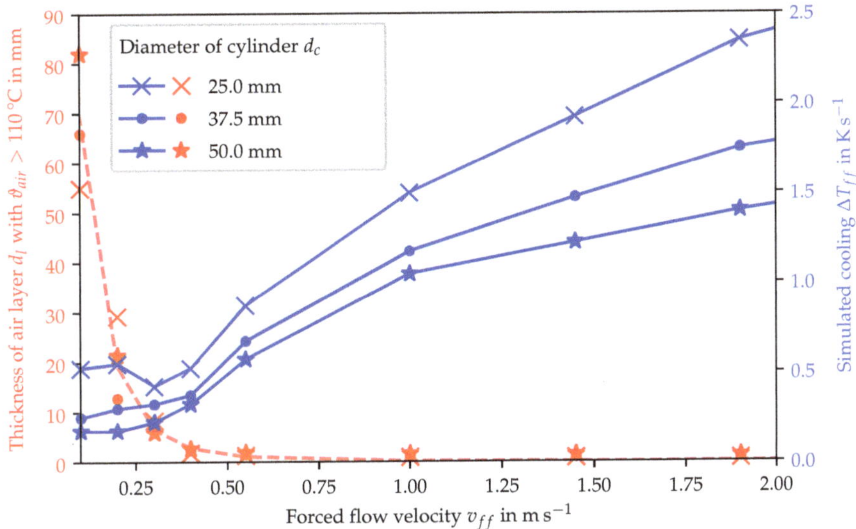

Figure 3. The thickness of the hot air layer d_l and the cooling of the component as functions of the cylinder diameter d_c and the forced air flow velocities v_{ff}. The initial velocity of the free convective flow v_{fc} varies and is shown in Table 1.

Table 1. Initial velocity of the free convective flow v_{fc} in relation to the cylinder diameter d_c.

d_c in mm	25.0	37.5	50.0
v_{fc} in $\mathrm{m\,s^{-1}}$	0.57	0.57	0.68

4.2. Simulation Conclusions

The simulations showed a direct connection between forced flow velocity v_{ff} to the layer thickness d_l and the cooling rate ΔT_{ff}. While a minimum layer thickness was desired, the reduction of the layer thickness for flow velocities $v_{ff} > 1\,\mathrm{m\,s^{-1}}$ was considered marginal compared to the increase of the cooling rate ΔT_{ff} through the forced air flow. Therefore, the flow velocities for the design of the actuator were chosen to be $1\,\mathrm{m\,s^{-1}} < v_{ff} < 4\,\mathrm{m\,s^{-1}}$ across the whole cylinder surface, assuming a linear increase in cooling ΔT_{ff} with further increasing v_{ff}.

To transfer the findings from the simulation to the design of a laboratory setup, a setup with three nozzles was implemented. The main aim was to achieve a mostly homogeneous velocity distribution over the area of a cylinder (diameter $d_c = 50\,\mathrm{mm}$; length $l_c = 250\,\mathrm{mm}$) at a distance of $500\,\mathrm{mm}$. The comparison between the simulated velocities and the measured velocities of the nozzle setup is shown in the following section 5. When considering the interaction between the light from the

FPS and the forced air flow, it was concluded that the measurement light would most be affected by the warmer areas under the hot object (light blue areas in Figure 2c,d). Therefore, the combined setup should include nozzles and FPS viewing the object from roughly the same direction.

5. Setup

The setup for the experiments is presented in this section. An image of the complete setup is shown in Figure 4. The measurement system followed the concept of multi-camera fringe projection profilometry and was based on the findings of Bräuer-Burchard et al. [16] and Reich et al. [17]. We explained the calibration and the main functions of the measurement system in a previous paper [15]. The forced air flow actuator consisted of three commercially available nozzles with a minimum inner diameter of $d_n = 0.8\,\text{mm}$ and a custom-made aluminum connection block. The design study examining different flow directions and nozzle setups resulting in the presented setup is shown in Appendix A. The connection block placed the nozzles at equidistant points on a line with 100 mm in between them and an angle to the outside of 5°. A magnetic valve and a manually operated pressure regulator connected the nozzles to the pressurized air container in the laboratory. An image of the nozzles is part of Figure 4 (upper left). The nozzles were set up above the measurement object to use the occurring forces to hold the measurement object in place. To ensure that the measurement light and the forced air flow took similar paths, the FPS was also mounted above the object. Depending on the air pressure p_{ff} set in the pressure regulator, the velocities of the air flow field were controlled. Figure 5 shows the designed and simulated velocity field in comparison to a measured field using an air pressure of $p_{ff} = 2.5\,\text{bar}$. The flow velocity measurements were conducted using a scale and a hand-held anemometer. The flow velocity field showed significant differences from the simulated velocity field in magnitude and in form. The main reasons for the different shapes lied in the lack of a concentric effect of screw threads, which were used to connect the nozzles and the connection block. The differences in velocity magnitude were mainly due to the low accuracy of the hand-held anemometer, as well as numerical dispersion and diffusion effects in the simulation.

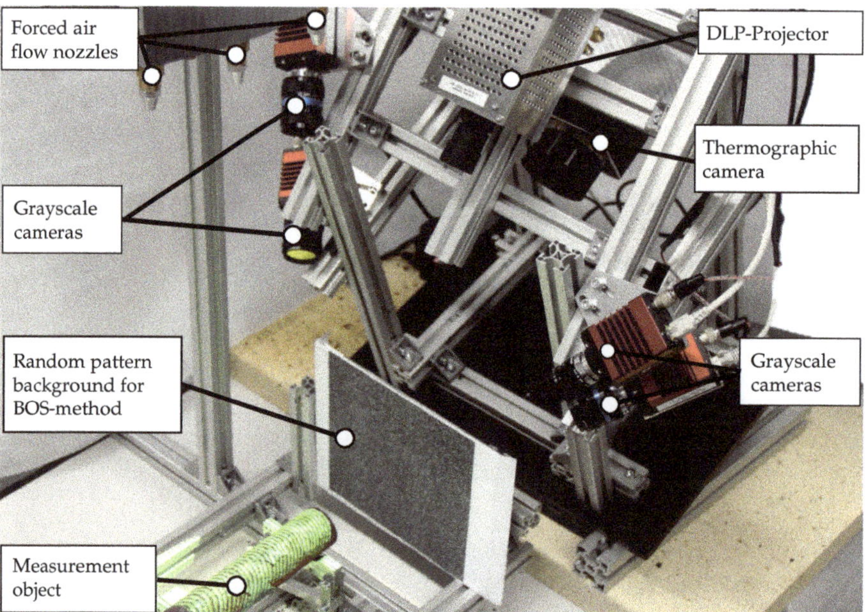

Figure 4. Image of the complete measurement setup.

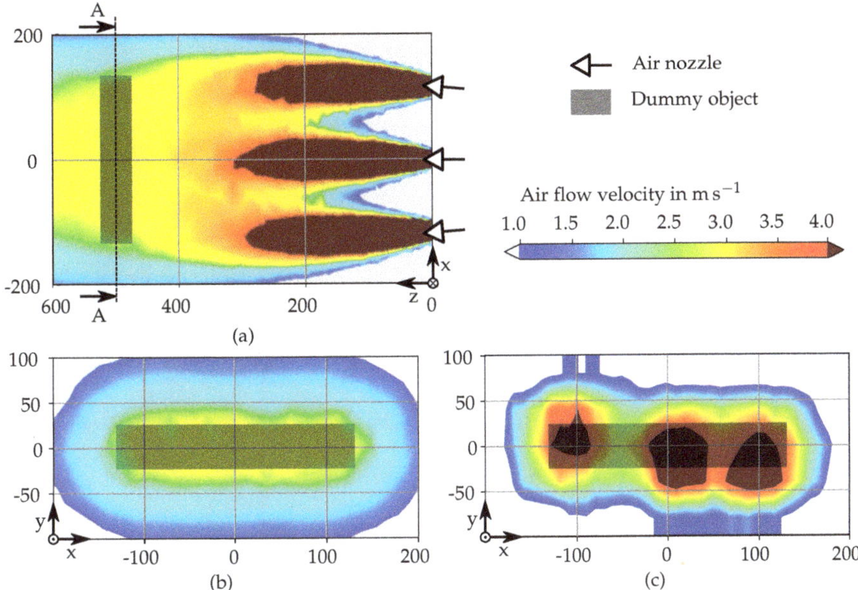

Figure 5. Comparison of the simulated and measured flow velocity fields of the nozzle setup. (**a**) Top view on the simulated air flow; (**b**) Section A-A through (**a**); (**c**) measurement of air flow velocities comparable to (**b**) in the implemented experimental setup. All given distances are in mm.

Even though there were velocity differences of up to 90 %, the overall design process was considered successful, since the aim of surrounding a cylinder ($d_c = 50\,\text{mm}, l_c = 250\,\text{mm}$) with an homogeneous air flow field with $1\,\text{m s}^{-1} < v_{ff} < 4\,\text{m s}^{-1}$ was achieved.

6. Experiments

For the experiments, the magnetic valve for the forced air flow was connected to the projecting sequence of the measurement system using a software trigger. To ensure a fully developed air flow field and account for triggering delays, the projection of the first image was delayed by $t_{d,0} = 0.3\,\text{s}$ compared to the opening of the magnetic valve. The measurement sequence started with a measurement without the air flow and was followed by a measurement with an active flow system to reduce the effect of cooling on the analysis and ensure a fully developed refractive index field during the measurement without forced air flow.

6.1. Experimental Plan

To evaluate the influence of the forced air flow on the light deflection effect, different experiments were conducted:

1. Measurement of the cooling effect of the forced air flow onto the cylinder,
2. using the pixel displacement from the BOS method to estimate the indirect influence of the air flow on the light deflection effect and
3. analysis of the direct influence by means of using the reconstruction error metric.

The cooling effect (1) was measured using the thermographic infrared (IR) camera (VarioCAM from InfraTec, Dresden, Germany, with a maximum frame rate of 10 Hz) mounted underneath the projector (see Figure 4). To this end, a red-hot cylinder was measured for a duration of 18 s at the maximum frame rate. During the temperature measurement, the magnetic valve was opened for 8 s.

The emission coefficient was set to one ($\varepsilon = 1$), and the mean temperature of an arbitrarily selected rectangle is plotted in relation to the passed time in Figure 6.

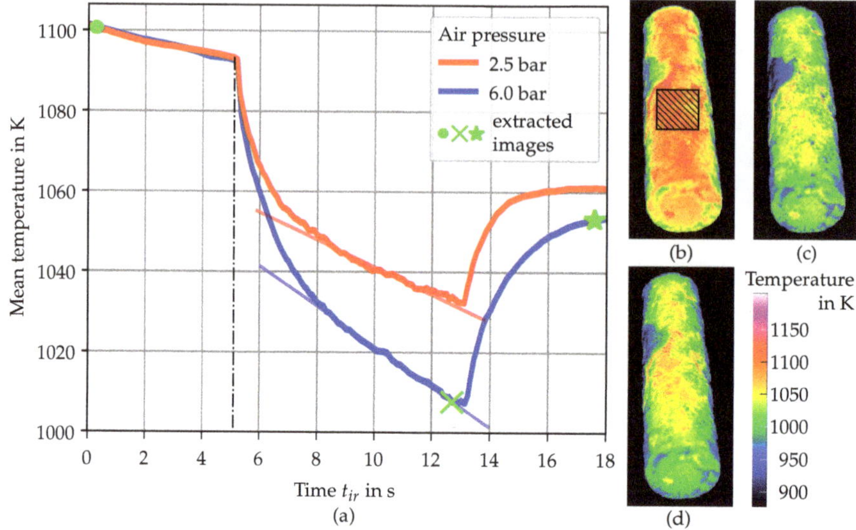

Figure 6. Thermographic images from the IR camera and the development of the mean temperature. (**b–d**) are taken at the specified points in time indicated by the green spots in (**a**). The window from which the mean temperature data in (**a**) is taken is marked by a crosshatched rectangle in (**b**).

To estimate the size of a refractive gradient field qualitatively (2), the BOS method was used. To this end, a video camera (not shown) was placed in front of the random pattern background without the measurement object in between both. The video capture was started before the hot cylinder was placed. After an undisclosed time delay, the magnetic valve was opened. The first image of the video was used as the reference image to calculate the results. Here, the pixel displacement was used as a means to quantify the light deflection relatively, since a reconstruction of valid refractive index values required extensive a priori knowledge about the setup and the measured field. This procedure was valid in the presented case, since the experiments were conducted without moving any parts of the setup during the video sequence.

To analyze the effect of the interaction of the forced air flow actuator and refractive index field on triangulation measurements (3), the multi-camera FPS was used. Here, the cylinder was placed, and a reference measurement without active air flow was conducted. This was directly succeeded by activating the air flow actuator, waiting a defined delay time t_d, and then, triggering the start of the measurement. This procedure reduced the uncertainties induced by moving or exchanging the objects, as discussed in [15]. There was a difference between the direct and the indirect method of estimation for the light deflection effect. This was mainly due to the different paths of light that were examined. When evaluating Figure 4, it is clear that the light for the FPS was traversing different paths compared to the view rays for the BOS method. Due to the 2D nature of the BOS measurement setup, the pixel-wise magnitude of the light deflection effect could not be correlated directly to the reconstructed 3D points. Additionally, the random pattern background used for BOS measurements was blocking the field of view of the lower cameras and of the IR camera. Therefore, the experiments were not carried out simultaneously.

6.2. Experimental Results

The results from the cooling analysis are shown in Figure 6. The thermographic images (Figure 6b–d) are excerpts from the full sequence of images at specific points in time (see the green spots in Figure 6a). To estimate the cooling rate ΔT_{ff}, linear functions were fitted into the sectors with constant gradients. The slope of these functions was the cooling rate and summarized in Table 2.

Table 2. Cooling rates before and during compensation.

Pressure p_{ff} on Air Flow Actuator	0.0 bar	2.5 bar	6.0 bar
Cooling rate	$1.6\,\mathrm{K\,s^{-1}}$	$3.4\,\mathrm{K\,s^{-1}}$	$5.0\,\mathrm{K\,s^{-1}}$
Time sector definition	$0\,\mathrm{s} < t_{ir} < 5\,\mathrm{s}$	$8\,\mathrm{s} < t_{ir} < 13\,\mathrm{s}$	$8\,\mathrm{s} < t_{ir} < 13\,\mathrm{s}$

The results from the BOS experiment are shown in Figure 7. The high displacement values in Figure 7b–d in the lower middle were a result of the hot cylinder not being in place when the reference image was shot. Therefore, the distance from a similar looking pixel group in the reference image was high.

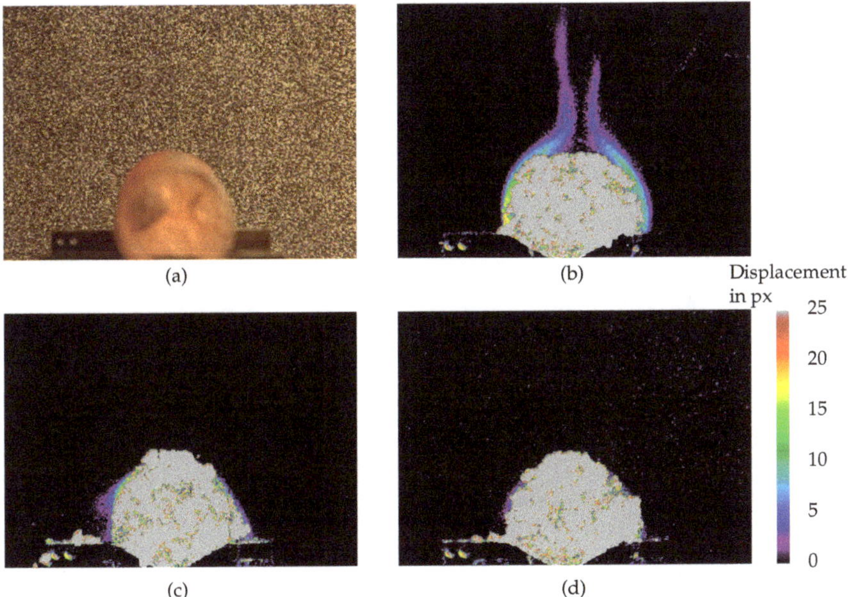

Figure 7. Results from the BOS experiment. (**a**) shows one of the regular images used to calculate the pixel displacement; (**b**) shows the result of the displacement calculation for an uninfluenced convection field; (**c**) the result for an air flow at $p_{ff} = 2.5\,\mathrm{bar}$; (**d**) the result for an air flow at $p_{ff} = 6.0\,\mathrm{bar}$.

The estimated influence of the refractive index field was observable in the comparison of the measurement of the cold object and the object in the red-hot state (see Figure 8) using the direct estimation method and the multi-camera FPS. The mean values of the shown histograms are found in Table 3 and Figure 9. The reconstruction error E_m in the reconstruction quality maps (Figure 8a–c) was not homogeneously distributed, with higher E_m values in the upper parts of the images, corresponding to points further away from the FPS. Both the reconstruction quality maps and the histograms showed only marginal differences between the experiments. An example of the reconstructed 3D datasets is shown in Appendix B.

Table 3. Mean deviation of the evaluated corresponding 3D points from the multi-camera fringe projection system.

Conditions	Deactivated Air Flow	Activated Air Flow
$T_c \approx 300\,\mathrm{K}; p_{ff} = 6.0\,\mathrm{bar}$	0.259 mm	0.253 mm
$T_c \approx 1300\,\mathrm{K}; p_{ff} = 2.5\,\mathrm{bar}; t_d = 0.3\,\mathrm{s}$	0.392 mm	0.645 mm
$T_c \approx 1300\,\mathrm{K}; p_{ff} = 2.5\,\mathrm{bar}; t_d = 3.3\,\mathrm{s}$	0.383 mm	0.316 mm
$T_c \approx 1300\,\mathrm{K}; p_{ff} = 6.0\,\mathrm{bar}; t_d = 0.3\,\mathrm{s}$	0.404 mm	0.688 mm
$T_c \approx 1300\,\mathrm{K}; p_{ff} = 6.0\,\mathrm{bar}; t_d = 3.3\,\mathrm{s}$	0.341 mm	0.281 mm

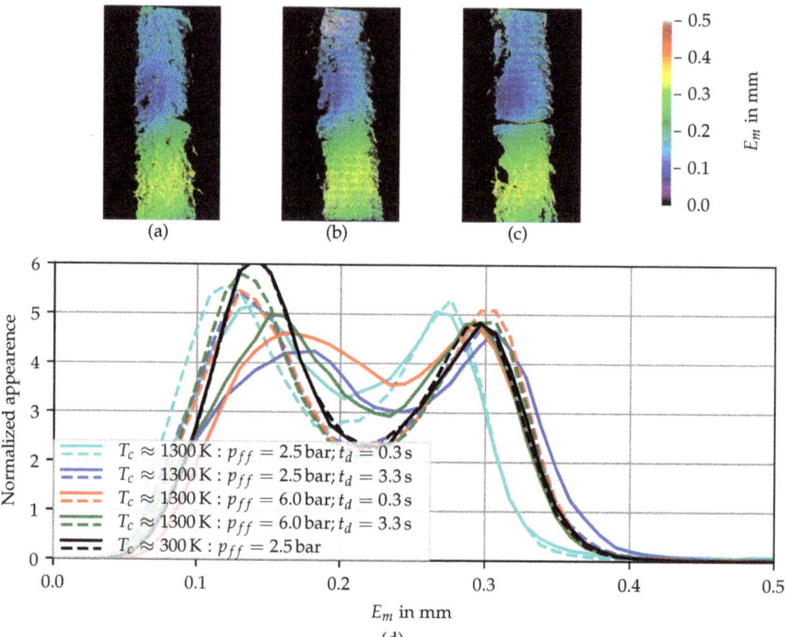

Figure 8. Evaluation of the results from the reconstructed corresponding points of the FPS. (**a–c**) show an excerpt from the full measurement with an activated forced flow actuator, in which the cylinder can be observed from all used cameras. (**a**) is taken at $T_c \approx 300\,\mathrm{K}; p_{ff} = 6.0\,\mathrm{bar}$; (**b**) is taken at $T_c \approx 1300\,\mathrm{K}; p_{ff} = 6.0\,\mathrm{bar}, t_d = 0.3\,\mathrm{s}$; (**c**) is taken at $T_c \approx 1300\,\mathrm{K}; p_{ff} = 2.5\,\mathrm{bar}, t_d = 3.3\,\mathrm{s}$; (**d**) shows the summary of the conducted experiments as histograms. Dashed lines indicate an activated flow actuator, while solid lines show no external influence on the refractive index field.

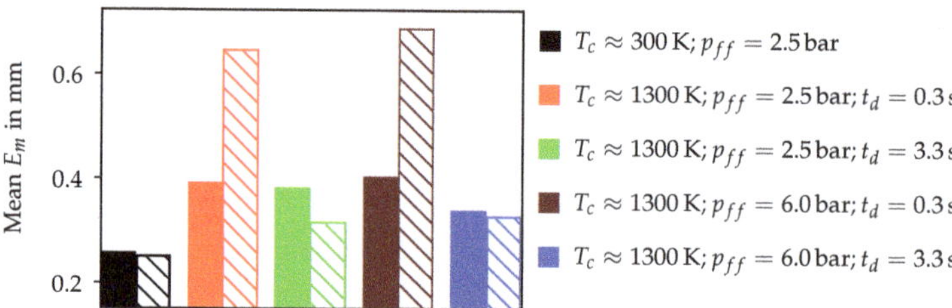

Figure 9. Summary of the measured reconstruction errors E_m (cf. Table 3) as a bar diagram of the average values. The solid area indicates a deactivated flow, while the hatched area is the result of an activated forced air flow.

Other observations during the experiment included a noticeable cooling of the component during the measurement as observable through a change of color. The applied force of the air field did not cause any visible vibration or movement of the object.

7. Discussion

Overall, the direct effect of the forced air flow on the development of a refractive index field around a hot cylinder was shown using the BOS method. The qualitative influence was similar in the simulation and measurement (compare Figures 5 and 7).

However, the effect was less distinguishable in the FPS measurements (see Figure 8 and Table 3). Here, the activation of the air flow yielded higher reconstruction errors for the immediate use ($t_d = 0.3\,\text{s}$) of the forced air flow compared to the uninfluenced measurement (compare the dashed and solid red and cyan lines in Figure 8d). This indicated a contradiction: the size of the light-deflecting refractive index field was reduced, but the reconstruction error was increased. In the presented setup, the change of the radiation color of the specimen might have an increasing influence on the reconstruction quality. A cooling of the object resulting in a lowering of the radiation through self-emission was observed during the experiment and could be concluded from the thermographic measurement (see Figure 6). To investigate the stated matter, the necessary time delay between the activation of the air flow and the start of the projection sequence was estimated to be between the start of the flow activation (dotted-dashed line) and the start of the linear temperature development at $t_{ir} \approx 8.0\,\text{s}$. Adding the activation time $t_{d,0}$ resulted in an improved time delay of $t_d = 3.3\,\text{s}$. The result of the implementation of this improved delay is shown in Figure 8 (blue/green lines). It is obvious that the improved time delay resulted in an improvement in the estimated reconstruction quality compared to the initial time delay $t_d = t_{d,0} = 0.3\,\text{s}$, as well as the measurement without the forced air flow.

Any concerns about an induced vibration through the forced air flow and a resulting decrease of reconstruction quality could be dispelled through the comparison of the measurements at $T_c \approx 300\,\text{K}$. If any vibration would have been induced, a significant increase in the reconstruction error metric would have been noticed.

8. Conclusions

Overall, the influence of the forced air flow on the light deflection effect was shown. The BOS measurement showed a significant reduction in pixel displacement, which was considered to be proportional to the light deflection. However, the results from the reconstruction quality experiment showed an only marginal improvement through the forced air flow. The reasons for this were already stated in the previous section and seemed plausible considering the results from the cooling experiment and the improved time delay. However, the reconstruction quality for this setting was lower compared to the experiment at $T_c \approx 300\,\text{K}$. Therefore, the change of radiation color through cooling needs to be investigated as an additional influence on the optical geometry measurement of hot objects.

This indicated a superposition of effects: on the one hand, the reduced light deflection resulted in a reduced reconstruction error metric while there needed to be an additional effect to then increase the error metric again. The underlying effect here might be the change of the surface color during the air flow induced cooling. The fringe projection reconstruction technique based on Peng's multi-frequency phase-shift method [18] used requires a constant ambient illumination strength. This ambient light influence is compensated using the temporal average of the phase shift sequence of the highest frequency. A change of background illumination may cause an error in phase reconstruction for each camera and, therefore, a change in the reconstruction error metric. However, a further investigation of this matter is beyond the scope of this paper and is therefore omitted.

9. Summary and Future Work

In this paper, a setup was presented to reduce the light deflection effect when using optical triangulation systems to measure the geometry of a red-hot object. A simulation was used to estimate the velocity of the free convection around the hot object. The same tool simulated the interaction of a forced convection flow with the free convection to gain insight into the necessary flow velocities to suppress the development of the refractive index field around the hot object. That knowledge was used to simulate and implement a forced air flow actuator to be used in combination with a fringe projection system. The effect of this actuator on the light deflection effect was indirectly estimated using a BOS-based method and directly measured using a reconstruction error metric. While the indirect method showed a minimization of the refractive index field, the direct method showed a reduction of the reconstruction error while not reaching the quality level of the measurement of a cold object. Different reasons for this behavior were discussed, but not further investigated.

Overall, the aim of the developed setup (reduction of reconstruction error) was reached while not being sufficient in comparison with a standard measurement. While the design method resulted in a successful development of the setup, it may not be generalizable due to changing boundary conditions when varying the geometry of the measurement object. The developed setup in the present configuration may be used to inspect shaft-like components.

In the future, an investigation of the changing surface color due to the cooling effect of the forced air flow needs to be conducted. Findings of this investigation can then be used to compensate for a changing background color, if the effect is found to be significant.

Author Contributions: Conceptualization, L.Q., C.R. and R.B.; Data curation, L.Q. and C.R.; Formal analysis, L.Q. and R.B.; Funding acquisition, M.K. and E.R.; Investigation, L.Q. and C.R.; Methodology, L.Q.; Project administration, M.K. and E.R.; Resources, L.Q.; Software, L.Q.; Supervision, M.K. and E.R.; Validation, L.Q.; Visualization, L.Q. and C.R.; Writing—original draft, L.Q.; Writing—review & editing, C.R. and M.K. All authors have read and agreed to the published version of the manuscript.

Funding: This research was funded by Deutsche Forschungsgemeinschaft (DFG); Collaborative Research Centre 1153 (CRC) Process Chain to Produce Hybrid High-performance Components Through Tailored Forming (252662854); Subproject C5 Multiscale Geometry Inspection of Joining Zones.

Conflicts of Interest: The authors declare no conflict of interest.

Abbreviations

The following abbreviations are used in this manuscript:

BOS   Background-oriented Schlieren
FPS   Fringe projection system
IR   Infrared

Appendix A. Simulated Case Study

The results from the simulated case study are shown in Figure A1 regarding the direction of the forced air flow and in Figure A2 for the design of the nozzle setup. Analyzing the direction of the air flow focuses on areas that exhibit low air flow velocities ($u < 0.36\,\mathrm{m\,s^{-1}}$). The air stays in the same place in these areas and is consequently heated by the hot cylinder. Therefore, there will be a refractive index gradient in these areas at a magnitude comparable to the gradient without air flow. For the given reasons, the air flow should be applied comparable to Case (a), i.e., perpendicular to the cylinder main axis. In all cases, the measurement system should be placed near the forced air flow actuator to avoid areas of low air flow velocities interfering with the measurement light.

Focusing on the nozzle setup, a case study was conducted using ANSYS Discovery AIM. The examined cases are shown in Figure A2. The results are comparable to the ones in Figure 5 while showing a slightly different excerpt from the simulation field. Analyzing the case study, an homogeneous velocity profile of 250 mm at a distance of 500 mm to the nozzles (cf. dashed lines in Figure A2) is considered to be desirable (cf. Section 5). A sectional view through the simulated velocities fields is shown in Figure A3. The simulated profiles are symmetrical to $x = 0\,\mathrm{mm}$. The setup using three nozzles and an angle of 5° show a nearly homogeneous profile at approximately $3.0\,\mathrm{m\,s^{-1}}$ for $-125\,\mathrm{mm} < x < 125\,\mathrm{mm}$. This is the chosen nozzle configuration for the suppression device.

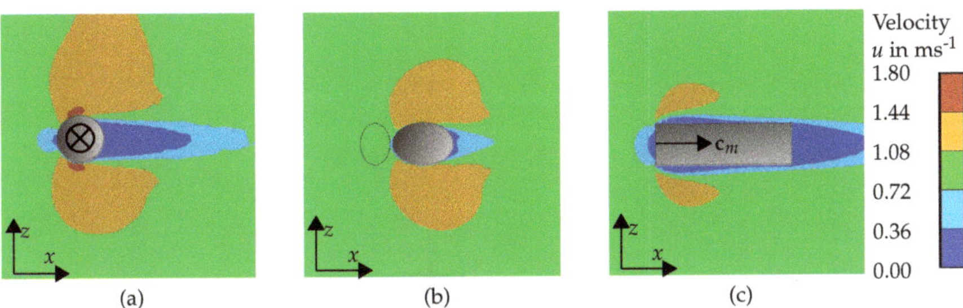

Figure A1. Results from the simulated case study examining different forced air flow $v_{ff} = 1.0\,\mathrm{m\,s}^{-1}$ directions in relation to a cylinder. The direction is varied by tilting the object and is described by the cylinder main axis $\mathbf{c}_m$ in world coordinates. The vector for the direction of the forced air flow is $(1,0,0)^T$. (**a**) $\mathbf{c}_m = (0,1,0)^T$; (**b**) $\mathbf{c}_m = (1,1,0)^T$; (**c**) $\mathbf{c}_m = (1,0,0)^T$.

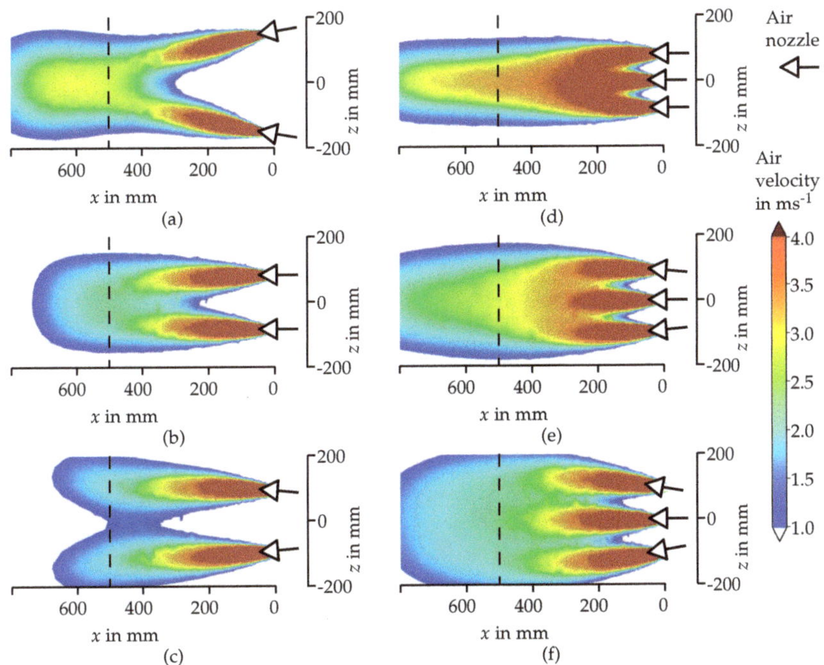

Figure A2. Results from the simulated case study using multiple nozzles. A positive angle value indicates nozzles pointing towards each other, and a negative value indicates nozzles pointing away from each other. The dashed line depicts the design distance, in which an homogeneous field is desired. Air flow velocities $v_{ff} < 1.0\,\mathrm{m\,s}^{-1}$ are masked and depicted in white. (**a**) Two nozzles 400 mm apart with an angle of $-10°$; (**b**) two nozzles 200 mm apart with an angle of $0°$; (**c**) two nozzles 200 mm apart with an angle of $5°$; (**d**) three nozzles with an angle of $0°$; (**e**) three nozzles with an angle of $5°$; (**f**) three nozzles with an angle of $10°$.

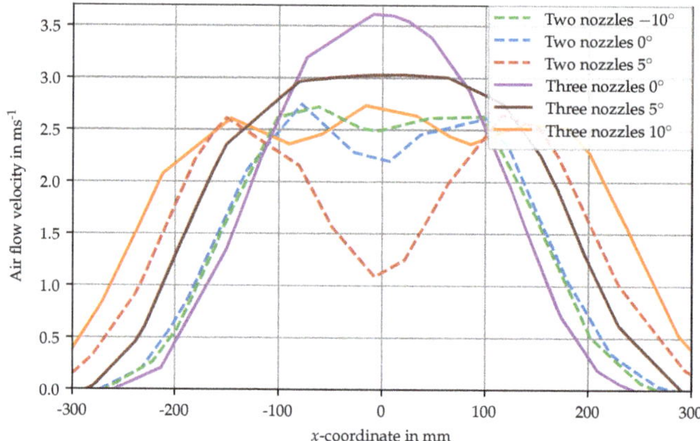

Figure A3. Sectional view through the air flow velocities profiles of Figure A2 (dashed lines).

Appendix B. 3D Geometry Reconstructions

An example set of reconstructed height maps $z(\mathbf{u}_p)$ is shown in Figure A4. These height maps are the fundamentals to calculate the deviations maps $E_m(\mathbf{u}_p)$. There is one height map for each camera-camera and camera-projector pair. The height maps reconstructed from cameras on the same side of the measurement setup are omitted (explained in detail in [15]).

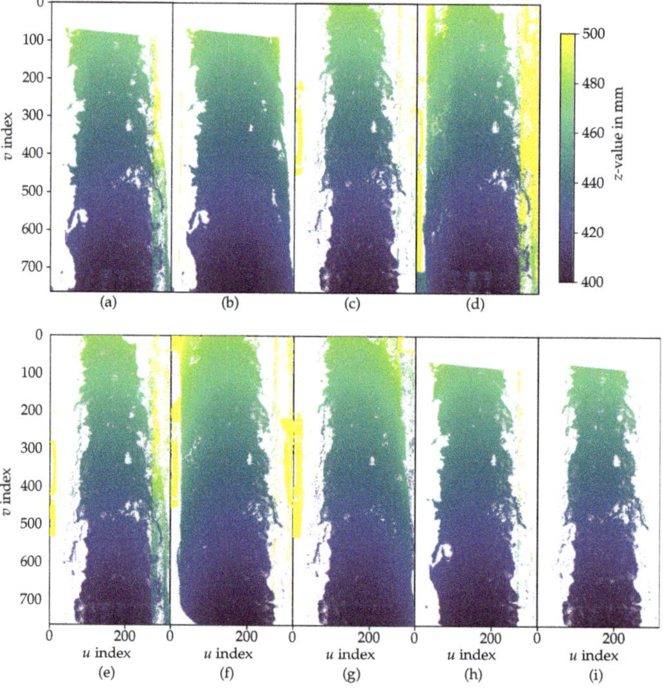

Figure A4. Example of a set of reconstructed height maps, depicted as the map of the corresponding z-values. (**a–h**) Height maps from each camera-camera pair and each camera-projector pair, respectively; (**i**) average height map of all available sets. All points are masked that are not present in each height map. Masked points are not considered when calculating the reconstruction quality metric.

References

1. Behrens, B.A.; Chugreev, A.; Selinski, M.; Matthias, T. Joining zone shape optimisation for hybrid components made of aluminium-steel by geometrically adapted joining surfaces in the friction welding process. *AIP Conf. Proc.* **2019**, *2113*, 040027.
2. Behrens, B.A.; Chugreev, A.; Matthias, T.; Poll, G.; Pape, F.; Coors, T.; Hassel, T.; Maier, H.J.; Mildebrath, M. Manufacturing and evaluation of multi-material axial-bearing washers by tailored forming. *Metals* **2019**, *9*, 232. [CrossRef]
3. Uhe, J.; Behrens, B.A. Manufacturing of Hybrid Solid Components by Tailored Forming. In *Production at the Leading Edge of Technology*; Springer: Berlin/Heidelberg, Germany, 2019; pp. 199–208.
4. Böhm, C.; Kruse, J.; Stonis, M.; Aldakheel, F.; Wriggers, P. Virtual Element Method for Cross-Wedge Rolling during Tailored Forming Processes. *Procedia Manuf.* **2020**, *47*, 713–718. [CrossRef]
5. Kruse, J.; Jagodzinski, A.; Langner, J.; Stonis, M.; Behrens, B.A. Investigation of the joining zone displacement of cross-wedge rolled serially arranged hybrid parts. *Int. J. Mater. Form.* **2019**, 1–13. [CrossRef]
6. Hawryluk, M.; Ziemba, J.; Sadowski, P. A review of current and new measurement techniques used in hot die forging processes. *Meas. Control* **2017**, *50*, 74–86. [CrossRef]
7. Han, L.; Cheng, X.; Li, Z.; Zhong, K.; Shi, Y.; Jiang, H. A Robot-Driven 3D Shape Measurement System for Automatic Quality Inspection of Thermal Objects on a Forging Production Line. *Sensors* **2018**, *18*, 4368. [CrossRef]
8. Mejia-Parra, D.; Sánchez, J.R.; Ruiz-Salguero, O.; Alonso, M.; Izaguirre, A.; Gil, E.; Palomar, J.; Posada, J. In-Line Dimensional Inspection of Warm-Die Forged Revolution Workpieces Using 3D Mesh Reconstruction. *Appl. Sci.* **2019**, *9*, 1069. [CrossRef]
9. Beermann, R.; Quentin, L.; Pösch, A.; Reithmeier, E.; Kästner, M. Light section measurement to quantify the accuracy loss induced by laser light deflection in an inhomogeneous refractive index field. In Proceedings of the SPIE Optical Metrology, Optical Measurement Systems for Industrial Inspection X, Munich, Germany, 25–29 June 2017; Volume 10329, p. 103292T.
10. Quentin, L.; Beermann, R.; Pösch, A.; Reithmeier, E.; Kästner, M. 3D geometry measurement of hot cylindric specimen using structured light. In Proceedings of the SPIE Optical Metrology, Optical Measurement Systems for Industrial Inspection X, Munich, Germany, 25–29 June 2017; Volume 10329, p. 103290U.
11. Beermann, R.; Quentin, L.; Reithmeier, E.; Kästner, M. Fringe projection system for high-temperature workpieces–design, calibration, and measurement. *Appl. Opt.* **2018**, *57*, 4075–4089. [CrossRef] [PubMed]
12. Richard, H.; Raffel, M. Principle and applications of the background oriented schlieren (BOS) method. *Meas. Sci. Technol.* **2001**, *12*, 1576. [CrossRef]
13. Farnebäck, G. Two-frame motion estimation based on polynomial expansion. In *Scandinavian Conference on Image Analysis*; Springer: Berlin/Heidelberg, Germany, 2003, pp. 363–370.
14. Beermann, R.; Quentin, L.; Pösch, A.; Reithmeier, E.; Kästner, M. Background oriented schlieren measurement of the refractive index field of air induced by a hot, cylindrical measurement object. *Appl. Opt.* **2017**, *56*, 4168–4179. [CrossRef] [PubMed]
15. Quentin, L.; Beermann, R.; Kästner, M.; Reithmeier, E. Development of a reconstruction quality metric for optical three-dimensional measurement systems in use for hot-state measurement object. *Opt. Eng.* **2020**, *59*, 064103. [CrossRef]
16. Bräuer-Burchardt, C.; Möller, M.; Munkelt, C.; Heinze, M.; Kühmstedt, P.; Notni, G. Determining exact point correspondences in 3D measurement systems using fringe projection—Concepts, algorithms, and accuracy determination. *Appl. Meas. Syst.* **2012**, 211–228. [CrossRef]
17. Reich, C.; Ritter, R.; Thesing, J. 3-D shape measurement of complex objects by combining photogrammetry and fringe projection. *Opt. Eng.* **2000**, *39*, 224–231. [CrossRef]
18. Peng, T.; Gupta, S.K.; Lau, K. Algorithms for constructing 3-D point clouds using multiple digital fringe projection patterns. *Comput.-Aided Des. Appl.* **2005**, *2*, 737–746. [CrossRef]

Article

Microstructural Evolution and Mechanical Properties of Hybrid Bevel Gears Manufactured by Tailored Forming

Bernd-Arno Behrens [1], Anna Chugreeva [1,*], Julian Diefenbach [1], Christoph Kahra [2], Sebastian Herbst [2], Florian Nürnberger [2] and Hans Jürgen Maier [2]

[1] Institut für Umformtechnik und Umformmaschinen (Forming Technology and Machines), Leibniz Universität Hannover, 30823 Garbsen, Germany; behrens@ifum.uni-hannover.de (B.-A.B.); diefenbach@ifum.uni-hannover.de (J.D.)

[2] Institut für Werkstoffkunde (Materials Science), Leibniz Universität Hannover, 30823 Garbsen, Germany; kahra@iw.uni-hannover.de (C.K.); herbst@iw.uni-hannover.de (S.H.); nuernberger@iw.uni-hannover.de (F.N.); maier@iw.uni-hannover.de (H.J.M.)

* Correspondence: chugreeva@ifum.uni-hannover.de; Tel.: +49-51176218280

Received: 17 September 2020; Accepted: 8 October 2020; Published: 13 October 2020

Abstract: The production of multi-metal bulk components requires suitable manufacturing technologies. On the example of hybrid bevel gears featuring two different steels at the outer surface and on the inside, the applicability of the novel manufacturing technology of Tailored Forming was investigated. In a first processing step, a semi-finished compound was manufactured by cladding a substrate using a plasma transferred arc welding or a laser hotwire process. The resulting semi-finished workpieces with a metallurgical bond were subsequently near-net shape forged to bevel gears. Using the residual heat after the forging process, a process-integrated heat treatment was carried out directly after forming. For the investigations, the material combinations of 41Cr4 with C22.8 (AISI 5140/AISI 1022M) and X45CrSi9-3 with C22.8 (AISI HNV3/AISI 1022M) were applied. To reveal the influence of the single processing steps on the resulting interface, metallographic examinations, hardness measurements and micro tensile tests were carried out after cladding, forging and process-integrated heat treatment. Due to forging and heat-treatment, recrystallization and grain refinement at the interface and an increase in both, hardness and tensile strength, were observed.

Keywords: hybrid components; bevel gears; hot forging; tailored forming; process-integrated heat treatment; air-water spray cooling; self-tempering

1. Introduction

The material choice for conventional monolithic components, commonly used in bulk metal forming at industrial scale, represents a compromise between material properties and economical requirements [1]. In parts with locally varying mechanical, thermal or chemical operating conditions, it is rather challenging to identify a single material, which fulfils the individual specifications. Increasing demands on technical components due to current trends towards resource and energy efficiency render this issue even more difficult and have triggered the development of alternative technological solutions to overcome the material-specific restrictions. In this context, multi-material (hybrid) designs combining the benefits of different materials in a single component offer a great potential for creating high-performance components with extended functionality and resource efficiency [2]. While this concept is well-established in the sheet metal industry (tailored blanks [3] or clad rolling [4,5]), further research is required in bulk metal forming, e.g., for hot forging processes. Specifically, the development of novel technologies for manufacturing metallic multi-material parts is required. The Collaborative Research Centre 1153 investigates the innovative process chain of Tailored Forming to manufacture hybrid metallic high-performance components by hot or semi-hot forming of prior joined preforms. Combinations of different joining and forming processes are evaluated on the example of technical parts such as bearing washers [6], bearing bushings [7] and transmission shafts [8]. In the following, the hot forging of hybrid metal bevel gears from cladded preforms made of two different steel grades combined with a subsequent process-integrated heat treatment are presented. The investigations involve an analysis of the interface using metallographic examinations before and after forming as well as after the integrated heat treatment. For a quantitative characterization of the bond quality, tensile tests were conducted using micro tensile samples and hardness values were measured.

2. State of the Art

Forging is a key technology for manufacturing technical components with complex geometries and provides for continuous fiber flow and excellent mechanical properties. Depending on the initial state of the applied semi-finished parts, there are two methods of forming multi-material components: compound and hybrid forging. In compound forging, raw parts assembled without a metallurgical bond are used. The joining of these parts takes place during forming to the end geometry. The challenge in joining by forming is to create a metallurgical bond between the different materials. In compound forging, the focus is often on parameter case studies in order to achieve a sufficient bond quality, as the latter depends on specific process conditions such as temperature, contact pressures and relative displacement between the materials. For example, Sun et al. studied a hot isothermal compression bonding process of two different steel grades (Q235 and 316 L) using experimental and numerical investigations. They found that the element diffusion distance in the near interface zone grows with increasing deformation temperatures, effective strains and holding times. However, increasing effective strain rates have a negative impact on bond quality [9]. Investigations of Kong et al. on the forge welding of steel-aluminum compounds (AISI 316 L/6063 aluminum) reveal that the forming temperature has the highest influence on the resulting bond quality and the tensile strength of the joint [10]. Wohletz and Groche studied a joining process combining forward and cup extrusion for manufacturing steel-aluminum parts (AISI 1015/6082 T6 aluminum) [11]. They observed that an increased formation of oxide scale on the contact surfaces at elevated temperatures has a negative influence on the resulting bond quality. Due to the above-mentioned restrictions, which result in a narrow range of suited possible process parameters, it is difficult to ensure a homogeneous bond quality when forming hybrid parts with complex geometry, process-related variation in strains and non-uniform material flow. In such a case, the application of previously joined semi-finished parts featuring a metallurgical bond is advantageous for achieving uniform characteristics at the interface. This approach, called hybrid forging, aims at improving the microstructure and the mechanical properties of the material compound. For instance, Förster et al. investigated a two-step forging process of aluminum-enclosed magnesium work pieces, which had previously been joined by co-extrusion [12]. After forging, the magnesium core was crack-free and fully enclosed by aluminum, even at the front ends of the prior extruded, initially aluminum-free profile sections. Though in these regions no metallurgical bond was observed, the bonding achieved by extrusion was maintained. Domblesky et al. conducted axial and side compression tests of friction-welded workpieces with a serial arrangement made from the same materials or material combinations (copper, steel, aluminum) [13]. All material combinations demonstrated a good workability during the forming stage. For combinations of identical materials, the tensile tests showed uniform deformation and material flow similar to that of monolithic materials. In the case of dissimilar material combinations, most of the deformation and subsequent fracture occurred in the softer metal. Klotz et al. performed an isothermal forging of bi-metal gas turbine discs made of two different Ni-based superalloys from hot isostatically pressed billets [14]. They found that bond quality after forming depends on the initial state of the hot isostatically pressed preforms. The formed specimens showed a refined microstructure due to recrystallization at hot forming temperatures.

If the hot-forged steel parts feature a near-net-shape geometry and the temperature after forming is in the austenite regime, a quenching and tempering can be carried out directly after hot forming [15]. Such an integration of the heat treatment into the hot-forming process is used to reduce process costs and times. By applying a controlled cooling with an air-water spray, locally adapted cooling rates can be achieved ranging from low cooling rates as in gas quenching up to high cooling rates as in immersion cooling in water [16]. In addition, the quenching process can be interrupted in order to self-temper the hardened surface using the residual heat remaining in the core of the component. By spray cooling, a continuous and gentle hardness transition is achievable, which can be favorable regarding stress distribution compared to parts featuring a steep hardness gradient [17]. Hence, the near-net shape Tailored Forming of bevel gears in combination with spray cooling offers the potential to manufacture parts with a high fatigue strength and extended service life.

3. Materials and Methods

3.1. Initial Geometry

Hybrid designs can be beneficial for all technical components that experience different loads throughout the part. Accordingly, it is favorable to place high-performance materials in highly stressed areas, while the remaining structural areas can be made of lightweight or low-cost materials. In the following, the Tailored Forming of a bevel gear combining two different steel grades is discussed. For the contacting tooth flanks exposed to rolling loads under operating conditions, a high-strength steel 41Cr4 (1.7035–AISI 5140) and a martensitic valve steel X45CrSi9-3 (1.4718–AISI HNV3) were used. The core of the bevel gear consisted of the low-alloyed steel C22.8 (1.0460–AISI 1022M). According to the assumed load collective, the coaxially arranged hybrid workpieces were produced by two different types of weld cladding: plasma transferred arc deposition welding (PTAW) and laser hotwire cladding (LHC). The corresponding cladded preforms are shown in Figure 1a,b.

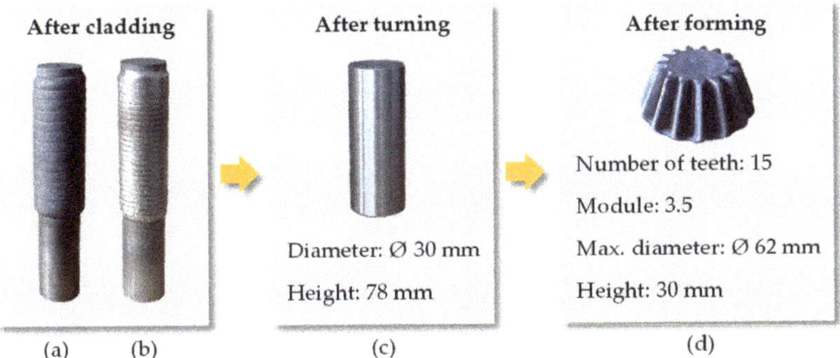

Figure 1. Cladded preforms produced with plasma transferred arc deposition welding (PTAW) (**a**) and laser hotwire cladding (LHC) (**b**) using a substrate with a diameter of 27 mm; geometry of the turned workpiece (**c**) and bevel gear (**d**).

For both methods, the cladding was welded onto a rotating cylinder of the C22.8 substrate. Due to the simultaneous axial movement of the cylinder, the weld seams formed a spiral pattern, which resulted in a wavy contour of the joining zone and non-uniform thickness of the clad layer as shown in a longitudinal cross section in Figure 2. Its thickness depends on the width of the sealing seams and the welding penetration depth, which differ according to the cladding method. Based on recommended thickness values determined in prior investigations on the compound forging of cylindrical gears, the diameter of the substrate cylinders was set to Ø 27 and Ø 28 mm for PTAW [18]. A smaller diameter is not suitable due to the high heat input during cladding. For these specimens, the material combination C22.8 (substrate)/41Cr4 (cladding) was used. For the LHC process, a substrate diameter of Ø 27 mm was favorable to achieve the minimal required thickness of the clad layer but prevent a cladding with several overlapping layers. In this case, a combination of C22.8 (substrate)/X45CrSi9-3 (cladding) was investigated. In order to ensure a constant diameter of the part, the cladded workpieces were additionally turned to a diameter of Ø 30 mm and shortened to a length of 78 mm (Figure 1c). The geometry of the hybrid bevel gear is illustrated in Figure 1d).

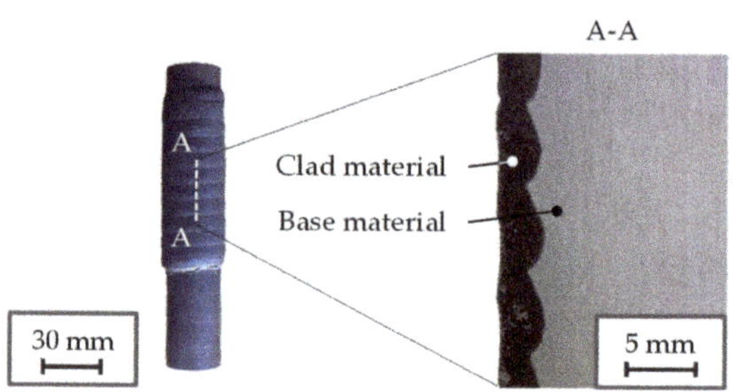

Figure 2. Material distribution in plasma transferred arc (PTA) cladded workpiece for a 41Cr4/C22.8 combination with an initial substrate diameter of Ø 27 mm along the cutting line A-A (longitudinal cross section).

3.2. Die Forging

The hybrid workpieces were used to form the near-net-shaped geometry of the bevel gear. The forging experiments were carried out on a screw press Lasco SPR 500 (LASCO Umformtechnik GmbH, Coburg, Germany) with a maximum capacity of 40 kJ per single step. The corresponding forming tool system depicted in Figure 3 was designed modularly. The lower die and the upper geared die are the main components, creating the mold. The geared die is located in the upper part of the tool. This ensures an appropriate detaching of the final forgings by their own weight. The height of the bevel gear is defined by limit stops. Before forming, the forging dies were heated up to a temperature of 200 °C. The maximum forging force was approx. 510 kN.

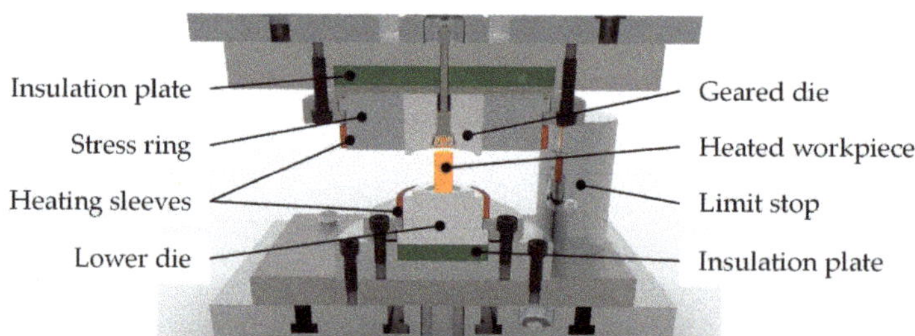

Figure 3. Forging tool system for manufacturing hybrid bevel gears.

The heating of the workpieces was carried out inductively. For an optimal die filling, an axially inhomogeneous heating profile was used (Figure 4). Details about the heating strategy are given in [19]. An axial temperature gradient of ca. 200 °C was achieved by off-center positioning of the workpiece inside the induction coil. By means of the electromagnetic end effect occurring in the workpiece areas located close to the edge of the induction coil, it was possible to set lower temperatures in the lower part of the workpiece [20]. In order to ensure high reproducibility of the forging tests, the pre-heated workpieces were automatically transferred from the induction coil to the forging tool system. After forging, the bevel gears were directly transported to the air-water spray cooling station and subsequently processed in an integrated heat treatment step. In the reference process route without heat treatment, the bevel gears were cooled in still air.

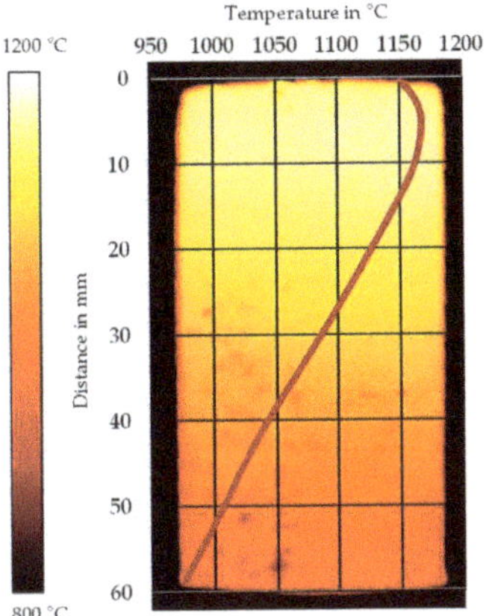

Figure 4. Thermographic image and axial temperature profile measured at the workpiece surface before forming.

3.3. Heat Treatment

The forming process was followed by a process-integrated surface hardening by quenching using an air-water spray cooling (Figure 5). The residual heat in the bevel gear after forging was employed for process-integrated heat treatment. By applying a short intensive cooling, a martensitic surface layer was formed; by interrupting the intensive cooling after a given time, a subsequent self-tempering of this surface layer was realized. This self-tempering of the martensitic surface layer occurs when residual heat from the core of the not fully quenched bevel gear flows to the surface.

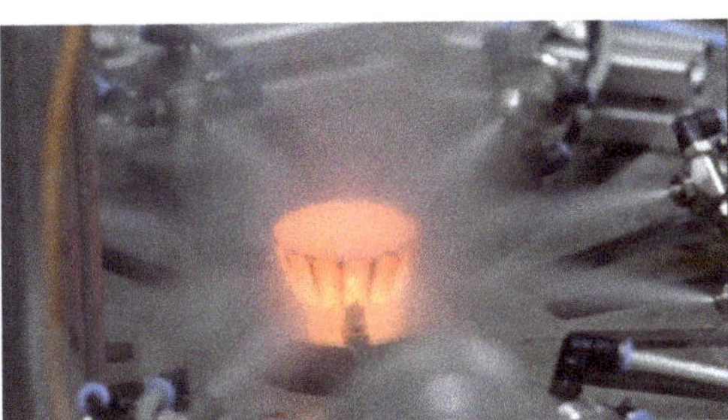

Figure 5. Quenching of a rotating bevel gear in an air-water spray.

The spray cooling system consisted of eight air-water spray nozzles (Internal Mix Nozzles, SUJ12, Spraying Systems Co®, Wheaton, IL, USA) annularly arranged around a rotating mount for the bevel gear. The nozzles were aligned at a distance of 100 mm from the workpiece surface. By varying air and water inlet pressures, the cooling rate was adjusted. A martensitic surface layer in the toothing area of the bevel gears was produced by a short quenching phase with high cooling rates. Subsequently, the residual heat remaining in the core of the bevel gear was employed to temper the surface layer. By means of employing active cooling, the self-tempering temperature could be controlled. A pyrometer was used to monitor the set self-tempering temperature. The pyrometer recorded the temperature at the tooth tip during the self-tempering phase. Since air-water spray was employed during self-tempering, measurement was only possible between two air-water spray pulses. Hence, short pulses of air-water spray were employed automatically every time the surface temperature exceeded the desired self-tempering temperature. To ensure a uniform heat treatment start temperature after transporting the forged bevel gears from the forging press to the spray cooling system, a second pyrometer measured the temperature on the top side of the bevel gear. By means of numerical simulations of process-integrated surface hardening and tempering, the spray parameters (inlet pressures), the duration of the quenching phase and the self-tempering temperature were determined; see Table 1. The heat transfer coefficients for the simulation were estimated by prior cooling tests on bevel gears. To compute the cooling curves by numerical simulations of the quenching process, boundary conditions were adapted as described in [21].

Table 1. Heat treatment parameters of the different hybrid bevel gears.

Material Combination	41Cr4, 41Cr4/C22.8, Ø 27 and Ø 28 mm	X45CrSi9-3/C22.8, Ø 27 mm
Start temperature	950 °C	1000 °C
Duration 1st phase	10 s	8 s
Cooling medium 1st phase	Air-water spray	Air-water spray
Tempering temperature 2nd phase	300 °C	750 °C
Cooling medium 2nd phase	Air-water spray	Air-water spray

Figure 6 shows the time-temperature curves recorded during heat treatment of the bevel gears with a first phase of quenching followed by the self-tempering phase. The temperatures were measured at the tooth flanks with a pyrometer and used to control the self-tempering process. If the specified self-tempering temperature of 300 °C in the bevel gears with the material combination 41Cr4/C22.8 was exceeded, the air-water spray pulse was activated for a short period to reduce surface temperature. This resulted in temperature oscillations (green and orange curves) during the tempering phase starting at about 15 s. Due to the higher recommended self-tempering temperatures of up to 750 °C for the steel X45CrSi9-3, no air-water spray cooling was required in this heat treatment phase; hence, for the material combination X45CrSi9-3/C22.8, no such cooling related oscillations occurred. Instead, only oscillations with lower amplitude are visible caused by the rotation of the bevel gears during heat treatment. The system was manually stopped when an active control of the tempering temperature was no longer required, so that the total self-tempering times seen in Figure 6 vary. The sudden temperature drop at the end of each plot is caused by stopping the measurement. After self-tempering, the bevel gears were placed outside of the spray cooling arrangement and cooled down in still air.

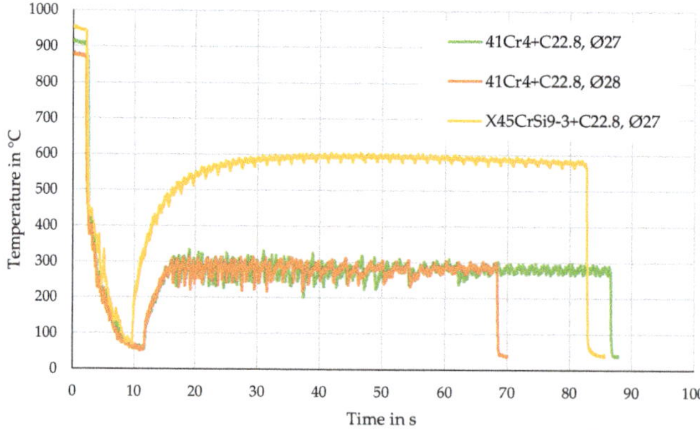

Figure 6. Temperature curves measured with a pyrometer at the tooth flank of the bevel gear.

3.4. Investigation of the Joining Zone

To characterize the microstructural evolution after each process step, cross-sections were extracted from the hybrid workpieces and the bevel gears. Due to deviating forming temperature and strain distribution in the upper and in the lower part of the bevel gear, two sampling positions A and B were employed (Figure 7a). After metallographic preparation, the specimens were etched with 5% nitric acid solution. To reveal the martensitic microstructure of the X45CrSi9-3 steel, an etching with Beraha II reagent was applied to the cladded and forged samples, and a V2A etchant was used on the heat-treated specimens. A detailed microstructural analysis of the combination X45CrSi9-3/C22.8 is given in [19]. Hardness measurements according to Vickers (HV0.5) were carried out for both the cladding layer and the substrate close to the interface [22]. In the bevel gears, the hardness was examined in the tooth tip area. The average values and the standard deviations given in Table 2 were calculated based on 10 indentations each in the cladding layer and in the substrate.

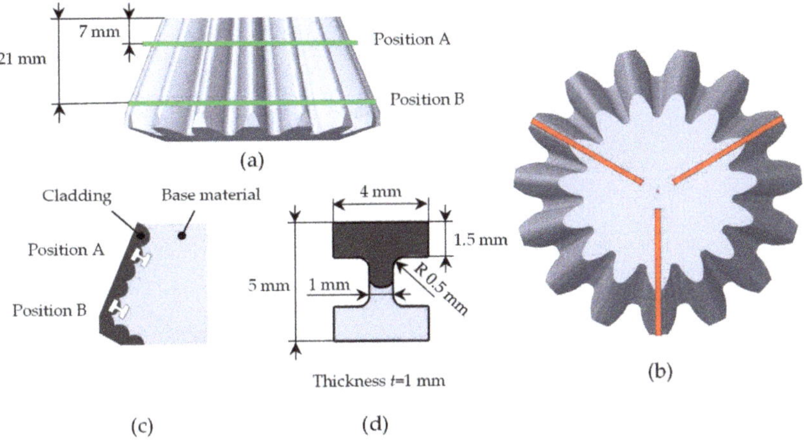

Figure 7. Sampling positions of the cross-sectional micrographs for metallographic examination and hardness measurements (**a**), extraction position of the sample plates (**b**), extraction position of the tensile specimens from the sample plates (**c**), geometry of the micro tensile specimen (**d**).

Table 2. Average hardness values (HV 0.5) of cladding and substrate after cladding, forging and heat treatment.

Material Combination			41Cr4/C22.8, Ø27 mm	41Cr4/C22.8, Ø28 mm	X45CrSi9-3/C22.8, Ø27 mm
Cladded workpiece		Cladding layer	299 ± 16	289 ± 13	379 ± 13
		Substrate	164 ± 7	165 ± 13	146 ± 3
After forging	Position A	Cladding layer	221 ± 9	262 ± 8	462 ± 77
		Substrate	170 ± 3	166 ± 4	148 ± 4
	Position B	Cladding layer	224 ± 7	238 ± 3	276 ± 4
		Substrate	156 ± 8	152 ± 4	140 ± 2
After heat treatment	Position A	Cladding layer	513 ± 7	539 ± 9	589 ± 23
		Substrate	410 ± 6	413 ± 6	268 ± 5
	Position B	Cladding layer	503 ± 6	534 ± 7	732 ± 7
		Substrate	345 ± 12	340 ± 14	188 ± 3

For a mechanical characterization of the interface, tensile test specimens with the geometry given in Figure 7d were cut from the longitudinal cross section of the cladded workpieces and the hybrid bevel gears using wire electrical discharge machining (EDM). The samples from the bevel gears were cut from the tooth tip area according to the sampling position depicted in Figure 7b. At first, three thin plates with a thickness of 1 mm were eroded from each section and etched with $FeCl_3$ reagent to reveal the material distribution. Thus, the tensile specimen from both positions A and B could be prepared precisely by means of wire EDM featuring the joining zone in the center of the specimen as shown in Figure 7c. Overall, six specimens were extracted from each bevel gear and each workpiece.

The experiments were carried out using a tensile testing machine Zwick Retro (Line ZwickRoell, Ulm, Germany) with a maximum capacity of 10 kN. Prior to testing, the specimens were pre-stressed with a load of 5 N. The stress increase rate was set to 30 MPa/s according to standard EN ISO 6892-1 [23].

4. Results

The results of the microstructural examination are summarized in Figure 8. In both types of PTAW-cladded workpieces made of 41Cr4/C22.8 (Figure 8a,b), the clad material (marked as CM) shows a pearlitic microstructure with ferrite along the former austenitic grain boundaries. The base material (marked as BM) has a ferritic-pearlitic microstructure with a prevailing ferrite fraction typical for low eutectoid steels. Close to the joining zone, a needle-shaped Widmanstätten structure is visible, which is induced by the high cooling rates after the cladding process. The forging process has a positive influence by completely transforming the coarse-grained weld microstructures. A smooth transition between cladding and substrates is visible. Grain refinement by recrystallization due to thermomechanical treatment took place at both investigated positions A and B (Figure 8d–h).

In the LHC workpieces combining X45CrSi9-3 and C22.8, the cladding material X45CrSi9-3 mainly consists of martensite and nodular-shaped pearlite. The substrate contains a mixture of ferrite and pearlite (Figure 8c). However, the Widmanstätten structure, which appears in the substrate close to the joining zone after cladding, is less pronounced compared to the 41Cr4/C22.8 workpieces. This can be explained by lower heat input during LHC in comparison to the PTAW, resulting in a lower thickness of the heat-affected zone (HAZ). Probably, there was also some difference in the cooling rates as well as in the start temperature of the transformation. In the transition zone, a pearlite interlayer with a thickness up to 50 μm can be observed, which is retained after forging.

Contrary to the 41Cr4/C22.8 workpieces, the micrographs of the combination X45CrSi9-3-C22.8 show some microstructural differences at positions A and B (Figure 8f–i). While the substrate shows grain refinement in both cases, the former austenitic grains of the cladding material have a coarser structure in position A than in position B. This means that the cladding of X45CrSi-9 is more sensitive to the temperature gradient between lower and upper part of the bevel gear (Figure 4) and the strain differences than the cladding of 41Cr4. The higher forming temperature at position A can result in intensive grain growth, which cannot be compensated for by recrystallization due to the lower strain at this position.

The heat treatment showed the desired impact on the microstructure in all cases. Due to the high cooling rates during spray quenching, a fully martensitic microstructure free of pearlite was achieved in both claddings. The subsequent self-tempering, applied to reduce brittleness and residual stresses in the hardened microstructure, tempers the martensite in the cladding layer as depicted in Figure 8j–o. A similar microstructure is also present in the partially heat-affected substrate close to the joining zone, providing a smooth transition between cladding and substrate. Moreover, the pearlitic interlayer observed in the material combination X45CrSi9-3/C22.8 was fully suppressed by the heat treatment.

The average hardness values given in Table 2 are in line with the microstructural features described above. The hardness values of X45CrSi9-3 were in all cases higher than those of 41Cr4, since at least some fraction of martensite forms even for the low cooling rates when cooling in still air. For both investigated material combinations, the hardness of the substrate and of the cladding material remained at the same level after forging as after cladding. However, the heat treatment resulted in a substantial increase in the material strength, and the tempered martensite in the cladding layer showed hardness values above 500 HV 0.5. For X45CrSi9-3, a maximum hardness of approx. 730 HV 0.5 was achieved at position B and of approx. 590 HV 0.5 at position A. This variance can be attributed to the microstructural differences observed between positions A and B.

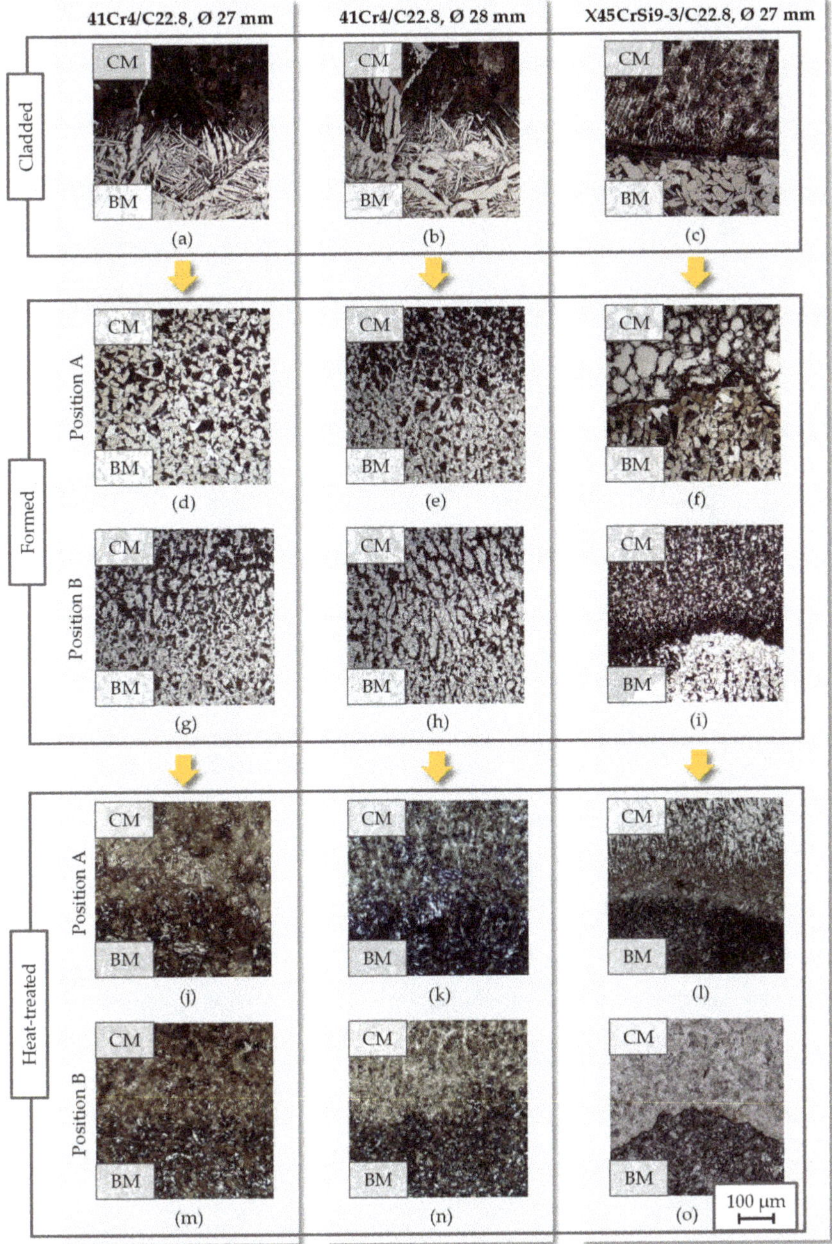

Figure 8. Microstructural evolution in the material combinations 41Cr4/C22.8 and X45CrSi9-3/C22.8 after cladding (**a–c**), forming (**d–i**) and heat-treatment (**j–o**); etched with 5% nitric acid solution excluding (**c,f,i**)—etched with Beraha II reagent—and (**l,o**)—Etched with V2A etchant.

The average tensile strength values for all investigated conditions are summarized in Figure 9. In all specimens, fracture in the hybrid samples occurred within the gauge section on the side of the base material with the lower strength. Thus, the measured values are not labeled as bond strength but as tensile strength. In analogy to the hardness values, the strength values for the conditions after cladding and after forming show no substantial differences (Figure 9a–c), and the strength values at positions A and B are similar. The values for mono-material specimens made of 41Cr4 (marked grey) are considerably higher after forming than those of the hybrid specimens. This can be attributed to the elevated mechanical properties of the steel 41Cr4 in comparison to the C22.8 substrate, in which all hybrid specimens fractured.

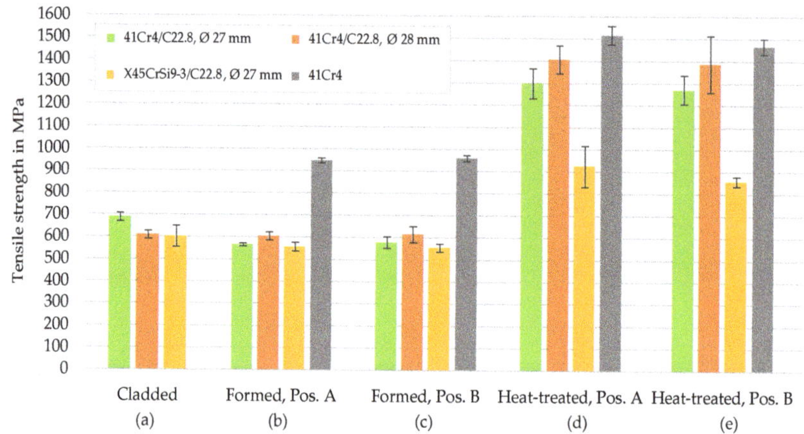

Figure 9. Tensile strength values of cladded workpieces (**a**), forged bevel gears at position A (**b**) and B (**c**), heat-treated bevel gears at position A (**d**) and B (**e**).

The integrated heat treatment increased the material strength as depicted in Figure 9d,e. In comparison to the condition after forging, the strength values of the samples of 41Cr4/C22.8 (marked green and orange) doubled and were close to those of the steel 41Cr4. The samples of X45CrSi9-3/C22.8 show a lower strength. This can be attributed to higher self-tempering temperatures, and thus a lower hardness of the substrate and an increased hardness difference between cladding and substrate compared to 41Cr4/C22.8. Despite the microstructural differences at position A and B, the corresponding strength values do not differ substantially from each other.

5. Discussion

The previous methods for production of hybrid gears are generally represented by shrink fitting, friction welding [24] or bi-metal casting [25]. In the field of forging, there are only few investigations on the compound forging of straight bevel gears, where forming and the joining are combined in a single stage. The key challenge in joining by forming is the creation of the metallurgical bond between raw parts. Politics et al. studied the material flow behavior of different material combinations (steel-aluminum, steel-lead, steel-copper and copper-lead) depending on tooth ring thickness as well as friction properties in the interface zone and in the contact area between forging tool and workpiece [26,27]. Wu et al. carried out similar research for a steel-aluminum-combination focusing on gap size and height difference between ring and core [28]. In this context, the Tailored Forming technology using pre-joined workpieces represents an innovative approach for production of hybrid gears.

Each process step in the process chain of Tailored Forming contributes to the final quality of the bevel gear. The application of previously joined cladded workpieces ensures continuous joints and easy handling of multi-material workpieces throughout the whole process chain. The forging process after cladding facilitates a grain refinement of the coarse weld microstructure, which positively affects the mechanical properties of the parts. The heat treatment contributes to the functionality of the bevel gears under operating conditions by increasing strength and hardness values of the tooth flanks. The micrographs in Figure 8 illustrate the microstructural evolution along the process chain.

As revealed by the hardness measurements, heat treatment has the most decisive influence on the strength of cladding and substrate. The hardness increase in the substrate can be attributed to the fact that the investigated area was located in the heat-affected zone (HAZ) during the heat treatment (Figure 10). Due to different outer diameters of the bevel gears, tooth sizes and axial temperature gradients, the depths of the HAZ at positions A and B differ. For instance, the higher heat input to the gear core at position B results in prolonged surface cooling. Therefore, a shallower HAZ is observed at this position. Independent of the initial substrate diameter, the HAZ is of a similar thickness at both positions A (Figure 10a,c) and B (Figure 10b,d) for the material combination 41Cr4/C22.8. In the case of material combination X45CrSi9-3/C22.8, the size of the HAZ is reduced in comparison with 41Cr4/C22.8 as a result of the different heat treatment strategy with higher self-tempering temperatures. At position B (Figure 10f), this effect is even more pronounced than at the position A due to the higher heat emanating from the bevel gear core.

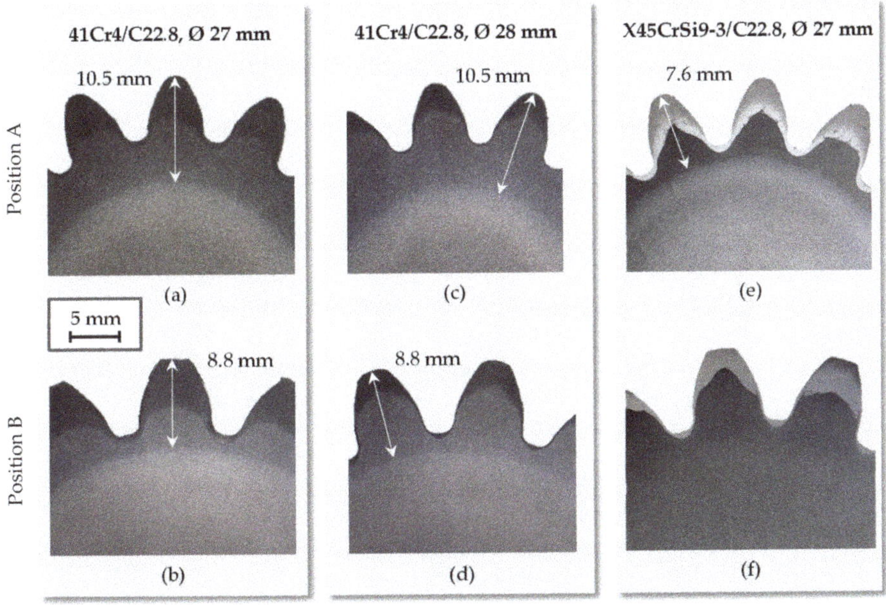

Figure 10. Heat-affected zones after quenching and self-tempering in bevel gears of the material combinations 41Cr4/C22.8 (**a**–**d**) and X45CrSi9-3/C22.8 (**e,f**) at position A (**a,c,e**) and B (**b,d,f**).

Comparing the tensile strength values of the bevel gears of material combination 41Cr4/C22.8, the bevel gears with a substrate diameter of 28 mm show a higher strength after forging and heat treatment (Figure 9). Since the tensile specimens were prepared with the joining zone always located in the center of the specimens (Figure 7d), their extraction position from the bevel gears slightly differed due to a varying cladding layer thickness for Ø 27 mm and Ø 28 mm. The tensile specimens for workpieces with an initial substrate diameter of Ø 27 mm are slightly displaced in the direction of the tooth tip core. This results in a lower tensile strength compared to the Ø 28 mm specimens (Figure 9b,c). With respect to heat treatment, the larger initial substrate diameter and the corresponding thinner cladding layer resulted in an increased hardening depth in the part of the tensile test specimens that corresponded to the substrate. Thus, a slightly higher tensile strength for the Ø 28 mm specimens after quenching and self-tempering was observed (Figure 9d,e). The lower tensile strength of the X45CrSi9-3/C22.8 bevel gears compared to those consisting of 41Cr4/C22.8

results from the increased self-tempering temperatures, reducing strength in the C22.8 part of the tensile test specimens (cf. Figure 6). The heat in the component is therefore not dissipated as quickly as in the 41Cr4/C22.8 bevel gears. Here, the increased self-tempering temperatures are intended to foster a secondary hardening by the formation of chromium carbides [29]. Self-tempering at higher temperatures and the lower heat dissipation result in a decreased hardness of the substrate (188/268 HV 0.5 for X45CrSi9-3/C22.8 compared to 340/410 HV 0.5 for 41Cr4/C22.8); cf. Table 2. Accordingly, the tensile strength of such specimens is reduced.

In future studies, the functionality of the investigated bevel gears will be tested under operating conditions. This will require a grinding of the tooth flanks in order to achieve a high meshing of the forged pinion with the mating gear. These experiments will provide information on the service life behavior of hybrid bevel gears of different material combinations and with varying cladding thicknesses and heat-treatment conditions.

6. Conclusions

In the present study, the process chain of Tailored Forming was used for producing high-performance hybrid bevel gears. The main results can be summarized as follows:

- The forming process has a great influence on the microstructure refinement. Due to the thermomechanical processing during forging, the initially coarse weld microstructure recrystallizes.
- The cooling strategy has the most important influence on the resulting values of hardness and tensile strength. High cooling rates by air-water spray quenching are the main strengthening factor. In air-cooled condition, hardness and strength remain almost unaffected.
- A process-integrated heat treatment by air-water spray quenching and self-tempering allows tailoring hardness and tensile strength in the cladding and the substrate.
- Since fracture of the hybrid tensile samples occurred in the lower strength substrate, a high quality of the bond at the interface was obtained.
- Hybrid bevel gears of the material combination 41Cr4/C22.8 manufactured by Tailored Forming feature a tensile strength similar to bevel gears of 41Cr4 mono-material.
- In comparison, bevel gears of the material combination X45CrSi9-3/C22.8 show a higher hardness in the cladding layer but a lower hardness in the substrate. While the increased hardness in the cladding is attributed to the formation of chromium carbides, the lower hardness in the substrate is caused by the increased self-tempering temperatures recommended for the steel grade X45CrSi9-3.
- The combination of Tailored Forming and a process-integrated heat treatment allows producing components with adapted hardness gradients by employing different materials and customized time-temperature profiles.

Author Contributions: Conceptualization, B.-A.B., A.C., J.D., C.K., S.H., F.N. and H.J.M.; investigation, A.C., J.D., C.K. and S.H.; writing—original draft preparation, B.-A.B., A.C., J.D., C.K., S.H., F.N. and H.J.M.; writing—review and editing, B.-A.B., A.C., J.D., C.K., S.H., F.N. and H.J.M.; visualization, A.C., J.D., C.K. and S.H.; supervision, B.-A.B., F.N. and H.J.M. All authors have read and agreed to the published version of the manuscript.

Funding: This research was funded by the Deutsche Forschungsgemeinschaft (DFG, German Research Foundation) grant number 252662854.

Acknowledgments: The results presented in this paper were obtained within the Collaborative Research Centre 1153 "Process chain to produce hybrid high-performance components by Tailored Forming" in subprojects A2 and B2. The authors would like to thank the subproject A4 (Alexander Barroi and Maximillian Mildebrath) for supplying cladded hybrid workpieces and the German Research Foundation (DFG) for the financial support of this project.

Conflicts of Interest: The authors declare no conflict of interest.

References

1. Ashby, M.F. Multi-objective optimization in material design and selection. *Acta Mater.* **2000**, *48*, 359–369. [CrossRef]
2. Martinsen, K.; Hu, S.J.; Carlson, B.E. Joining of dissimilar materials. *CIRP Ann.* **2015**, *64*, 679–699.
3. Merklein, M.; Johannes, M.; Lechner, M.; Kuppert, A. A review on tailored blanks—Production, applications and evaluation. *J. Mater. Process. Technol.* **2014**, *214*, 151–164. [CrossRef]
4. Wang, S.; Liu, B.X.; Chen, C.X.; Feng, J.H.; Yin, F.X. Microstructure, mechanical properties and interface bonding mechanism of hot-rolled stainless steel clad plates at different rolling reduction ratios. *J. Alloys Compd.* **2018**, *766*, 517–526. [CrossRef]
5. Carradò, A.; Sokolova, O.; Ziegmann, G.; Palkowski, H. Press joining rolling process for hybrid systems. *Key Eng. Mater.* **2010**, *425*, 271–281. [CrossRef]
6. Behrens, B.-A.; Chugreev, A.; Matthias, T.; Poll, G.; Pape, F.; Coors, T.; Hassel, T.; Maier, H.J.; Mildebrath, M. Manufacturing and evaluation of multi-material axial-bearing washers by tailored forming. *Metals* **2019**, *9*, 232. [CrossRef]
7. Behrens, B.-A.; Sokolinskaja, V.; Chugreeva, A.; Diefenbach, J.; Thürer, S.; Bohr, D. Investigation into the bond strength of the joining zone of compound forged hybrid aluminium-steel bearing bushing. *AIP Conf. Proc.* **2019**, *2113*, 040028.
8. Behrens, B.-A.; Bonhage, M.; Bohr, D.; Duran, D. Simulation Assisted Process Development for Tailored Forming. *Mater. Sci. Forum* **2019**, *949*, 101–111. [CrossRef]
9. Sun, C.Y.; Li, L.; Fu, M.W.; Zhou, Q.J. Element diffusion model of bimetallic hot deformation in metallurgical bonding process. *Mater. Des.* **2016**, *94*, 433–443. [CrossRef]
10. Kong, T.F.; Chan, L.C.; Lee, T.C. Qualitative study of bimetallic joints produced by solid state welding process. *Sci. Technol. Weld. Join.* **2008**, *13*, 679–682. [CrossRef]
11. Wohletz, S.; Groche, P. Temperature influence on bond formation in multi-material joining by forging. *Procedia Eng.* **2014**, *81*, 2000–2005. [CrossRef]
12. Förster, W.; Binotsch, C.; Awiszus, B. Process Chain for the Production of a Bimetal Component from Mg with a Complete Al Cladding. *Metals* **2018**, *8*, 97. [CrossRef]
13. Domblesky, J.; Kraft, F.; Druecke, B.; Sims, B. Welded preforms for forging. *J. Mater. Process. Technol.* **2006**, *171*, 141–149. [CrossRef]
14. Klotz, U.E.; Henderson, M.B.; Wilcock, I.M.; Davies, S.; Janschek, P.; Roth, M.; Gasser, P.; McColvin, G. Manufacture and microstructural characterisation of bimetallic gas turbine discs. *Mater. Sci. Technol.* **2005**, *21*, 218–224.
15. Bach, F.W.; Krause, C.; Behrens, B.-A.; Dahndel, H.; Huskic, A. Integration of heat treatment in precision forging of gear wheels. *Arab. J. Sci. Eng. Sect. B Eng.* **2005**, *30*, 103–112.
16. Puschmann, F.; Specht, E. Transient measurement of heat transfer in metal quenching with atomized sprays. *Exp. Therm. Fluid Sci.* **2004**, *28*, 607–615. [CrossRef]
17. Krause, C.; Wulf, E.; Nürnberger, F.; Bach, F.W. Wärmeübergangs-und Tropfencharakteristik für eine Spraykühlung im Temperaturbereich von 900–100 °C. *Forsch. IM Ing.* **2008**, *72*, 163–173. [CrossRef]
18. Behrens, B.-A.; Bistron, M.; Küper, A. Investigation of load adapted gears and shafts manufactured by compound-forging. *J. Adv. Manuf. Syst.* **2008**, *7*, 175–182. [CrossRef]
19. Behrens, B.-A.; Diefenbach, J.; Chugreeva, A.; Kahra, C.; Herbst, S.; Nürnberger, F. Tailored Forming of Hybrid Bevel Gears with Integrated Heat Treatment. *Procedia Manuf.* **2020**, *47*, 301–308.
20. Rudnev, V.; Loveless, D.; Cook, R.L. *Handbook of Induction Heating*; CRC Press: London, UK, 2017.
21. Herbst, S.; Steinke, K.F.; Maier, H.J.; Milenin, A.; Nürnberger, F. Determination of heat transfer coefficients for complex spray cooling arrangements. *Int. J. Microstruct. Mater. Prop.* **2016**, *11*, 229–246. [CrossRef]
22. *EN ISO 6507-1: Metallic Materials—Vickers Hardness Test—Part 1: Test Method*; International Organization for Standardization: Geneva, Switzerland, 2005.
23. *EN ISO 6892-1: Metallic Materials—Tensile Testing—Part 1: Method of Test at Room Temperature*; International Organization for Standardization: Geneva, Switzerland, 2005.
24. Wadleigh, A.S. Multi-Metal Composite Gear/Shaft—Produced by Consecutively Forming, Friction Welding, Reforming and Friction Welding, then Machining Three Elements of Dissimilar Metals. U.S. Patent No. 5492264, 20 February 1996.

25. Miller, W.H. Gear Blanks. U.S. Patent No. 3847557, 12 November 1974.
26. Politis, D.J.; Lin, J.; Dean, T.A. Investigation of material flow in forging bi-metal components. *Steel Res. Int.* **2012**, *231*, 234.
27. Politis, D.J.; Lin, J.; Dean, T.A.; Balint, D.S. An investigation into the forging of Bi-metal gears. *J. Mater. Process. Technol.* **2014**, *214*, 2248–2260. [CrossRef]
28. Wu, P.; Wang, B.; Lin, J.; Zuo, B.; Li, Z.; Zhou, J. Investigation on metal flow and forming load of bi-metal gear hot forging process. *Int. J. Adv. Manuf. Technol.* **2017**, *88*, 2835–2847. [CrossRef]
29. Deng, H.X.; Shi, H.J.; Tsuruoka, S.; Yu, H.C.; Zhong, B. Influence of Heat Treatment on Characteristic Behavior of Co-Based Alloy Hardfacing Coatings Deposited by Plasma Transferred Arc Welding. *Key Eng. Met.* **2011**, *462*, 593–598. [CrossRef]

Article

Applying Membrane Mode Enhanced Cohesive Zone Elements on Tailored Forming Components

Felix Töller [1],*, Stefan Löhnert [2] and Peter Wriggers [1]

[1] Institute of Continuum Mechanics, Leibniz University Hannover, An der Universität 1, 30823 Garbsen, Germany; wriggers@ikm.uni-hannover.de

[2] Institute of Mechanics and Shell Structures, Technische Universität Dresden, August-Bebel-Straße 30, 01219 Dresden, Germany; stefan.loehnert@tu-dresden.de

* Correspondence: toeller@ikm.uni-hannover.de; Tel.: +49-511-7621-7579

Received: 31 August 2020; Accepted: 25 September 2020; Published: 5 October 2020

Abstract: Forming of hybrid bulk metal components might include severe membrane mode deformation of the joining zone. This effect is not reflected by common Traction Separation Laws used within Cohesive Zone Elements that are usually applied for the simulation of joining zones. Thus, they cannot capture possible damage of the joining zone under these conditions. Membrane Mode Enhanced Cohesive Zone Elements fix this deficiency. This novel approach can be implemented in finite elements. It can be used within commercial codes where an implementation as a material model is beneficial as this simplifies model preparation with the existing GUIs. In this contribution, the implementation of Membrane Mode Enhanced Cohesive Zone Elements as a material model is presented within MSC Marc along with simulations showing the capabilities of this approach.

Keywords: tailored forming; membrane mode enhanced cohesive zone elements; damage; joining zone

1. Introduction

The requirements for technical products are rising. Fulfilling these becomes more and more difficult using mono material components. A hybrid component combines different materials and can be tailored to meet specific requirements that might not be possible with a single material. The joining process to produce hybrid components from two or more mono material parts is usually at the end of the production chain. If a forming process is included, joining is either during or after the forming process. Doing the joining process before forming allows to treat the joining zone during forming geometrically and thermomechanically. This provides more flexibility to tailor the hybrid component regarding its requirements; this process is called Tailored Forming (Figure 1).

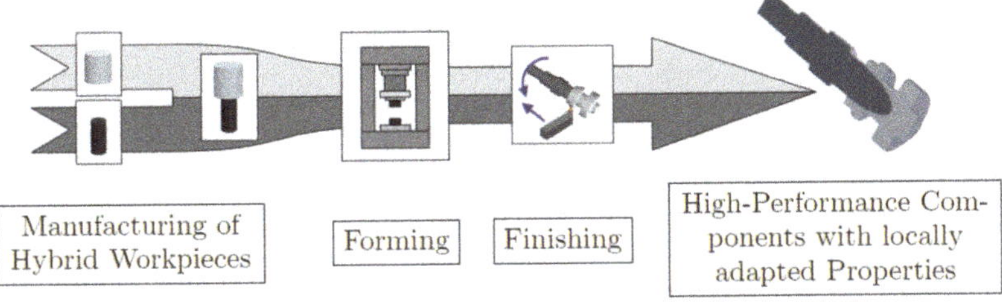

Figure 1. Tailored forming process chain [1].

The joining zone is intended to be firmly bonded. In the case of aluminium and steel, intermetallic compounds often arise. Their existence shows that a firm bond is obtained during joining. Subsequent heat treatments cause thickening of intermetallic compounds to about 2 µm [2]. Though practical investigations indicate advantages of joining zones without or with a non-measurable intermetallic compound thickness during forming [3].

Thin layers are often modelled using cohesive zone elements. Standard formulations use traction separation laws for the constitutive relation which are not sufficient here. Forming of hybrid bulk metal components might include severe membrane deformations that are additional deformation modes not describable by separations. Internal Thickness Extrapolation (InTEx) [4] and Membrane Mode Enhanced Cohesive Zone Elements (MMECZE) [5] are two novel approaches that consider additional loading directions. Internal Thickness Extrapolation aims on layers that are thin but not completely flat, reconstructs a volumetric geometry from a flat element topology and applies bulk material models. Membrane Mode Enhanced Cohesive Zone Elements are focused on flat joining zones and combine Traction Separation Laws for the separation loading directions with a bulk material capturing the damage contribution from membrane mode deformation of the joining zone. Here, the latter approach is more convenient due to the flatness of the joining zone.

2. Membrane Mode Enhanced Cohesive Zone Elements

Membrane Mode Enhanced Cohesive Zone Elements are extensively described in [5]. Here, key points are summarised that are necessary for the implementation and application.

The Continuum Damage Mechanics framework from Lemaitre [6] is a concept that can describe damage in various bulk material models. Damage ($0 \le D \le 1$) reduces the effective surface area for load transmission. Stress is distributed on a smaller portion of the material; the elastic and the plastic response are affected. For Membrane Mode Enhanced Cohesive Zone Elements the framework is applied at an interface. Instead of tensorial stresses, traction vectors transmit the load through the material. Figure 2 shows the joining zone with partial defects (∂S_D) that reduce the surface area for load transmission (t). The effective traction due to a reduction of the load transmitting surface area is

$$\widetilde{t} = \frac{t}{1-D} \tag{1}$$

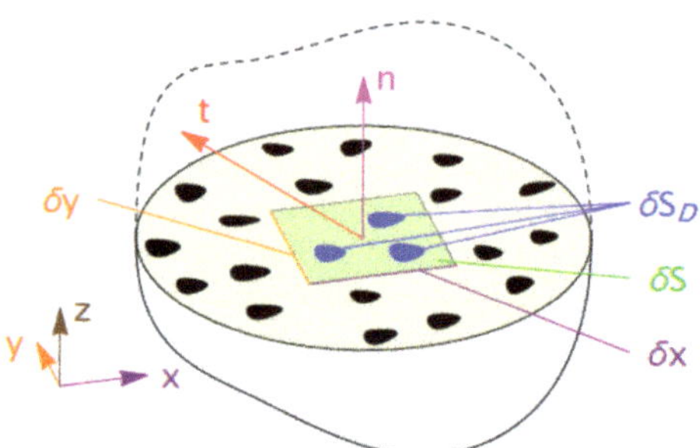

Figure 2. Damage of an interface [5].

The change of damage is related to the rate $\dot{D}$ which depends on the combination of the plastic multiplier $\dot{\lambda}$ describing the amount of deformation and the strain energy density release rate Y. The latter is the amount of elastic energy stored in the material that is released due to damage softening

$$\dot{D} = \dot{\lambda}\,\frac{1}{1-D}\left(\frac{Y}{S}\right)^{s} \tag{2}$$

$$\text{with} \quad Y = \frac{1}{2}\frac{1}{(1-D)^2}\left(\frac{t_t^2}{E} + \frac{t_s^2}{G}\right) \tag{3}$$

where s and S are damage related material parameters, t_t and t_s are the tension and shearing part of the traction, and E and G are the Young's and shearing modulus.

Traction Separation Laws (TSLs) are defined on the cohesive surface (black dashed in Figure 3) that describes the midplane of the initially coincident faces of the upper and lower component

$$x_{cs} = \frac{x_u + x_l}{2} \tag{4}$$

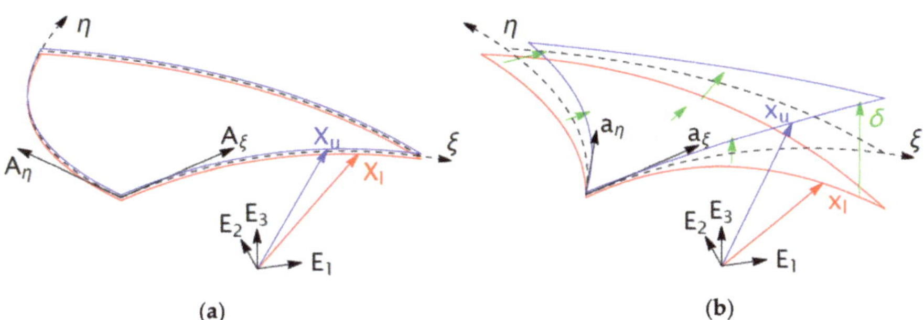

Figure 3. Separation of a joining zone; initial configuration (**a**) and current configuration (**b**) [5].

TSLs describe the traction over separation relation. They are often curves with a nonlinear load increase that levels off smoothly for a rising displacement until failure. This might be realistic for the application of modelling adhesives. Within Tailored Forming brittle failure of an infinitesimal thin joining zone has to be described. The theoretically infinitesimal stiffness with sharp load release after failure is modelled using a large but finite artificial stiffness k to circumvent numerical problems

$$t = k\,\delta \tag{5}$$

where t is the traction resulting from the separation δ (see green vectors in Figure 3). This traction separation relation induces a reaction force in the direction contrary to the separation.

The membrane mode deformation can be modelled, see Figure 3, as angular and length changes of the vectors A_ξ and A_η in the initial configuration to a_ξ and a_η in the current configuration. Describing these vectors in a two-dimensional local coordinate system that follows the cohesive surface (two dimensional quantities with ˆ in the following) allows to compute a two-dimensional deformation gradient

$$F = \frac{\partial \hat{x}}{\partial \hat{X}} = \frac{\partial \hat{x}}{\partial \hat{\xi}} \cdot \frac{\partial \hat{\xi}}{\partial \hat{X}}$$
$$\text{with } \frac{\partial \hat{x}}{\partial \hat{\xi}} = \left(\begin{array}{cc} \hat{A}_\xi & \hat{A}_\eta \end{array} \right) \text{ and } \frac{\partial \hat{\xi}}{\partial \hat{X}} = \left(\begin{array}{cc} \hat{a}_\xi & \hat{a}_\eta \end{array} \right)^{-1} \tag{6}$$

Based on the deformation gradient a variety of material models can be applied; here the Lemaitre model is used. The plastic multiplier as amount of (plastic) deformation can be computed (right in Figure 4). Together with tractions from the traction separation relation (left) a damage rate ($\dot{D}$) results. Accumulation of the damage D (or certainly high tractions t_n and t_s) might then cause interruption of the cohesion.

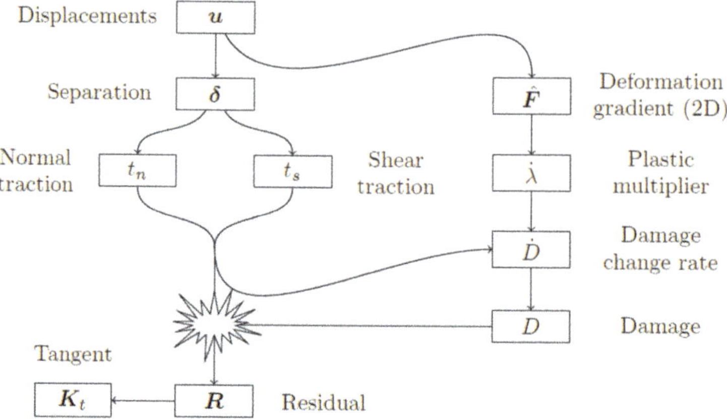

Figure 4. Merging cohesive zone modes and membrane modes [5].

The Membrane Mode Enhanced Cohesive Zone Element concept does not necessitate this, but failure is updated explicitly here. This means the load transmission is cut off in the time step after failure occurred. Such procedure is numerically very advantageous as failure induced instabilities are inhibited and the nonlinearity of the computation is limited to the nonlinearity of the forming process itself.

3. Cohesive Modelling in MSC Marc

Marc is a nonlinear finite element solver from MSC Software (MSC Marc Mentat 2019 Feature Pack 1, Newport Beach, CA, US). The pre- and postprocessor Mentat, also from MSC Software, allows to setup simulations for Marc. Linear and quadratic Cohesive Zone Elements are provided by the software package [7]. Figure 5 shows the connectivity (a) and Gauss integration points (b) of element 188. The connectivity corresponds to a brick element, though the thickness in one direction is (initially) zero so the geometry complies with a quadrilateral element. As integration is only executed in the cohesive surface (dashed midplane, Figure 5a), a Gauss integration scheme contains only $2 \times 2 = 4$ points like a quadrilateral element. Wedge or triangle like element types and Newton Cotes integration schemes are also available. The degrees of freedom are displacements in three directions. In case of a thermomechanical analysis a weak coupling is realized with a staggered approach, i.e., a separate thermal element is used additionally to the mechanical element (e.g., the thermal element 222 corresponds to the mechanical element 188).

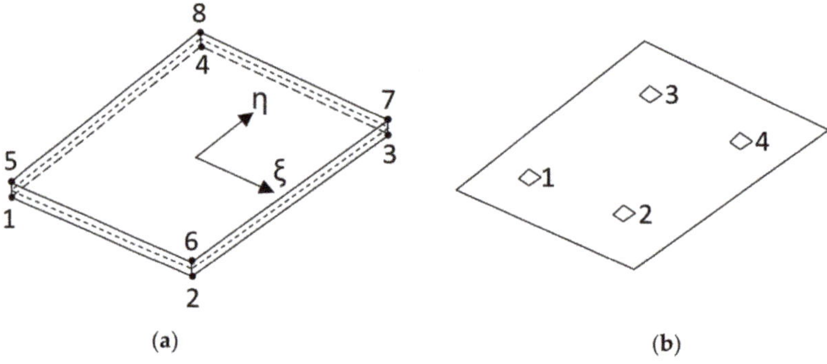

Figure 5. MSC Marc element 188: connectivity (a) and Gauss integration points (b) [7].

Modelling of cohesive zones is realised as a subsequent step of solid modelling. Existing nodes are doubled and the Cohesive Zone Element is placed. Cohesive elements can either immediately be inserted (Toolbox/General/Matching Boundaries) or when a certain load criterion is reached (Toolbox/Fracture mechanics/Delamination). Here, the direct insertion is beneficial to track membrane changes even if the load remains small. The inserted cohesive elements can use user subroutines to define the material behaviour.

Figure 6 shows the analysis path from pre- to postprocessing; element and material subroutines are drawn as sub steps of processing to discuss the implementation that interacts here. During processing, coordinates and displacements are handed over to the element subroutine. In the case of cohesive elements, separations are computed and the material subroutine is called. The material subroutine returns a tangent and stress. This is processed in the element subroutine to an element stiffness matrix and a residual.

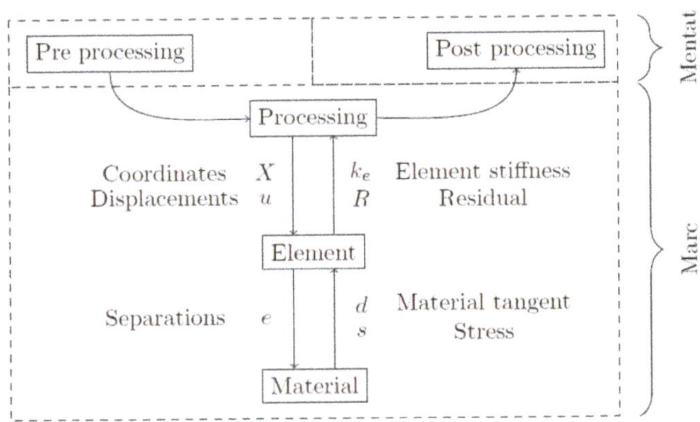

Figure 6. Processing.

4. Membrane Mode Enhanced Cohesive Zone Element Technology as a Material Subroutine

A cohesive formulation based on separations does not need to know more about the deformation than the separations. Though Membrane Mode Enhanced Cohesive Zone Elements need information about the membrane deformation. Therefore, in [5] they are implemented as a user element. To comfortably use this new technology within MSC Marc, it has to be implemented as a material model. The missing information has to be gathered and the differing output has to be discussed as well.

4.1. Gathering Information about the Membrane Deformation

For the sake of simplicity, Figure 6 only shows the separation as most important information that is transferred from the element to the material. Some further data are the external element number m(1) and the integration point number nn. However, the required information about the membrane deformation is not transferred. Though Marc provides some functions that help to gather the required information about the membrane deformation [8]. The (internal) element nodes nodes and the number of nodes num can be determined using

```
call elnodes(ielint(m(1)),num,nodes)
```

where `ielint(m(1))` converts the external to an internal element number. The number of nodes is used to identify the current element type. The internal node numbers have to be converted to external node numbers

```
nodeid = nodext(nodes(i))
```

as these can be used to obtain nodal values with

```
call nodvar(icod,nodeid,valno,nqncomp,nqdatatype)
```

where, e.g., `icod=0` gives coordinates and `icod=1` displacements.

The requested data is returned in `valno`; `nqncomp` and `nqdatatype` are the size and type of the returned data. With these the initial and current cohesive surface geometry can be constructed; e.g. the corner points for a linear quad like element in initial configuration are

```
xs(:,1) = (xl(:,1) + xl(:,5))/2
xs(:,2) = (xl(:,2) + xl(:,6))/2
xs(:,3) = (xl(:,3) + xl(:,7))/2
xs(:,4) = (xl(:,4) + xl(:,8))/2,
```

where `xl` are the node coordinates. For the current configuration the displacements have to be added. All this happens within a material subroutine that is called in one certain integration point. The location of integration point number nn for a linear quad with Gauss integration is

```
select case (nn)
  case (1)
    xi  = - 1.d0/dsqrt(3.d0)
    eta = - 1.d0/dsqrt(3.d0)
  case (2)
    xi  = + 1.d0/dsqrt(3.d0)
    eta = - 1.d0/dsqrt(3.d0)
  case (3)
    xi  = - 1.d0/dsqrt(3.d0)
    eta = + 1.d0/dsqrt(3.d0)
  case (4)
    xi  = + 1.d0/dsqrt(3.d0)
    eta = + 1.d0/dsqrt(3.d0)
end select  .
```

With standard bilinear ansatz functions and derivatives

```
N(1)  = 0.25d0 * ( 1.d0 - xi) * ( 1.d0 - eta)
N(2)  = 0.25d0 * ( 1.d0 + xi) * ( 1.d0 - eta)
N(3)  = 0.25d0 * ( 1.d0 + xi) * ( 1.d0 + eta)
N(4)  = 0.25d0 * ( 1.d0 - xi) * ( 1.d0 + eta)
Nxi(1)  = 0.25d0 *                 (-1.d0 + eta)
Nxi(2)  = 0.25d0 *                 ( 1.d0 - eta)
Nxi(3)  = 0.25d0 *                 ( 1.d0 + eta)
Nxi(4)  = 0.25d0 *                 (-1.d0 - eta)
Neta(1) = 0.25d0 * (-1.d0 + xi)
Neta(2) = 0.25d0 * (-1.d0 - xi)
Neta(3) = 0.25d0 * ( 1.d0 + xi)
Neta(4) = 0.25d0 * ( 1.d0 - xi)
```

the 3D tangential vectors in initial and current configuration can be determined. E.g., for the quadrilateral like element with the number of cohesive surface points num2 the tangential vectors in the initial configuration are

```
axi3Di = (/ 0.d0, 0.d0, 0.d0 /)
aet3Di = (/ 0.d0, 0.d0, 0.d0 /)
do i = 1, num2
  axi3Di(1)  = axi3Di(1) + xs(1,i) * Nxi(i)
  axi3Di(2)  = axi3Di(2) + xs(2,i) * Nxi(i)
  axi3Di(3)  = axi3Di(3) + xs(3,i) * Nxi(i)
  aeta3Di(1) = aeta3Di(1) + xs(1,i) * Neta(i)
  aeta3Di(2) = aeta3Di(2) + xs(2,i) * Neta(i)
  aeta3Di(3) = aeta3Di(3) + xs(3,i) * Neta(i)
```

end do

and accordingly, in the current configuration. With the procedure in [5] the 2D tangential vectors and finally the 2D deformation gradient can be computed that describes the membrane deformation.

4.2. Returning Stress and Material Tangent

The stress returned from a standard Traction Separation Law as a material routine for interfaces contains three components: normal direction (N) and two shearing directions ($S1$ and $S2$)

$$s = \begin{pmatrix} s_N \\ s_{S1} \\ s_{S2} \end{pmatrix} \tag{7}$$

The corresponding tangent considers the changes with respect to normal and (two) shearing separations

$$d = \begin{pmatrix} \partial s_N/\partial\delta_N & \partial s_N/\partial\delta_{S1} & \partial s_N/\partial\delta_{S2} \\ \partial s_{S1}/\partial\delta_N & \partial s_{S1}/\partial\delta_{S1} & \partial s_{S1}/\partial\delta_{S2} \\ \partial s_{S2}/\partial\delta_N & \partial s_{S2}/\partial\delta_{S1} & \partial s_{S2}/\partial\delta_{S2} \end{pmatrix} \tag{8}$$

Taking all directions into account, two in plane stretching ($IP1$ and $IP2$) directions and the out of plane shearing direction (OPS) have to be added

$$s = \begin{pmatrix} s_N \\ s_{S1} \\ s_{S2} \\ s_{IP1} \\ s_{IP2} \\ s_{OPS} \end{pmatrix} \tag{9}$$

$$d = \begin{pmatrix} \partial s_N/\partial\delta_N & \partial s_N/\partial\delta_{S1} & \partial s_N/\partial\delta_{S2} & \partial s_N/\partial\delta_{IP1} & \partial s_N/\partial\delta_{IP2} & \partial s_N/\partial\delta_{OPS} \\ \partial s_{S1}/\partial\delta_N & \partial s_{S1}/\partial\delta_{S1} & \partial s_{S1}/\partial\delta_{S2} & \partial s_{S1}/\partial\delta_{IP1} & \partial s_{S1}/\partial\delta_{IP2} & \partial s_{S1}/\partial\delta_{OPS} \\ \partial s_{S2}/\partial\delta_N & \partial s_{S2}/\partial\delta_{S1} & \partial s_{S2}/\partial\delta_{S2} & \partial s_{S2}/\partial\delta_{IP1} & \partial s_{S2}/\partial\delta_{IP2} & \partial s_{S2}/\partial\delta_{OPS} \\ \partial s_{IP1}/\partial\delta_N & \partial s_{IP1}/\partial\delta_{S1} & \partial s_{IP1}/\partial\delta_{S2} & \partial s_{IP1}/\partial\delta_{IP1} & \partial s_{IP1}/\partial\delta_{IP2} & \partial s_{IP1}/\partial\delta_{OPS} \\ \partial s_{IP2}/\partial\delta_N & \partial s_{IP2}/\partial\delta_{S1} & \partial s_{IP2}/\partial\delta_{S2} & \partial s_{IP2}/\partial\delta_{IP1} & \partial s_{IP2}/\partial\delta_{IP2} & \partial s_{IP2}/\partial\delta_{OPS} \\ \partial s_{OPS}/\partial\delta_N & \partial s_{OPS}/\partial\delta_{S1} & \partial s_{OPS}/\partial\delta_{S2} & \partial s_{OPS}/\partial\delta_{IP1} & \partial s_{OPS}/\partial\delta_{IP2} & \partial s_{OPS}/\partial\delta_{OPS} \end{pmatrix} \tag{10}$$

The resistance of a zero thickness surface regarding in plane stretching and out of plane shearing is assumed to be zero. This can be justified considering a very thin layer where the reaction force depends on the integral over the thickness. Without thickness, the reaction force becomes zero. Hence in-plane stretching and out-of-plane shearing are assumed not to contribute to the residual and thus Membrane Mode Enhanced Cohesive Zone Elements build their element residual solely based on the separation modes. The other directions are still considered for damage calculations but can be omitted in the stress and consequently also in the tangent

$$s = \begin{pmatrix} s_N \\ s_{S1} \\ s_{S2} \\ 0 \\ 0 \\ 0 \end{pmatrix} \tag{11}$$

$$d = \begin{pmatrix} \partial s_N/\partial\delta_N & \partial s_N/\partial\delta_{S1} & \partial s_N/\partial\delta_{S2} & \partial s_N/\partial\delta_{IP1} & \partial s_N/\partial\delta_{IP2} & \partial s_N/\partial\delta_{OPS} \\ \partial s_{S1}/\partial\delta_N & \partial s_{S1}/\partial\delta_{S1} & \partial s_{S1}/\partial\delta_{S2} & \partial s_{S1}/\partial\delta_{IP1} & \partial s_{S1}/\partial\delta_{IP2} & \partial s_{S1}/\partial\delta_{OPS} \\ \partial s_{S2}/\partial\delta_N & \partial s_{S2}/\partial\delta_{S1} & \partial s_{S2}/\partial\delta_{S2} & \partial s_{S2}/\partial\delta_{IP1} & \partial s_{S2}/\partial\delta_{IP2} & \partial s_{S2}/\partial\delta_{OPS} \\ 0 & 0 & 0 & 0 & 0 & 0 \\ 0 & 0 & 0 & 0 & 0 & 0 \\ 0 & 0 & 0 & 0 & 0 & 0 \end{pmatrix} \tag{12}$$

At this point, it can be noticed that all necessary stress components for Membrane Mode Enhanced Cohesive Zone Elements can be returned and processed in the standard Cohesive Zone Element manner. Though the influence of in plane modes in the tangent cannot be captured in this way what might have an influence on the convergence. In [5], failure is updated explicitly. For this reason, the influence of membrane mode deformations during one load step cancels out

$$d = \begin{pmatrix} \partial s_N/\partial\delta_N & \partial s_N/\partial\delta_{S1} & \partial s_N/\partial\delta_{S2} & 0 & 0 & 0 \\ \partial s_{S1}/\partial\delta_N & \partial s_{S1}/\partial\delta_{S1} & \partial s_{S1}/\partial\delta_{S2} & 0 & 0 & 0 \\ \partial s_{S2}/\partial\delta_N & \partial s_{S2}/\partial\delta_{S1} & \partial s_{S2}/\partial\delta_{S2} & 0 & 0 & 0 \\ 0 & 0 & 0 & 0 & 0 & 0 \\ 0 & 0 & 0 & 0 & 0 & 0 \\ 0 & 0 & 0 & 0 & 0 & 0 \end{pmatrix} \tag{13}$$

and the tangent can also be processed in the standard Cohesive Zone Element manner.

4.3. Thermal and Mechanical Parameters

In [5], several simulations are carried out to fit parameters for the material model. The maximum allowable tensile and shear traction are determined to be $t_{t,max} = 365$ MPa and $t_{s,max} = 300$ MPa. The Lemaitre damage parameters are $s = 1.0$ and $S = 0.34$.

The thermal conductivity of the joining zone considered as bulk material within an InTEx element is investigated in [9]. For undamaged ($D = 0$) or fully damaged ($D = 1$) material, the conductivity is

$$\kappa(D = 0) = 68 \,\frac{W}{mK} \quad , \quad \kappa(D = 1) = 0.01 \,\frac{W}{mK} \quad . \tag{14}$$

This can be converted to a heat transfer coefficient h

$$h(D = 0) = 6800000 \,\frac{W}{m^2K} \quad , \quad h(D = 1) = 1000 \,\frac{W}{m^2K} \quad , \tag{15}$$

with the corresponding bulk material thickness of $t = 10\ \mu m$. It is assumed that h linearly depends on D

$$h(D) = (6800000\,(1 - D) + 1000\,D) \,\frac{W}{m^2K} \quad . \tag{16}$$

5. Simulation of a Transverse Link

The transverse link is an advanced demonstrator component for Tailored Forming. All processes and also the final geometry are currently under development. The raw hybrid part is produced by extrusion; a forging process and afterwards machining is used to generate the final geometry. According to the current state, the raw part is an L formed steel profile (32 mm × 32 mm, width = 7 mm) that is filled with aluminium (Figure 7a). The final geometry is flattened, contains an indentation in the middle and has three holes for mounting (Figure 7b).

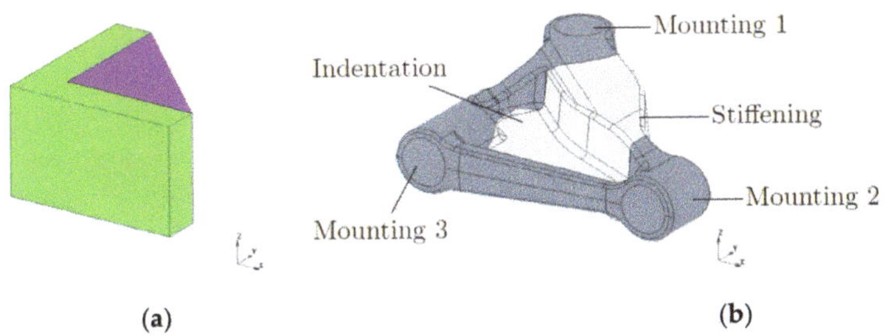

Figure 7. Transverse link; raw (**a**) and final geometry (**b**).

The forging process is simulated here to test Membrane Mode Enhanced Cohesive Zone Elements in a new load case especially with the implementation in MSC Marc. The simulations might also give an idea of the impact of forming on this specific part, though the validity is limited as no material data for this certain joining zone is available; the parameters in Section 4.3 stem from a different process. For the bulk materials temperatures of $\Theta_{st} = 750\,°C$ and $\Theta_{al} = 480\,°C$ and yield stresses of $\sigma_{y,st} = 270\,\text{MPa}$ and $\sigma_{y,al} = 40\,\text{MPa}$ [10] are assumed.

The component has a symmetry in thickness direction that is utilised during the simulations. In a first step, the forming tool is modelled by subtracting the final geometry (except for the holes) from a solid block. To prevent burr formation, the tool is extended in forming direction, see Figure 8; practically this can be realised in a closed die process. Simulations show that the indentation in the middle induces a severe loading of the joining zone; the slope at the edge of the indentation is large and coincides with the joining zone. This causes failure of the joining zone in a very early stage. Figure 9a shows a damage contour plot of the slightly formed component. Nodal averaging is activated to visualise the damage also in the bulk elements around. Even if damage ranges from 0 to 1, here 0.5 is used as upper limit as interface elements contribute a damage and bulk elements do not. In the case of a tetraeder mesh, or also if the corner here would be damaged from both sides, the averaging result can only be utilised to indicate that damage occurs. The extent is influenced as the share of damage contributing elements varies. Figure 9b only contains the interface elements, nodal averaging is deactivated and the damage can be seen directly on the interface elements using a scale from 0 to 1.

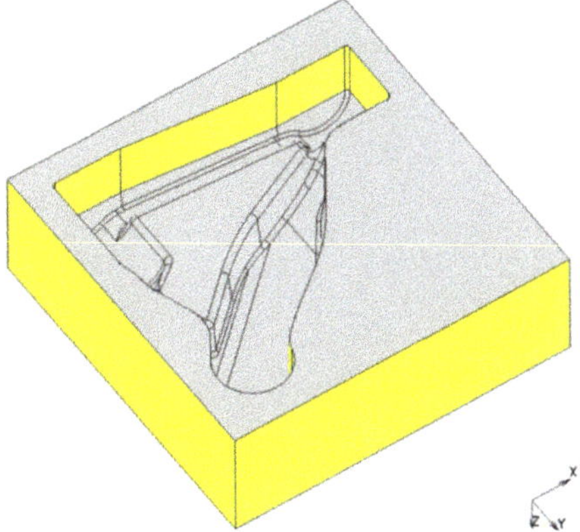

Figure 8. Forming tool.

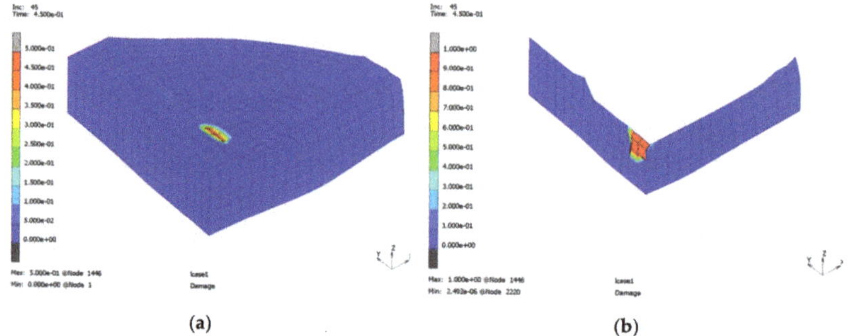

Figure 9. Forming process using the original geometry; whole geometry (**a**) and joining zone (**b**): joining zone damage.

Next, the simplified geometry in Figure 10 is used that does not contain the indentation in the middle. The results strongly depend on the starting position of the specimen in the form. In the case depicted in Figure 11a, a crack arises after the aluminium gets in contact with the radius of the form. With a slightly different position (1 mm shift in x and y direction), three positions get cracks, though later (Figure 11b). The inner corner is strongly shear loaded. On the backside (Figure 12) two shear induced cracks arise that are both close to the stiffening between mounting 1 and 2 (Figure 7b). The strong dependency on the position of the specimen in the form necessitates a holder or special geometry of the form that allows reliable and repeatable positioning. To further reduce loading, the form is again simplified by dropping the stiffening.

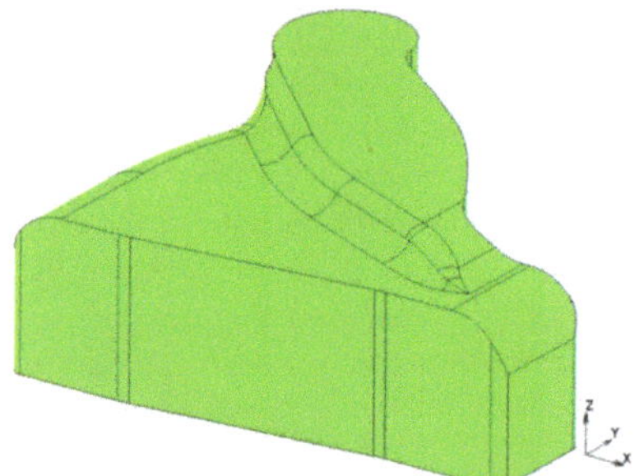

Figure 10. Simplified geometry.

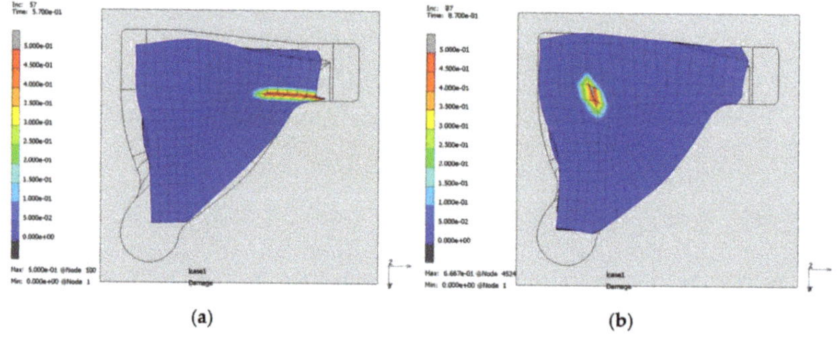

Figure 11. Forming process using a simplified geometry with two different initial positions (**a**) and (**b**) of the specimen in the form: joining zone damage.

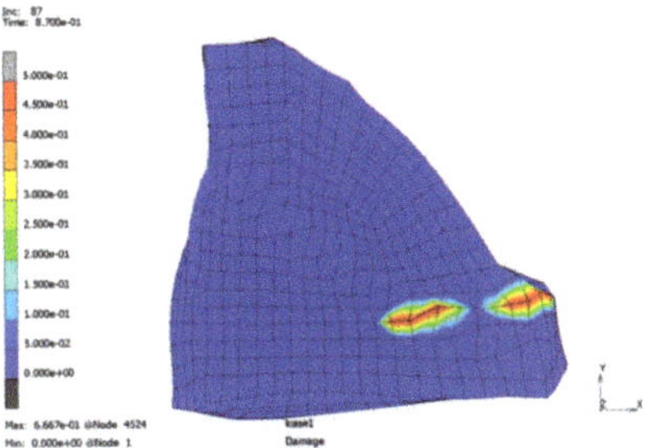

Figure 12. Forming process using a simplified geometry (backside of Figure 11b): joining zone damage.

The process can now be executed further. Though filling the edge around mounting 1 (Figure 7b) causes severe difficulties. The stronger steel and a thin portion of the weaker aluminium enter the groove and come into contact on both sides (Figure 13a). Further forming induces compression, the aluminium starts yielding normal to the compression direction and decohesion is caused by shearing (Figure 13b). The decohesion induces a reduced heat flow that results in a temperature jump, see Figure 14. Using an adapted temperature range (Figure 14b) the effect can be noticed better compared to the plot with the whole temperature range (Figure 14a).

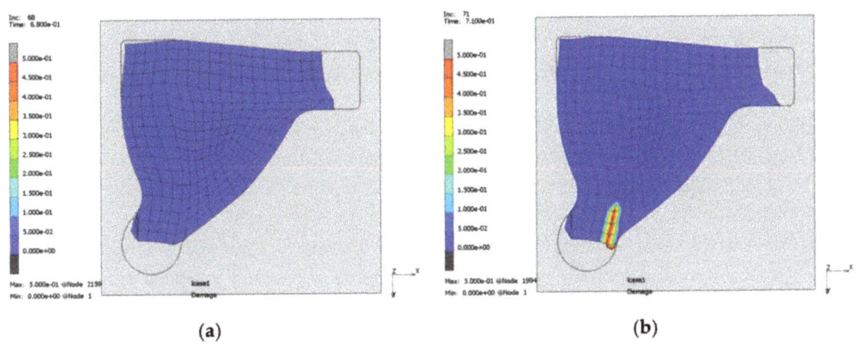

(a)   (b)

Figure 13. Forming process using a further simplified geometry; before (**a**) and after failure initiation (**b**): Joining zone damage.

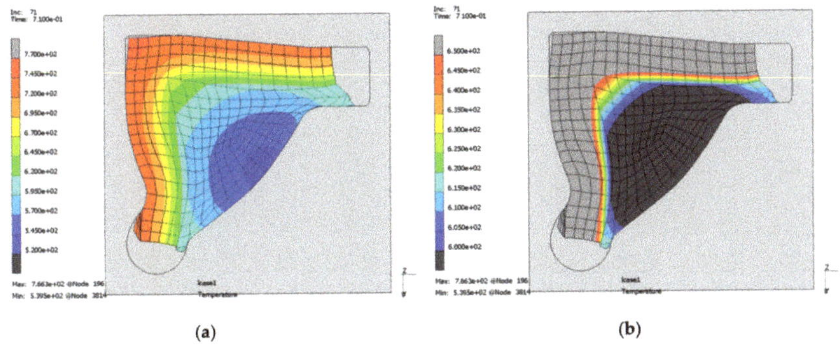

(a)   (b)

Figure 14. Forming process using a further simplified geometry; whole (**a**) and adapted temperature range (**b**): temperature distribution.

Using a tetraeder mesh (tetraeder for the bulk, triangle or wedge like elements for the interface) instead of a hexaeder mesh (hexaeder for the bulk, quadrilaterial or hexaeder like elements for the interface) leads to very similar results. Figure 15 shows the same increment as Figure 13b and the specimen exhibits failure in the same position.

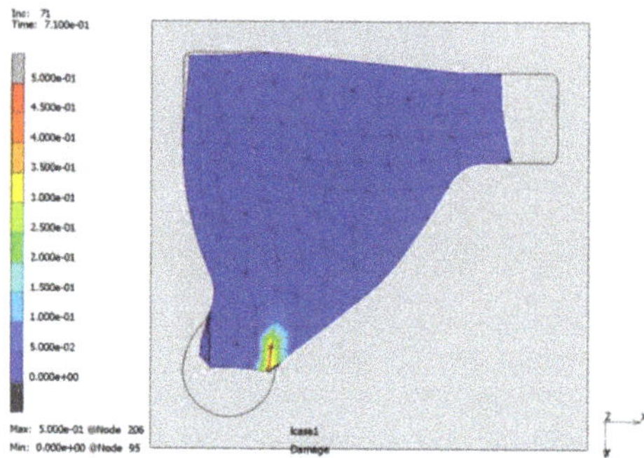

Figure 15. Forming process using tetraeder elements: joining zone damage.

6. Indentation Test

There are several standard tests for Cohesive Zone Elements, e.g. the double cantilever beam test or the end notch flexure test. However, they focus on shear or normal separation induced delamination. Therefore, an indentation test is used here that is influenced by membrane modes as well. A sandwich of 20 mm × 4 mm steel (upper part) and 20 mm × 4 mm aluminium (lower part) is compressed by a cylinder with a radius of 6 mm. The simulation is carried out in 3D but boundary conditions are applied to enforce plane strain conditions. Furthermore, symmetry conditions are utilised. The parameters for the joining zone are adopted from Section 4.3, for the bulk materials room temperature is assumed and accordingly yield stresses of $\sigma_{y,st} = 600$ MPa and $\sigma_{y,al} = 400$ MPa [10].

Figure 16 shows some increments of the simulation using a fine mesh. Predamaging, failure, and successive delamination can be observed. The contact algorithm adopted from [5] maintains a certain stiffness in normal direction only for compression. Therefore, penetration after failure is prevented but tangential sliding occurs.

The simulations are carried out with 10, 20, 30, 40, 50, and 60 elements in horizontal direction. Table 1 contains the increment and position of failure initiation for the different meshes. The three finer discretisations lead to very similar results regarding failure initiation.

Table 1. Location and increment of failure initiation.

Number of Elements in Horizontal Direction	Failure in Increment	Failure Position from Left Corner
10	59	0.21 mm
20	47	1.40 mm
30	43	0.93 mm
40	40	1.20 mm
50	39	1.16 mm
60	37	1.13 mm

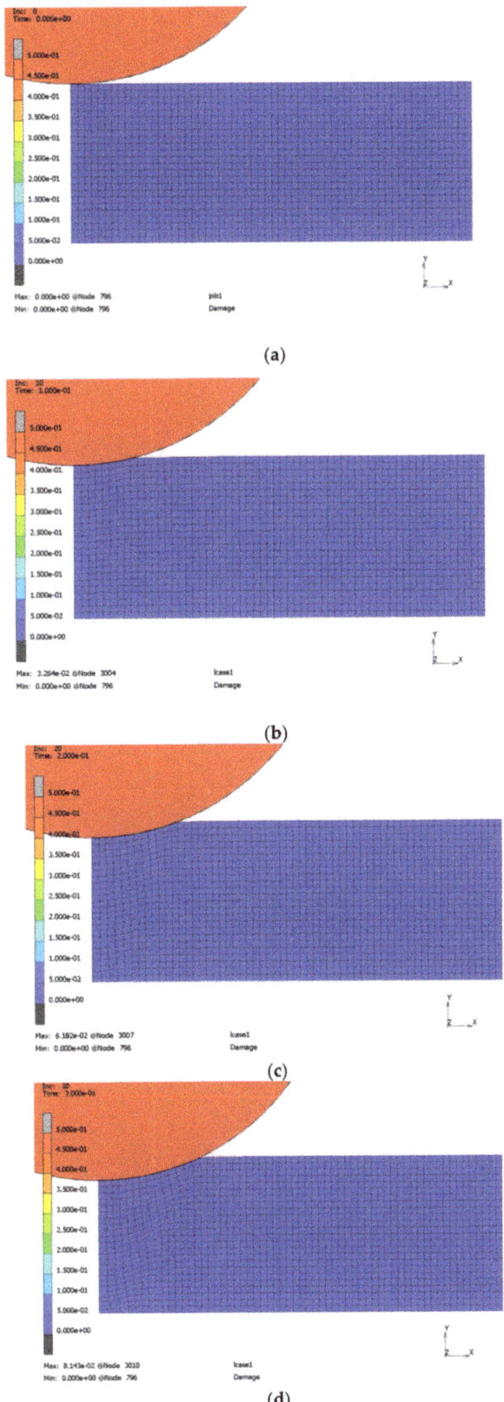

(**a**)

(**b**)

(**c**)

(**d**)

Figure 16. *Cont.*

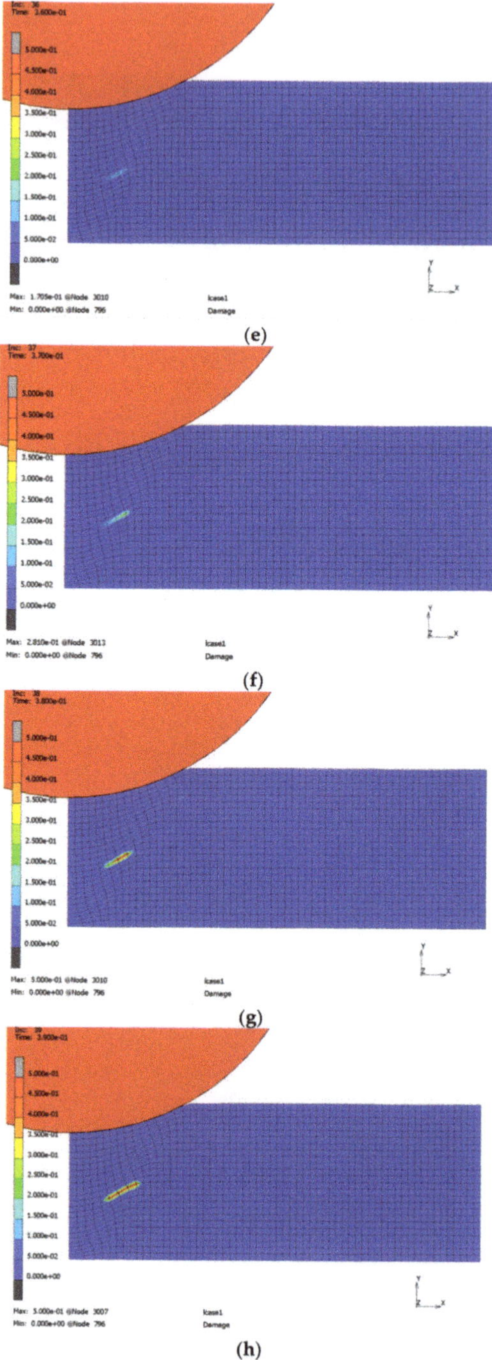

(e)

(f)

(g)

(h)

Figure 16. *Cont.*

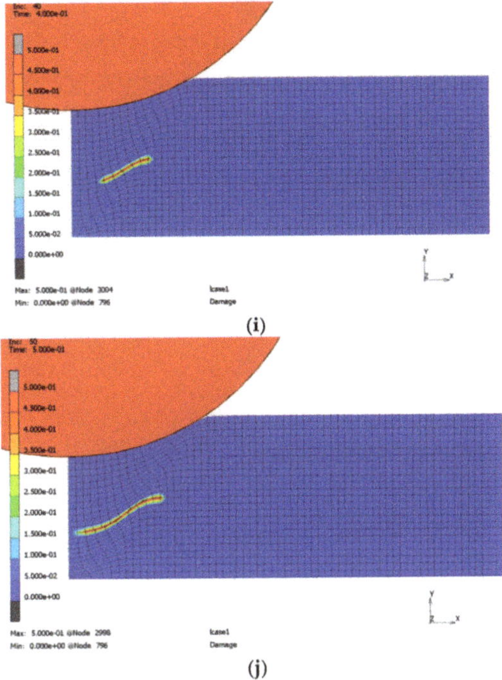

Figure 16. Indentation test: Joining zone damage: increment 0 (**a**), 10 (**b**), 20 (**c**), 30 (**d**), 36 (**e**), 37 (**f**), 38 (**g**), 39 (**h**), 40 (**i**), 50 (**j**).

7. Conclusions

Interfaces can be modelled using Cohesive Zone Elements. However, standard formulations do not consider membrane deformations that occur during forming of hybrid bulk material components. Membrane Mode Enhanced Cohesive Zone Elements capture these effects. An implementation of this technology in the nonlinear finite element solver MSC Marc is carried out. The new technology can be used for thermomechanical simulations in combination with all capabilities that Marc provides for usual Cohesive Zone Elements as it is implemented as a material subroutine in the provided standard interface element.

Simulations of a transverse link are carried out to test the technology. The material parameters stem from a different joining process and the development of the transverse link itself is not finished yet. Though some general findings might be valid:

(1)   The intended geometry contains severe deformation affecting the joining zone.
(2)   A more similar geometry of the raw component and the formed component would reduce the probability of failure initiation here.
(3)   The starting position in the form significantly influences the process and should be fixed.
(4)   A closed die process with burr prevention should be considered to prevent zones with very large deformation where a crack initiation might be triggered.

An indentation test is suited to investigate mesh dependency due to a simpler geometry. The load increment and also the position of failure initiation change from coarse to fine mesh, but are stable for further mesh refinement.

Author Contributions: Conceptualization, F.T.; software, F.T.; investigation, F.T.; writing—original draft preparation, F.T.; writing—review and editing, S.L. and P.W.; visualization, F.T.; supervision, S.L. and P.W.; funding acquisition, S.L. and P.W. All authors have read and agreed to the published version of the manuscript.

Funding: Funded by the Deutsche Forschungsgemeinschaft (DFG, German Research Foundation)—CRC 1153, subproject C4 - 252662854.

Conflicts of Interest: The authors declare no conflict of interest.

References

1. Geplanter Sonderforschungsbereich 1153—Prozesskette zur Herstellung hybrider Hochleistungsbauteile durch Tailored Forming—Einrichtungsantrag. 2014.

2. Herbst, S.; Dovletoglou, C.N.; Nuernberger, F. Method for Semi-Automated Measurement and Statistical Evaluation of Iron Aluminum Intermetallic Compound Layer Thickness and Morphology. *Met. Microstruct. Anal.* **2017**, *6*, 367–374. [CrossRef]

3. Behrens, B.A.; Bonhage, M.; Bohr, D.; Duran, D. Simulation Assisted Process Development for Tailored Forming. *Mater. Sci. Forum* **2019**, *949*, 101–111. [CrossRef]

4. Töller, F.; Löhnert, S.; Wriggers, P. Bulk material models in Cohesive Zone Elements for simulation of joining zones. *Finite Elements Anal. Des.* **2019**, *164*, 42–54. [CrossRef]

5. Töller, F.; Löhnert, S.; Wriggers, P. Membrane mode enhanced cohesive zone elements. *Eng. Comput.* **2020**. Unpublished work.

6. Lemaitre, J.; Desmorat, R. *Engineering Damage Mechanics: Ductile, Creep, Fatigue and Brittle Failures*; Springer Science & Business Media: Berlin/Heidelberg, Germany, 2005.

7. MSC.Software Corporation. MSC.Marc Volume B: Element Library. version 2019.

8. MSC.Software Corporation. MSC.Marc Volume D: User Subroutines and Special Routines. version 2019.

9. Töller, F.; Löhnert, S.; Wriggers, P. Thermo-mechanical coupling for internal thickness extrapolation elements. In Proceedings of the 8th GACM Colloquium on Computational Mechanics: For Young Scientists from Academia and Industry, Kassel, Germany, 28–30 August 2019; Gleim, T., Lange, S., Eds.; Kassel University Press GmbH: Kassel, Germany, 2019.

10. Behrens, B.-A.; Chugreev, A.; Selinski, M.; Matthias, T. Joining zone shape optimisation for hybrid components made of aluminium-steel by geometrically adapted joining surfaces in the friction welding process. In *Proceedings of the AIP Conference Proceedings*; AIP Publishing: College Park, MD, USA, 2019; Volume 2113, p. 040027.

Article

Feeling Machine for Process Monitoring of Turning Hybrid Solid Components

Berend Denkena, Benjamin Bergmann and Matthias Witt *

Institute of Production Engineering and Machine Tools, Leibniz University Hannover, 30823 Garbsen, Germany; denkena@ifw.uni-hannover.de (B.D.); bergmann@ifw.uni-hannover.de (B.B.)
* Correspondence: witt@ifw.uni-hannover.de; Tel.: +49-511-762-18095

Received: 11 June 2020; Accepted: 8 July 2020; Published: 10 July 2020

Abstract: The realization of the increasing automation of production systems requires the guarantee of process security as well as the resulting workpiece quality. For this purpose, monitoring systems are used, which monitor the machining based on machine control signals and external sensors. These systems are challenged by innovative design concepts such as hybrid components made of different materials, which lead to new disturbance variables in the process. Therefore, it is important to obtain as much process information as possible in order to achieve a robust and sensitive evaluation of the machining. Feeling machines with force sensing capabilities represent a promising approach to assist the monitoring. This paper provides, for the first time, an overview of the suitability of the feeling machine for process monitoring during turning operations. The process faults tool breakage, tool wear, and the variation of the material transition position of hybrid shafts that were researched and compared with a force dynamometer. For the investigation, longitudinal turning processes with shafts made of EN AW-6082 and 20MnCr5 were carried out. The results show the feeling machine is sensitive to all kinds of examined errors and can compete with a force dynamometer, especially for roughing operations.

Keywords: turning; process monitoring; tailored forming; feeling machine; benchmark

---

1. Introduction

In production, the aim is to achieve full automation of manufacturing through autonomous machining processes. In this context, process monitoring systems are an important part of modern production plants. They protect machines and machine operators from damage, reduce downtime, and improve workpiece quality by eliminating chatter [1]. Rehorn et al. estimated, if the computer numerical control (CNC) machine tool is equipped with a monitoring system, downtime can be reduced by up to 20% while productivity can be raised by 50%. Machine utilization even increases by more than 40% [2]. The increasing complexity of series production represents a challenge for process monitoring. This also includes the use of new design strategies, e.g., by combining different materials in hybrid workpieces. These workpieces with locally adapted properties offer a promising approach for designing components with energy-efficient applications while reducing the use of high-alloy materials. This is the main focus of the Collaborative Research Center (CRC) 1153, where material combinations are joined and formed with different processes and then machined [3,4]. Due to different material properties and chemical compositions, the machining properties and chip formation mechanisms change during the cutting process and pose challenges for machining. In particular, the force gradient during material transition can lead to increased tool wear and the various material properties affect the workpiece geometry and surface quality. In addition, the trend to single part production or small batch sizes as well as the high demands on surface quality and manufacturing tolerances pose further challenges. For this reason, modern process monitoring systems must be provided with process information of the highest possible quality.

In this context, more monitoring approaches that combine different signals and features are being researched [5–7]. A process parameter that offers a high level of information regarding various production errors is the process force [2,8]. Balsamo et al. showed that a catastrophic tool failure during turning can be monitored by multi-sensor signal processing of force and acoustic emission [9]. Jie et al. successfully developed an approach for tool condition monitoring during the machining of titanium alloys, which is also based on the force and acoustic emission signal [10]. In addition, the development of a cloud-based framework for monitoring manufacturing processes for online process monitoring services takes into account process force information due to its high sensitivity [11]. In general, the machine tool does not provide detailed information about forces. Therefore, various approaches have been studied in the past to extract this information parallel to the machining. A method to obtain information about the process force is to extract the process components from control signals [12,13]. This is associated with a high modelling effort, which has to be carried out for each machine due to manufacturing and assembly tolerances. By using external sensors, process forces can be measured with high sensitivity [14,15]. These sensor systems require high acquisition costs, are limited in their flexibility, and restrict the installation space for workpieces [16]. Further approaches to determine process force information have focused on modelling forces by virtual processing [17]. The simulations reach their limits if tool wear and thermal effects are to be considered in real time. In addition, a high modeling effort is required to take the structural dynamic effects into account. To overcome these challenges, Denkena et al. developed the idea of the "feeling" machine tool. They integrated sensory machine components, which are placed close to the process and are located directly in the force flow. These sensory components are equipped with strain gauges that measure the structural strain caused by the process force. Forces are then reconstructed from the strains. With this approach, the static and dynamic behavior of the machine structure is generally maintained and a flexible force measurement is realized. In milling machines, the spindle slides have been modified to become a sensory machine component. Based on the force sensing properties of the slide, the static tool deflection was determined during the process. The workpiece quality was improved by online adaptation of the tool position using an axis offset [18]. The monitoring of geometrical errors of the workpiece was investigated by a feeling clamping system for a milling process. Therefore, holes were drilled into the workpiece to represent a material defect for a subsequent flank milling operation. A dynamometer was used as a reference. The errors could be detected by both measuring systems based on confidence limits. However, due to a lower signal-to-noise ratio, the feeling clamping system had a lower sensitivity to the error [19]. Denkena et al. integrated metallic strain gauges in four carriages of a linear guide of a spindle-driven position axis [20]. The load on the carriages could be measured in two directions orthogonally to the direction of travel. The quality of the recorded measurement signals was suitable for detecting parallelism errors in the linear guide. Based on the achieved force, sensitivity process monitoring could not be performed.

This paper focuses for the first time on the investigation of the suitability of a feeling machine for process monitoring of turning operations. In this context, the process force is measured based on a feeling turret during machining and is compared with the measured forces of a dynamometer. Individual process errors, which affect the process and workpiece quality, are investigated. These include tool breakage, tool wear, and the varying material transition position of hybrid workpieces. Based on the considered process errors and the measuring accuracy of the feeling turret, the sensitivity of the monitoring is discussed for various process parameters.

2. Materials and Methods

2.1. Materials

When researching the system, different investigations were performed with two different materials, which were presented in Table 1. The samples were shafts with a diameter of 30 or 40 mm and a length of 120 mm. Mono-material components were selected to investigate tool breakage and tool wear.

When studying the performance of material detection friction-welded specimens, two mono-material shafts were used.

Table 1. Material properties.

Material Properties	EN AW-6082	20MnCr5
density [g/cm^3]	2.70	7.75
hardness [HV]	100	268
ultimate tensile strength [MPa]	310	980–1280
Young's modulus [GPa]	70	210
thermal expansion coefficient [10^{-6}K^{-1}]	23.4	11.5
thermal conductivity [W/mK]	170–220	42

2.2. Machining and Data Acquisition

Experimental tests were carried out on the turning center Gildemeister CTX420 linear (Germany). An industrial personal computer (IPC) was connected to an open platform communication server and fieldbus interface of the Siemens Powerline 840D machine control (München, Germany) as well as to the CAN bus of the feeling turret. For the longitudinal turning, an indexable insert DNMG150404-FP5 by Walter AG (Tübingen, Germany) was used. The machining operations were all carried out without the use of cooling lubricant. The process force was measured by the sensory abilities of the turret, which is equipped with strain gauges (HBM, Darmstadt, Germanycountry). The strain gauges were set up as a Wheatstone-bridge and glued into the notch ground. The notches were designed by finite element simulation to increase the measured strain. A total of six strain gauges were set up as Wheatstone-bridges are placed on the turret. The signals are digitized by electronic devices close to the measuring position and transmitted to the IPC via a CAN-BUS [21]. According to the occurring strains during machining, the forces were calculated. The determined force resolution of the feeling turret is 64 N in the x-direction, 43 N in the y-direction, and 44 N in the z-direction. A dynamometer (Kistler 9129A, Winterthur, Switzerland) was used as a reference for the force measurement. Both systems have a sampling rate of 1000 Hz. The low-pass frequency for the feeling turret is 25 Hz and 30 Hz for the dynamometer. The passive force F_p acts in the x-direction, the cutting force F_c acts in the y-direction, and the feed force F_f acts in the z-direction. The numerical control (NC) provided the axis positions with a frequency of 83 Hz. The measurement setup is shown in Figure 1.

Within the evaluation, the process force was also simulated to investigate the process parameter range in smaller steps. The force model was used according to Kienzle [22]. The Kienzle parameters applied for the modelling were determined experimentally on the presented test set-up for the material and tool combinations. These parameters are shown in Table 2.

Table 2. Summary of Kienzle parameters.

Parameter	Symbol	Material	
		20MnCr5	EN AW-6082
specific cutting force	$k_{c1.1}$	2140 N/mm^2	742 N/mm^2
material exponent	m_c	0.250	0.236

Figure 1. Presentation of the measurement setup.

2.3. Simulation of Process Errors

A tool breakage can result in a sudden increase in process force. This can be caused by the broken cutting-edge jamming between the tool and workpiece. Afterward, the process force decreases because the gap between tool and workpiece must first be overcome by the travel movement. Then the remaining part of the tool collides with the workpiece, which increases the process forces again [23]. The described drop in force was simulated in the experiments by a material defect depicted in Figure 2a. The missing material results in an identical force characteristic as in the case of a tool breakage. The advantage of these simulated errors is the reproducibility and the fact that the machine tool components are not critically stressed. To investigate, if tool wear can be identified with the systems, the resulting force for machining with a new and a worn tool was measured. The worn tool exhibits a cutting-edge failure with a length of 465 µm, which is depicted in Figure 2b.

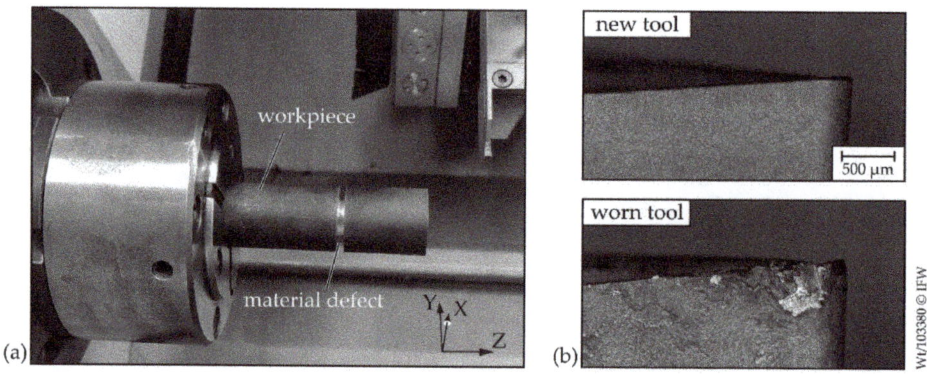

Figure 2. (a) Prepared workpiece and (b) presentation of the new and worn tool.

2.4. Monitoring

The confidence limits for monitoring tool breakage were calculated by performing 10 longitudinal turning operations. The measured cutting forces were used to calculate the upper L_{up} and lower L_{low} confidence limits. Thus, the confidence estimator approach according to Brinkhaus was used [24]. The confidence limits were updated after each process based on the expected mean value $\bar{x}$ standard deviation σ, and a safety value C_{saftey} for each running measurement value i. The safety value influences the distance between $\bar{x}$ and the confidence limits. A safety value of $C_{saftey} = 6$ was used, which is usually the default setting of a monitoring system.

$$L_{up,\,low}(i) = \bar{x}(i) \mp C_{saftey} \cdot \sigma(i) \tag{1}$$

In order to take account of temporal fluctuations, the confidence limit was calculated for a function $h(i)$. This function determines an envelope for the expected value $x(i)$, which allows a time lag for a specified time K. During the investigations, K was set to 100 ms. Based on this value, no false alarms occurred after four processes in the teach-in phase.

$$\begin{aligned} h_{up}(i) &= max(x(i-K), \ldots, x(i+K)) \\ h_{low}(i) &= min(x(i-K), \ldots, x(i+K)) \end{aligned} \tag{2}$$

According to the combination of Equations (1) and (2), the shift of time and amplitude was taken into account when calculating the confidence limits. The values of h and σ were calculated by a moving average to allow dynamic weighting of the different measurements.

$$\begin{aligned} L_{up}(i) &= \bar{h}_{up}(i) + C_{saftey} \cdot \sigma\!\left(h_{up}(i)\right) \\ L_{low}(i) &= \bar{h}_{low}(i) - C_{saftey} \cdot \sigma\!\left(h_{low}(i)\right) \end{aligned} \tag{3}$$

For detecting the currently machined material, the material-specific cutting force Mat_{Fc} was monitored [25]. In this approach, the cutting force F_c was normalized by the material removal rate Q_w. The material removal rate was calculated online using a dexel-based cutting simulation. Input parameters for the simulation were the axis positions from the machine control. A material compound was subsequently monitored on the basis of a defined boundary surface.

$$Mat_{Fc} = F_c / Q_w \tag{4}$$

3. Results and Discussion

3.1. Tool Breakage

In cutting operations, process errors typically lead to an increase or decrease of the process force. For this reason, the process force is well suited for a sensitive and robust monitoring in machining. To determine the sensitivity of force measuring systems, material defects can be applied to the workpiece. Similar to tool breakage, missing material lead to a drop in the process force. This characteristic was used in order to compare the cutting force $F_{c,SG}$, which was determined by the feeling turret, and the cutting force $F_{c,dyn}$, which was measured with the dynamometer. A longitudinal turning process was carried out with a new tool for a cutting speed of 300 m/min and a feed rate of 0.1 mm. During machining, the depth of cut was reduced stepwise from 1 to 0.1 mm. This process was repeated 10 times. Afterward, a shaft was machined, which was prepared with a groove that had a width of 3 mm. The missing material causes a decrease in cutting force at each depth of cut, depicted in Figure 3.

The comparison of the two force measurement systems showed that they have an identical time behavior with regard to the signal change. The amplitude of the decreasing cutting force is also identical for both systems. Therefore, it can be concluded that both systems are highly sensitive to a change in cutting force due to tool breakage. However, by comparing the signals of the 10 good processes, it becomes clear that the cutting force $F_{c,SG}$ has a lower signal-to-noise ratio (SNR).

$$SNR = \mu / \sigma \tag{5}$$

The SNR was determined for each depth of cut a_p by the mean value of the amplitude μ and the standard deviation σ of the signal. The results of the SNR for both measuring systems are summarized in Table 3.

Table 3. Signal-to-noise ratio for the measured cutting force by using the dynamometer and feeling turret.

Signal-to-Noise Ratio (SNR)	Depth of Cut a_p [mm]				
	1	0.5	0.3	0.2	0.1
cutting force by dynamometer $F_{c,dyn}$	46	149	78	90	70
cutting force by feeling turret $F_{c,SG}$	40	26	16	12	7

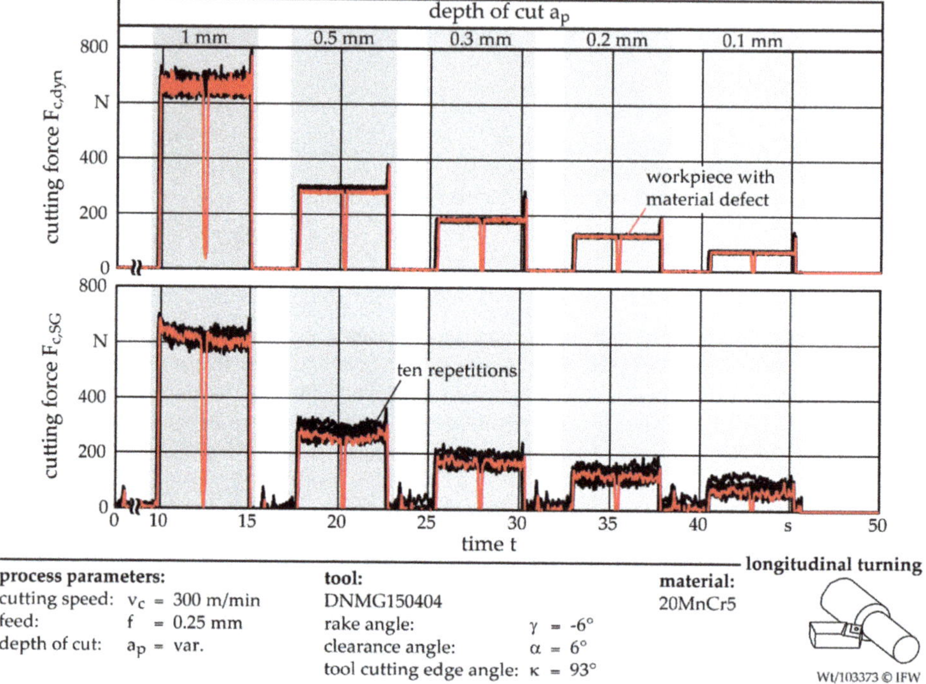

Figure 3. Longitudinal turning with and without a material defect for different depths of the cut.

The SNR was determined for each depth of cut a_p by the mean value of the amplitude μ and the standard deviation σ of the signal. At an a_p of 0.5 mm, $F_{c,dyn}$ exhibits an SNR of 149. With a decreasing a_p, the SNR also decreases further. For a_p = 0.1 mm, the SNR is still 70. The measured cutting force by the dynamometer has the lowest SNR for a_p = 1 mm with 46. The deviation to the other depths of cut is caused by different chip formation, which led to a higher vibration of the system. For a_p = 1 mm, the $F_{c,SG}$ shows a similar SNR with 40. However, with each reduction of a_p, the signal-to-noise ratio decreases for the signal $F_{c,SG}$. For the depth of cut of a_p = 0.1 mm, the SNR is 7 and, thus, lower by a factor of 10 compared to $F_{c,dyn}$. The repeatability of the individual measurements also shows a different quality for both signals. The evaluation of the 10 good processes demonstrates that the cutting force $F_{c,SG}$ has larger variations between the measurements than $F_{c,dyn}$. With a depth cut of a_p = 1 mm, the standard deviation σ_i for the mean amplitudes of the different workpieces is for both systems being almost identical with approximately 15 N. By changing the process parameters to a lower a_p, the standard deviation decreases for $F_{c,dyn}$ to σ_i = 1.7 N while σ_i of $F_{c,SG}$ is around 10 N.

If confidence limits are used to monitor the signal, the higher signal-to-noise ratio and variation of the mean amplitude leads to a widening of the statistical limits and, thus, to a decreasing monitoring sensitivity. In order to compare $F_{c,SG}$ and $F_{c,dyn}$ with regard to their qualification for confidence limit based monitoring, the feature of the normed bandgap (NB) was determined [26]. The normed bandgap indicates the range in which the normalized signal has to be changed to trigger an alarm. For this purpose, the confidence limit was calculated based on the distribution of the signal around its long-term average. After eight processes, the limits had become close to the process. A further approximation was only possible in very small steps due to the high standard deviation of the measured signal. It can be assumed that 10 measurements are sufficient to generate the confidence limit and calculate the NB.

$$\text{NB} = \frac{k}{\bar{x}} = \frac{L_{up} - \bar{x}}{\bar{x}} \tag{6}$$

To evaluate the monitoring quality, the gap between the mean value $\bar{x}$ and the upper confidence limit L_{up} was evaluated. The distance k between $\bar{x}$ and L_{up} was normalized to the average. Figure 4 shows the confidence limits after 10 processes for the cutting forces $F_{c,dyn}$ and $F_{c,SG}$ and the comparison of the normed bandgap (NB) for both measuring systems.

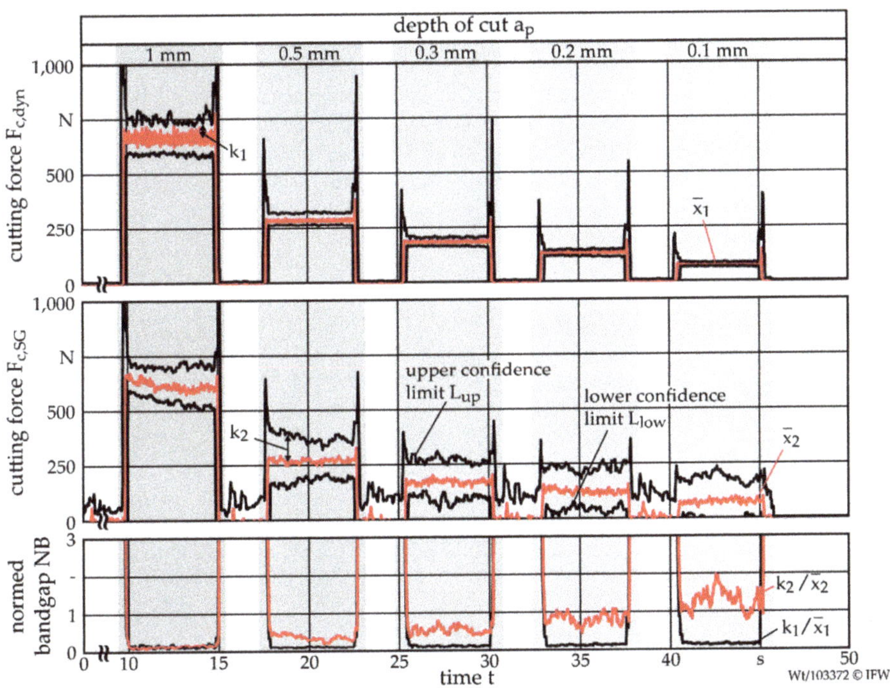

Figure 4. Confidence limits and normed bandgaps for $F_{c,dyn}$ and $F_{c,SG}$ during a longitudinal turning process with different depths of cut.

The sensitivity of the confidence limit, which is calculated based on the cutting force of the feeling turret, declines with decreasing force. This is reflected by the increasing normed bandgap. For an a_p of 1 mm, both signal sources achieve a similar monitoring performance. At a normed bandgap of 0.3, a process error must result in a signal change of 30% to trigger the alarm. The ratio remains between NB = 0.1–0.2 for all examined depths of cut if the cutting force is measured with the dynamometer (Kistler, Winterthur, Switzerland). If the feeling turret is used, the normed bandgap for $a_p = 0.3$ mm shows a maximum of NB = 0.8. The NB increases to 2 with a reduction of a_p to 0.1 mm. Thus, the sensitivity of the dynamometer is higher by a factor of 10 compared to the feeling turret. The fact that the feeling turret has a significantly lower sensitivity at low process forces is based on the lower measurement resolution of the system relative to the dynamometer. This is primarily due to a higher signal-to-noise ratio, which has an increasing impact on the stochastic distribution of the signal for lower cutting forces. A generalized statement about the required NB of a signal cannot be made since each error is represented differently in the signal. However, based on the investigations, it is considered when no finishing operations can be monitored by the feeling turret. For roughing operations, on the other hand, a similar monitoring quality as with the dynamometer can be achieved.

3.2. Tool Wear

Tool wear generally leads to a change in process force. With increasing flank wear, friction increases, which, in turn, leads to an increasing cutting force and cutting temperatures. If the wear is mainly caused by crater wear, the change in rake angle can lead to a decrease in cutting force. Consequently, the change in cutting force is also an indicator for monitoring the type of wear that will influence the process and the workpiece quality. Tool wear is usually monitored by a static limit. Therefore, the average measured cutting force with a new tool is determined for a predefined tool path. Based on empirical values, a maximum tolerable change in force is defined as the limit for the tolerable wear. If the limit is exceeded, the monitoring system initiates a tool change before an intolerable influence on the workpiece quality occurs. In contrast to the confidence limits, the influence of the signal-to-noise ratio on the monitoring quality is significantly lower since the average value is evaluated over a defined range of the monitored signal. For a sensitive and robust monitoring of tool wear, the repeat accuracy is mostly important. In addition to the 10 good processes with a new tool, three measurements were carried out with a tool, which has a high chipping length of 465 µm. During the machining, identical process parameters of a_p, f, and v_c were used as for the previous longitudinal turning operations, depicted in Figure 5.

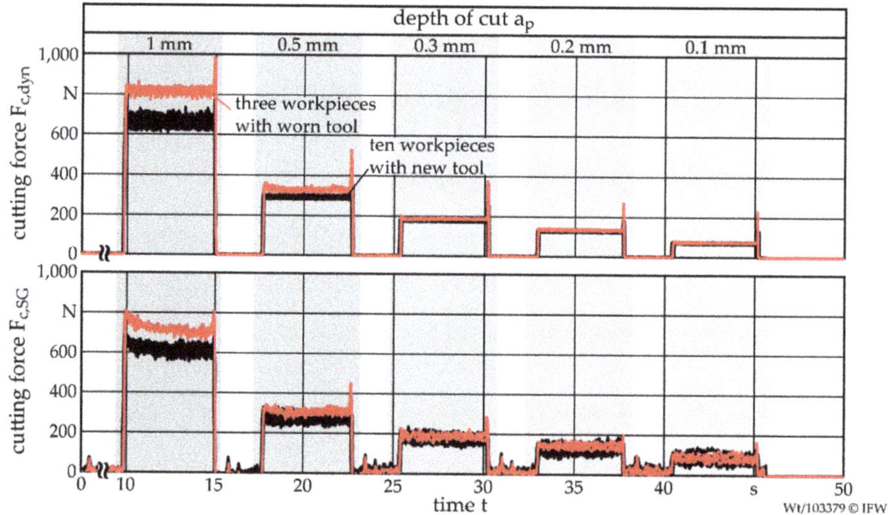

Figure 5. Cutting force characteristics for new and worn tool for different depths of cut.

Tool wear results in an increasing cutting force, which varies according to the depth of cut. At a depth of cut of $a_p = 1$ mm, an increase of 24% occurs with 158.5 N for $F_{c,dyn}$. With a reduction to $a_p = 0.5$ mm, the wear lead to an average increase in the cutting force of 38.6 N, which represents a percentage change of 13%. The triple standard deviation of the measurements with the new tool is $3\sigma = 9$ N for $F_{c,dyn}$. Assuming tool wear can be detected, if this value is lower than the resulting change in cutting force, monitoring can be performed for both a_p. The increase of $F_{c,dyn}$ at an a_p of 0.3 mm is, therefore, with 7.5 N too small to detect the current tool wear for the defined process parameters.

For the feeling turret, a 3σ for $F_{c,SG}$ of 53 N is determined at a depth cut of 1 mm. Thus, the tool wear can be robustly identified for $a_p = 1$ mm. For the remaining process parameters, the variation of the average cutting force is too low to realize tool wear monitoring.

As shown by the different depths of cut, no general statement can be made about the effect of tool wear on the percentage change in cutting force. Therefore, it is not possible to make a universal prediction of the ability to monitor tool wear. The influence of the wear mechanisms on the cutting force vary significantly based on the selected process parameters. For this reason, it is more relevant to compare what change in cutting force caused by tool wear can be theoretically identified with different signals. For this purpose, the cutting force was modelled according to Kienzle for a new tool and three wear-related force increases. The variation of the cutting speed has a smaller influence on the force compared to the feed and depth of cut. For this reason, the cutting speed was fixed for the research at $v_c = 300$ m/min. Based on the modelled cutting forces, an increase in force was determined for a combination of process parameters. It is assumed that the change can be monitored if the difference is higher than the triple standard deviation 3σ of the cutting force measurement. Based on the previous investigation, 3σ was set to $F_{c,dyn,3\sigma} = 9$ N and $F_{c,SG,3\sigma} = 53$ N. For both measuring systems, it process parameters were researched at which a force difference of 10%, 20%, and 30% results can be detected by the individual system, which was depicted in Figure 6. This investigation was carried out for a steel (20 MnCr5) and an aluminum (EN AW-6082).

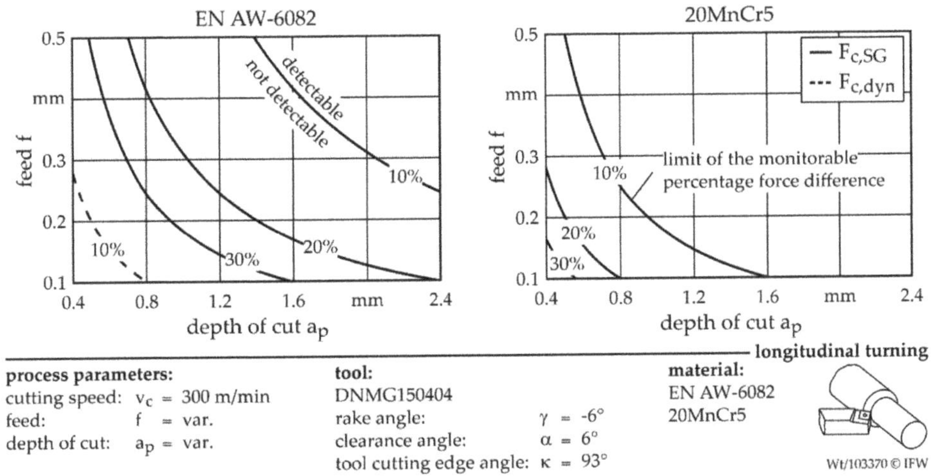

Figure 6. Representation of the monitorable percentage increase of the cutting force for EN AW-6082 and 20MnCr5 based on the feeling turret and dynamometer.

With the dynamometer, almost the total process parameter range considered for both materials can be monitored with regard to the three selected percentile cutting force changes. Only for aluminum, the theoretical detection is not possible for low depths of cut and low feeds. In this case, the cutting force difference caused by a wear-related force increase of 10% is too low. The evaluation of the cutting force measured by the feeling turret shows the potential for monitoring tool wear for a wide range of process parameters. Especially for steel, a large range is covered. For aluminum, higher a_p and f are required due to the different material properties. For the highest expected change of 30%, this includes process parameters of $a_p = 0.5$ mm and $f = 0.17$ mm. In the case of aluminum, large parts of the considered process parameter range can theoretically not be monitored, especially if the cutting force changes by only 10% due to tool wear. In general, higher a_p and f are used for machining aluminum than for steel. Then tool wear can also be monitored for aluminum during roughing operations.

3.3. Material Detection

The online detection of the machined material is needed to realize a material-specific machining of hybrid components. For this reason, the sensitivity of monitoring a material compound based on the material-specific cutting force was investigated. A longitudinal turning process of EN AW-6082 and 20MnCr5 was performed and the signals of the dynamometer and the feeling turret were compared by the statistical overlap factor (SOF). The SOF determines the degree of separation of the material-specific cutting force between the first and second material. If a signal shows a high degree of separation due to different material properties, the SOF increases. In addition to the resulting standard deviation when measuring the cutting force, the influence of the variance of the simulated material removal rate Q_w was considered. The determined triple standard deviation for the simulation of Q_w is 3–4%. With $Q_{w,min}$, the value at the lower end of the triple standard deviation is taken into account for the first material i. In contrast, $Q_{w,max}$ is assumed for the second material j. This results in the material-specific cutting force being calculated too low for steel and too high for aluminum in order to obtain the highest possible variation. To guarantee robust monitoring, an SOF greater than six is necessary to separate two materials during machining.

$$SOF_{ij} = \frac{\left| \dfrac{F_{c,i}}{Q_{w,min}} - \dfrac{F_{c,j}}{Q_{w,max}} \right|}{\sigma\left(\dfrac{F_{c,i}}{Q_{w,min}}\right) + \sigma\left(\dfrac{F_{c,j}}{Q_{w,max}}\right)} \tag{7}$$

With a constant a_p of 1 mm, the SOF was calculated for a varying feed rate of $f = 0.1$–0.4 mm and a cutting speed of $v_c = 200$–400 m/min, which are depicted in Figure 7. The influence of the simulation error of Q_w on the calculated SOF can be neglected. For the investigated aluminum-steel compound, the SOF decreases by a maximum of 5% if the highest possible variation of Q_w is considered instead of assuming a constant value for Q_w. The significant impact is, therefore, the measuring accuracy and the signal-to-noise ratio of the cutting forces.

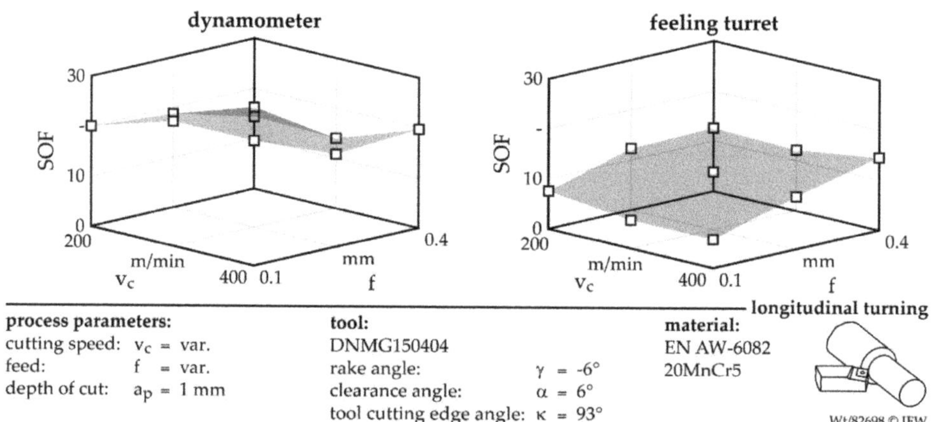

Figure 7. Material-specific cutting force for varying feed and cutting speed.

For the feeling turret, the sensitivity of the monitoring system improves if the the feed and, thus, the cutting force is increased. This is because the SNR is much more significant in the signal of the structure-integrated strain gauges compared to the signals of the dynamometer. Consequently, the SNR has a larger impact on the measurement of F_c and, subsequently, on the quality of the material identification at lower process forces. For a feed of $f = 0.1$ mm, the average measured SOF is 6.0, while, at $f = 0.4$ mm, the average SOF increases to 12.6. With the dynamometer, this effect does not appear due to the lower SNR. However, the SOF also varies for the material-specific cutting force measured by the dynamometer over the examined process parameter range. This is caused by the higher sensitivity to varying chip formation. Different chip forms results in changing peak2peak values of the signal and, consequently, in different standard deviations of the force signal. This effect is less significant for strain gauges due to the higher damping of the turret structure. In general, a higher SOF value and, thus, better sensitivity is achieved by using the dynamometer for all considered process parameters. The maximum statistical overlap factor is 23.9 while the minimum SOF is 13.5. Both measuring systems are, therefore, qualified to robustly monitor the material during the machining of the aluminium-steel compound for the examined process parameter range.

4. Conclusions

This paper presents the investigation of the application of a feeling machine for process monitoring. In order to replace the dynamometer, a feeling turret was used, which determines process forces based on the structure's integrated strain gauges. The monitoring quality was compared with the performance of a dynamometer regarding the detection of tool breakage, tool wear, and material transitions of hybrid components. Investigations have shown that, especially in roughing operations, which are characterized by higher process forces, the feeling turret has a good monitoring performance. At depths of cut of $a_p = 1$ mm, similar sensitivities were determined for monitoring with confidence limits as with a dynamometer. If steel is machined, tool wear can be detected with the cutting force of the feeling turret for roughing and finishing processes. For aluminum, on the other hand, only monitoring roughing processes can be realized. However, it must be assumed that the tool wear also leads to a significant change in cutting force of 10–30%. Similar to the dynamometer, the feeling turret is suitable for monitoring the materials of aluminium-steel compound (EN AW-6082/20MnCr5). For the examined process parameter range, the statistical overlap factor was higher than six. Consequently, a classification of the materials based on the material-specific cutting force can be achieved. In further steps, it will be investigated how the fusion of signals from the feeling machine and process data from the machine control and external sensors can be used to improve the monitoring quality.

Author Contributions: Conceptualization, M.W. and B.B.; funding acquisition, B.D. and B.B.; investigation, M.W.; methodology, M.W. and B.B.; supervision, B.D.; validation, B.B.; writing—original draft, M.W.; writing—review & editing, B.D. and B.B. All authors have read and agreed to the published version of the manuscript.

Funding: The results presented in this paper were obtained within the Collaborative Research Centre 1153 "Process chain to produce hybrid high performance components by Tailored Forming" in the subproject B5 (252662854). The authors would like to thank the German Research Foundation (DFG) for its financial and organisational support of this project.

Conflicts of Interest: The authors declare no conflict of interest.

References

1. Tönshoff, H.K.; Wulfsberg, J.P.; Jals, H.J.J.; König, W.; Van Luttervelt, C.A. Developments and Trends in Monitoring and Control of Machining Processes. *CIRP Ann.* **1988**, *37*, 611–622. [CrossRef]
2. Rehorn, A.G.; Jiang, J.; Orban, P.E. State-of-the-art methods and results in tool condition monitoring: A review. *Int. J. Adv. Manuf. Technol.* **2005**, *26*, 693–710. [CrossRef]
3. Behrens, B.A.; Chugreev, A.; Selinski, M.; Matthias, T. Joining zone shape optimisation for hybrid components made of aluminium-steel by geometrically adapted joining surfaces in the friction welding process. In *AIP Conference Proceedings, Proceedings of the 18th International Conference on Positron Annihilation, Vitoria-Gasteiz, Spain, 8–10 May 2019*; AIP Publishing: Melville, NY, USA, 2019. [CrossRef]
4. Behrens, B.A.; Chugreeva, A.; Mildebrath, M.; Diefenbach, J.; Barroi, A.; Lammers, M.; Hermsdorf, J.; Hassel, T.; Overmeyer, L. Manufacturing of High-Performance Bi-Metal Bevel Gears by Combined Deposition Welding and Forging. *Metals* **2018**, *8*, 898. [CrossRef]
5. Abellan-Nebot, J.V.; Subirón, F.R. A review of machining monitoring systems based on artificial intelligence process models. *Int. J. Adv. Manuf.* **2010**, *47*, 237–257. [CrossRef]

6. Teti, R. Advanced IT Methods of Signal Processing and Decision Making for Zero Defect Manufacturing in Machining. *Procedia CIRP* **2015**, *28*, 3–15. [CrossRef]

7. Siddhpura, A.; Paurobally, R. A review of flank wear prediction methods for tool condition monitoring in a turning process. *Int. J. Adv. Manuf.* **2013**, *65*, 371–393. [CrossRef]

8. Lauro, C.H.; Brandão, L.C.; Baldo, D.; Reis, R.A.; Davim, J.P. Monitoring and processing signal applied in machining processes—A review. *Measurement* **2014**, *58*, 73–86. [CrossRef]

9. Balsamo, V.; Caggiano, A.; Jemielniak, K.; Kossakowska, J.; Nejman, M.; Teti, R. Multi Sensor Signal Processing for Catastrophic Tool Failure Detection in Turning. *Procedia CIRP* **2016**, *41*, 939–944. [CrossRef]

10. Sun, J.; San, W.Y.; Soon, H.G.; Rahman, M.; Zhigang, W. Identification of Feature Set for Effective Tool Condition Monitoring-A Case Study in Titanium Machining. In Proceedings of the IEEE Conference on Automation Science and Engineering, Arlington, VA, USA, 23–26 August 2008; IEEE Xplore: Washington, DC, USA; pp. 273–278. [CrossRef]

11. Caggiano, A. Cloud-based manufacturing process monitoring for smart diagnosis services. *Int. J. Comput. Integr. Manuf.* **2018**, *31*, 612–623. [CrossRef]

12. Altintas, Y. Prediction of cutting forces and tool breakage in milling from feed drive current measurement. *J. Manuf. Sci. Eng.* **1992**, *114*, 386–392. [CrossRef]

13. Aslan, D.; Altintas, Y. Prediction of cutting forces in five-axis milling using feed drive current measurements. *IEEE/ASME Trans. Mechatron.* **2018**, *23*, 833–844. [CrossRef]

14. Möhring, H.C.; Nguyen, Q.P.; Kuhlmann, A.; Lerez, C.; Nguyen, L.T.; Misch, S. Intelligent tools for predictive process control. *Procedia CIRP* **2016**, *57*, 539–544. [CrossRef]

15. Brecher, C.; Eckel, H.M.; Motschke, T.; Fey, M.; Epple, A. Estimation of the virtual workpiece quality by the use of a spindle-integrated process force measurement. *Ann. CIRP* **2019**, *68*, 381–384. [CrossRef]

16. Teti, R.; Jemielniak, K.; Odonnell, G.; Dornfeld, D. Advanced Monitoring of Machining Operations. *Ann. CIRP* **2010**, *59*, 717–739. [CrossRef]

17. Altintas, Y.; Kersting, P.; Biermann, D.; Budak, E.; Denkena, B.; Lazoglu, I. Virtual process systems for part machining operations. *Ann. CIRP* **2014**, *63*, 585–605. [CrossRef]

18. Denkena, B.; Boujnah, H. Feeling machines for online detection and compensation of tool deflection in milling. *Ann. CIRP* **2018**, *67*, 423–426. [CrossRef]

19. Denkena, B.; Dahlmann, D.; Kiesner, J. Production Monitoring Based on Sensing Clamping Elements. *Procedia Technol.* **2016**, *26*, 235–244. [CrossRef]

20. Denkena, B.; Park, J.K.; Bergmann, B.; Schreiber, P. Force sensing linear rolling guides. In Proceedings of the Euspen's 18th International Conference & Exhibition, Venice, Italy, 4–8 June 2018.

21. Bergmann, B.; Witt, M. Feeling machine for material-specific machining. *Ann. CIRP* **2020**, *69*. [CrossRef]

22. Kienzle, O. Die Bestimmung von Kräften und Leistungen an spanenden Werkzeugmaschinen. *Zeitschrift des Vereines Deutscher Ingenieure* **1952**, *94*, 299–305.

23. Tlusty, J.; Andrews, G.C. A Critical Review of Sensors for Unmanned Machining. *Ann. CIRP* **1983**, *32*, 563–572. [CrossRef]

24. Brinkhaus, J.W. Statistische Verfahren zur selbstlernenden Überwachung spanender Bearbeitungen in Werkzeugmaschinen: (Statistical Methods for Self-Teaching Monitoring of Machining Operations in Machine Tools). Ph.D. Thesis, Leibniz University, Hannover, Germany, 18 December 2008.

25. Denkena, B.; Bergmann, B.; Witt, M. Automatic process parameter adaption for a hybrid workpiece during cylindrical operation. *Int. J. Adv. Manuf. Technol.* **2017**, *95*, 311–316. [CrossRef]

26. Denkena, B.; Dahlmann, D.; Damm, J. Self-adjusting process monitoring system in series production. *Procedia CIRP* **2015**, *33*, 233–238. [CrossRef]

Article

Production-Related Surface and Subsurface Properties and Fatigue Life of Hybrid Roller Bearing Components

Bernd Breidenstein [1], Berend Denkena [1], Alexander Krödel [1], Vannila Prasanthan [1,*], Gerhard Poll [2], Florian Pape [2] and Timm Coors [2]

[1] Institute of Production Engineering and Machine Tools, Leibniz University Hannover, An der Universität 2, 30823 Garbsen, Germany; breidenstein@ifw.uni-hannover.de (B.B.); denkena@ifw.uni-hannover.de (B.D.); kroedel@ifw.uni-hannover.de (A.K.)

[2] Institute of Machine Design and Tribology, Leibniz University Hannover, An der Universität 1, 30823 Garbsen, Germany; poll@imkt.uni-hannover.de (G.P.); pape@imkt.uni-hannover.de (F.P.); coors@imkt.uni-hannover.de (T.C.)

* Correspondence: prasanthan@ifw.uni-hannover.de; Tel.: +49-511-762-19091

Received: 25 August 2020; Accepted: 3 October 2020; Published: 7 October 2020

Abstract: By combining different materials, for example, high-strength steel and unalloyed structural steel, hybrid components with specifically adapted properties to a certain application can be realized. The mechanical processing, required for production, influences the subsurface properties, which have a deep impact on the lifespan of solid components. However, the influence of machining-induced subsurface properties on the operating behavior of hybrid components with a material transition in axial direction has not been investigated. Therefore, friction-welded hybrid shafts were machined with different process parameters for hard-turning and subsequent deep rolling. After machining, subsurface properties such as residual stresses, microstructures, and hardness of the machined components were analyzed. Significant influencing parameters on surface and subsurface properties identified in analogy experiments are the cutting-edge microgeometry, $\bar{S}$, and the feed, f, during turning. The deep-rolling overlap, u, hardly changes the residual stress depth profile, but it influences the surface roughness strongly. Experimental tests to determine fatigue life under combined rolling and rotating bending stress were carried out. Residual stresses of up to -1000 MPa, at a depth of 200 μm, increased the durability regarding rolling-contact fatigue by 22%, compared to the hard-turned samples. The material transition was not critical for failure.

Keywords: tailored forming; hybrid bearing; residual stresses; X-ray diffraction; rolling contact fatigue; bearing fatigue life

1. Introduction

One of today's challenges in mechanical engineering is the environmentally friendly and resource-saving production of components [1–3]. Moreover, in the fields of energy technology, medical technology, automotive engineering, and the aerospace industry, the requirements for high-performance solid components are increasing steadily [4–6]. In the automotive industry, for example, the aim is to reduce CO_2 emissions by reducing vehicle weight [7,8]. In addition, the development of components with a longer lifetime is one of the most important goals for the industry [9].

The choice of material is, therefore, always based on the requirements of the intended application of the component. Consequently, requirements such as a reduction in weight with a simultaneous increase in mechanical strength cannot be realized with the use of one material. Mono-material components are thus increasingly reaching their material and production-specific limits. One strategy for reducing the component weight of highly stressed components is to combine different materials in one component. This allows the design of components with different materials that are locally adapted to the respective requirements. In this way, components can be realized that combine partly contradictory properties, such as high mechanical strength and simultaneous weight reduction. This is the focus of the Collaborative Research Centre (CRC) 1153, in which the new process chain, known as "tailored forming", is investigated [10]. In existing production technologies for hybrid solid components, the joining of the components takes place during or after the forming process. In the tailored forming process chain, the components are joined from semi-finished products, at the beginning of the production route, and their properties are specifically influenced by the subsequent forming processes. The simple geometry of the components also facilitates the joining process. In addition, considerably more complex geometries can be produced by this strategy. Within the framework of the CRC, the entire process chain, from the joining of the components until the final machining of the hybrid components, is considered, and the operating behavior of these components is investigated.

In the process chain of solid-part production, machining plays a quality-determining role as the final step. Research results of the last decades show a significant influence of machining on the operating behavior of components due to surface and subsurface modification [11].

In particular, for rolling contacts with high contact stresses >2 GPa, a strong influence on the wear and fatigue life behavior as a function of the surface and subsurface properties is known [12]. In components like roller bearings, an alternating stress field during over-rolling can propagate fatigue-crack growth in the contact zone after high cycle numbers exceeding 10^7 revolutions [13,14]. Component failure occurs when the crack network expands toward the surface, causing spalling of the raceway [15]. Although the topography makes a significant contribution to the operating behavior of components, not all effects can be explained by it. Wear resistance and fatigue life of components are also significantly influenced by residual stresses [11,16]. Residual stresses also have an effect on magnetizability and chemical resistance [17]. In general, compressive residual stresses in the subsurface lead to a proportional increase in fatigue strength [18]. Due to plastic deformations in the subsurface area, the residual stress distributions are of considerable importance, as they have a strong influence on the fatigue limit and the crack initiation tendency of the material. For roller bearings, for example, an increase in fatigue life from 40% to 250% could be achieved by introducing residual compressive stresses [19–23].

However, it must be considered that a residual-stress-induced increase in fatigue strength depends largely on the strength of the material [18]. Denkena et al. were able to show that the probability of failure of roller bearings can be reduced by the targeted introduction of compressive residual stresses [24]. The influences of hard turning and deep rolling was investigated by Pape et al. in terms of bearing fatigue tests. Additionally, an FEM-based model, in combination with a calculation routine, was set up to compute the influence of residual stresses on bearing fatigue [25,26]. However, current research results show that too-high residual compressive stresses can also lead to a shorter service life [27,28]. A cause for this is not yet known and currently represents a research gap [29].

Surface and subsurface properties are significantly influenced by the choice of process parameters and tool microgeometry. The surface roughness is significantly influenced by the feed rate during turning. In addition, cutting speed and tool microgeometry also determine the final roughness of a component [30,31]. The mechanical and thermal effects during chip formation are the main factors influencing the subsurface condition. This can be specifically modified by subsequent processes, such as deep rolling [32]. In interaction with cutting speed and feed rate, the rounding of the cutting edge plays a dominant role in influencing the subsurface. Increasing rounding leads to a higher proportion of material being pressed under the cutting edge. This generally leads to compressive residual stresses below the surface. On the other hand, large cutting-edge radii lead to greater temperature development, which promotes the formation of tensile residual stresses [30,31]. The influence of process parameters and tool microgeometry on the surface and subsurface properties of hybrid components and their influence on the operating behavior is currently inadequately investigated.

The hybrid design offers the potential to combine contradictory requirements in one component. However, there is no knowledge about the application behavior of hybrid components in comparison to mono-material components. Hence, the aim of this study is to investigate the production-related surface and subsurface properties of tailored forming components in comparison to mono-material components. Research questions that need to be addressed are as follows:

1. Can the surface and subsurface properties be adjusted to a similar extent as for conventional components by varying the cutting parameters?
2. Do the components have the same operating behavior with regard to friction or deflection due to an alternating shear force?
3. Do the same damage mechanisms occur regarding rolling contact fatigue and possible structural failure?
4. Even if the joining zone is not directly loaded in the later application, is it still the weakest point in the component?

For this reason, the joining zone as a weak point of hybrid components is investigated in this work, in comparison to a reference sample set made from monolithic material. A possible influence of the application behavior of hybrid components by modifying the subsurface properties is further examined. A focus is set on the mechanical properties of finished components and their use case as hybrid machine elements under complex loads.

2. Materials and Methods

2.1. Materials

In the presented work, two different steel alloys, SAE5140 and SAE1020, were combined by friction welding. These materials were chosen in order to combine a high-strength steel for highly loaded part zones with a cheaper material for supporting structures. A shaft with material transition in axial direction serves as specimen geometry (see Figure 1a).

For the friction welding process, shafts with a diameter of $d = 30$ mm, a length of $l = 220$ mm for SAE1020, and $l = 150$ mm for SAE5140 were used. As process parameters, a friction speed of $n = 2000$ min^{-1}, a friction pressure of $p = 60$ MPa, and a friction path of $s = 4$ mm were applied. Finally, a compression of $p = 150$ MPa and a compression time of $t = 6$ s created the final bond. In the next step, the friction weld bead was removed. Afterward, the friction-welded shafts were impact extruded at a temperature of $T = 900$ °C, in order to change the bonding zone geometry. Hereafter the hybrid shafts were sawn to a total length of $l = 146.5$ mm and centered on both sides. The machining process used is divided in two steps. First, the hybrid shafts were roughly turned to an oversized diameter. Then the bearing surface of the pre-turned hybrid shafts was locally hardened at the largest diameter, at a temperature of 860 °C, by an induction hardening process. The heating time was $t = 0.5$ s, with $P = 90$ kW. Air and water with a pressure of $p = 3$ bar each were used for cooling by means of a spray nozzle. The cooling time was $t = 15$ s. The properties of the materials after the heat treatment are displayed in Figure 1b).

The difference in the hardness of both materials after the heat treatment is evident. SAE5140 is significantly harder than SAE1020 due to the higher carbon content. Therefore, in the finishing operation, different machining behavior of the respective materials is expected. Knowledge about the influence of different material properties on the machining behavior is consequently important in order to comply with shape, dimension, and position tolerances of the hybrid components. In the second and thus final machining step, the hybrid shafts are then turned to size with a required surface quality of Ra = 0.14 μm and with adapted process parameters. With higher roughness values, the load-carrying capacity of the bearings would not be fully utilized, since contacting roughness peaks in the mixed friction area can lead to surface-induced early failures.

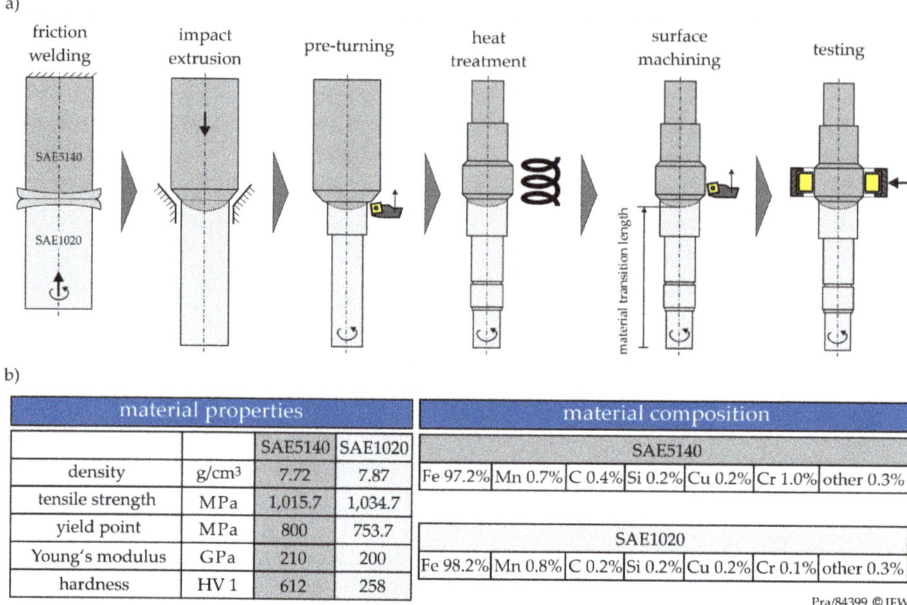

material properties				material composition						
		SAE5140	SAE1020	SAE5140						
density	g/cm³	7.72	7.87	Fe 97.2%	Mn 0.7%	C 0.4%	Si 0.2%	Cu 0.2%	Cr 1.0%	other 0.3%
tensile strength	MPa	1,015.7	1,034.7							
yield point	MPa	800	753.7	SAE1020						
Young's modulus	GPa	210	200	Fe 98.2%	Mn 0.8%	C 0.2%	Si 0.2%	Cu 0.2%	Cr 0.1%	other 0.3%
hardness	HV 1	612	258							

Pra/84399 © IFW

Figure 1. (**a**) Process chain for the production of hybrid shafts, and (**b**) material properties and composition.

2.2. Machining

Machining is one of the most important steps in the process chain of manufacturing hybrid components because it determines the final surface and subsurface properties of a component. Therefore, it is essential to understand the influences of the machining parameters on the surface and subsurface properties of hybrid components. In this work, the influence of turning and deep-rolling parameters on these properties were investigated. At first, analogy studies were carried out on simplified component geometries, with friction-welded hybrid shafts of SAE5140 and SAE1020. With the help of analogy studies, correlations and interactions between process parameters and subsurface properties were investigated. Subsequently, with the knowledge of the machining behavior and its influence on the surface and subsurface properties, hybrid shafts with different manufacturing strategies were produced. The operating behavior of these hybrid shafts was then investigated by roller-bearing tests. The results were compared to conventionally manufactured shafts as a reference.

2.2.1. Turning

Analogy experiments with a three-step variation of cutting speed, v_c, and feed, f, were carried out on cylindrical shafts with a material transition in axial direction. Coated indexable cemented tungsten carbide inserts DNMA150612 WAK20 with a symmetric cutting-edge rounding of S_α/S_γ = 30/30 µm, as well as S_α/S_γ = 75/75 µm, were used. Depth of cut, a_p, and cutting direction were kept constant. Process parameters from the analogy investigations that have a significant influence on the subsurface properties were then used for the production of hybrid shafts for fatigue tests. All turning experiments were conducted on a Gildemeister CTX420 linear machine tool (DMG Mori Aktiengesellschaft, Bielefeld, Germany). A rake angle of $\gamma = 3°$ was chosen for the longitudinal turning process. Analogy experiments with different feed directions (machining from hard–soft or soft–hard) on friction welded shafts revealed worse quality for the feed direction from SAE5140 (hard) into SAE1020 (soft) in the machining process. The reason is that clamping of the shaft with the SAE1020 side is inconvenient due to the lower strength of this material. This leads to slight chatter vibrations during machining. Therefore, in the following investigations, only the feed direction SAE1020 into SAE5140 was selected. The cutting parameters are summarized in Table 1. The choice of cutting parameters for the production of shafts for fatigue-life investigations is based on the lowest achieved roughness.

Table 1. Process parameters of the cutting operation.

Process Parameters	Unit	Analogy Experiments	Final Component
cutting speed, v_c	(m/min)	50–200	180
feed, f	(mm)	0.05–0.2	0.05
depth of cut, a_p	(mm)	0.1	0.1
cutting-edge geometry, S_α/S_γ	(μm)	30/30 and 75/75	30/30 and 75/75
cutting direction,	(-)	SAE1020→SAE5140	SAE1020→SAE5140

2.2.2. Deep Rolling

Additionally, deep rolling experiments were conducted on a CTX420 linear Gildemeister machine tool. For this, the indexable insert of the turning tool was replaced by a hydrostatic rolling tool HG6 with a ball diameter of d_b = 6.35 mm, manufactured by the company ECOROLL AG. Analogy studies were also initially carried out, and the findings were then transferred to the final machining. Before deep rolling, the shafts were turned with the process parameters of the final component, as shown in Table 1, with a cutting-edge geometry of S_α/S_γ = 30 μm. During the deep rolling process, the rolling pressure and the deep rolling speed were kept constant at p_r = 40 MPa and v_w = 180 m/min. The rolling overlap, u, was varied in three steps. Just as for turning, the parameters for fatigue life investigations for deep rolling experiments were also selected according to the best achieved surface quality. The deep rolling direction did not show any significant influence on the surface results. Therefore, the machining direction was chosen analogous to turning. The process parameters for deep rolling are summarized in Table 2.

Table 2. Process parameters of the deep-rolling operation.

Process Parameters	Unit	Analogy Experiments	Final Component
deep rolling speed, v_w	(m/min)	180	180
overlap, u	(%)	50–99	85
deep rolling pressure, p_r	(bar)	400	400
ball diameter, d_b	(mm)	6.35	6.35
deep rolling direction	(-)	SAE1020→SAE5140	SAE1020→SAE5140

2.3. Residual Stress Measurement

Residual stress measurements were carried out via X-ray diffraction (Agfa NDT Pantak Seifert GmbH & Co KG, Ahrensburg, Germany), using the $\sin^2\psi$-method described by Macherauch and Müller [33]. In order to get a non-destructive and depth-resolved measurement of residual stresses, the energy-dispersive measurement method is applied. Here, the determination of lattice strain is also based upon the well-established Bragg's law:

$$n{\cdot}\lambda = 2{\cdot}d_{hkl}{\cdot}sin\theta_{hkl} \tag{1}$$

where n is a natural number indicating the diffraction order, λ is the X-ray wavelength, d_{hkl} is the interatomic lattice spacing, and θ_{hkl} is the diffraction angle (Bragg angle). For the energy-dispersive measurement, white X-radiation from a tungsten anode tube on a Seifert XRD Space Universal diffractometer is used (Figure 2, left). High depth resolution is achieved at an acceleration voltage of U_a = 50 kV and an anode current of I_a = 60 mA. The Bragg angle, θ, remains constant throughout the measurement, at 20°. Thus, all interference lines are simultaneously determined in one diffraction spectrum (Figure 2, right). The peaks determined by this method represent a function of the wavelength,

λ, or photon energy, E_{ph}. Since the different interference lines are distinguished by different energy levels in the spectrum, they can be referred to different depth information. The more peaks in the direction of increasing photon energy can be evaluated, the more depth information about residual stresses is obtained (Figure 2, right). The collimator used has a diameter of 2 mm. Peak position was analyzed by the center of gravity method. The main difference between the well-established angle dispersive and the energy dispersive methods is that, with the energy-dispersive method, the wavelength or photon energy, λ, is varied at a constant Bragg angle, θ, and in the angle-dispersive method, the Bragg angle, θ, is varied at a constant wavelength, λ. The attainment of residual stress depth information up to 35 μm in the hybrid transition zone of steel is non-destructively possible with a single measurement [34]. In order to get a higher information density in the depth direction, electrolytic removal of material was additionally used. Measurements were done parallel and transverse to the feed direction.

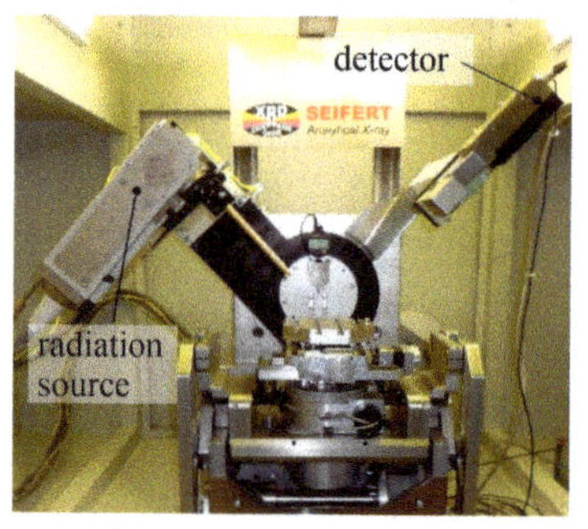

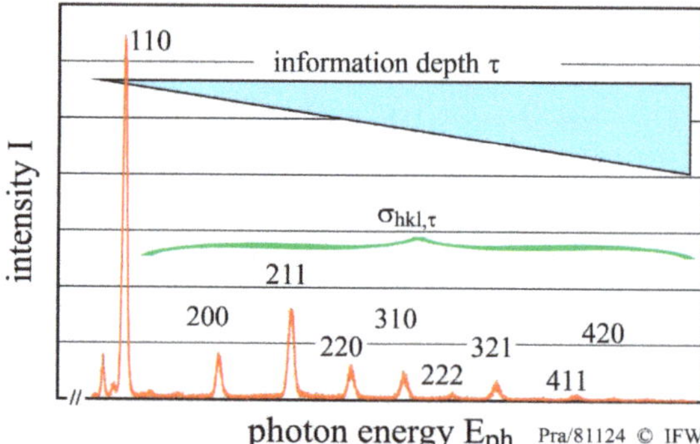

Figure 2. Energy-dispersive residual-stress measurement.

2.4. Fatigue Life Testing

After machining, bearing-fatigue tests were carried out on a test bench, according to Figure 3. The shaft with material transition in axial direction was later used to transmit torque and additional rolling contact stresses of a cylindrical roller bearing with an inner ring integrated into the shaft. Here, the specimen is mounted with two conventional deep groove ball bearings of type 6305 as supporting bearings. The SAE5140-part of the shaft acts as an inner ring for a cylindrical roller bearing (CRB) type RNU204. On the CRB, a radial preload of F = 2 kN is applied through a disc spring assembly. The resulting Hertzian contact stress at the inner race contact is p_H = 1.9 GPa. The operating conditions with approximately 2 GPa pressure correspond to medium to high loads for a roller bearing and thus represent the usual operating conditions for shortened life tests. The bearings were lubricated by a temperature-controlled circulating oil lubrication system with a constant volumetric flow rate of $\dot{V}$ = 0.3 L/min per bearing. Additional test parameters are shown in Figure 3, on the right. The radial loading leads to a superposition of rolling-contact stresses and rotating-bending stresses in the contacting surface between the shaft and the rolling element. Within the path of load, measurements using piezoelectric vibration transducers were carried out. In this way, bearing failure during over-rolling of surface spalls was detected by a self-designed condition monitoring system. The tests were automatically stopped if a critical vibration threshold of 150% of the steady state signal was exceeded. A statistical evaluation of the probability of failure of the series was carried out with these lifetimes. After completion of each test, the damage resulting from a surface chipping in the middle of the raceway on the hybrid shaft was documented. The damage and the microstructure were further analyzed by micrographs.

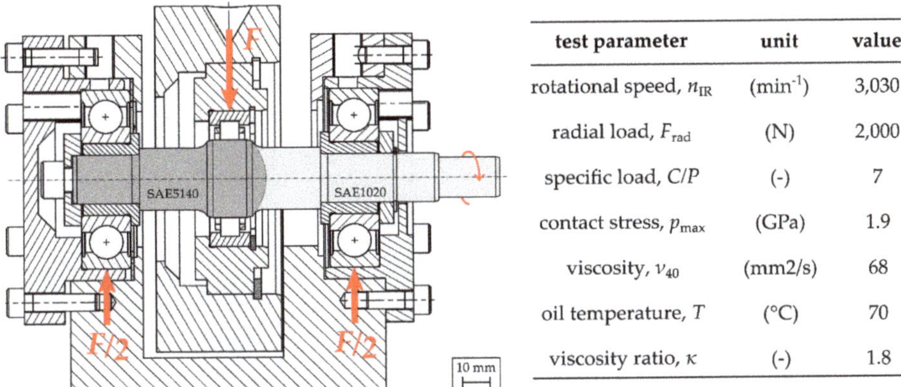

test parameter	unit	value
rotational speed, n_{IR}	(min⁻¹)	3,030
radial load, F_{rad}	(N)	2,000
specific load, C/P	(-)	7
contact stress, p_{max}	(GPa)	1.9
viscosity, v_{40}	(mm2/s)	68
oil temperature, T	(°C)	70
viscosity ratio, κ	(-)	1.8

Figure 3. Fatigue life test in three-bearing arrangement (**left**), and test parameters (**right**).

3. Results and Discussion

In this section, the surface and subsurface properties of the hybrid shafts and the results of the bearing fatigue tests are presented. In turning investigations, only the cutting-edge microgeometry was varied, because it has a decisive influence on the subsurface properties. The cutting-edge microgeometry is defined by the parameters S_γ, S_α, and the form factor κ. At this, S_γ and S_α are the distance between the separation point of the cutting-edge rounding and the tool tip of an ideal sharp cutting at rake face and flank face, respectively. In the presented study, symmetrical cutting-edge roundings, as described in Section 2.2.1, were used. Machining parameters were selected which provide a surface finish with Ra < 0.14 µm as minimum requirement. The knowledge of this was drawn from the analogy studies described above.

3.1. Surface Measurement

Surface roughness plays a decisive role when it comes to the later operating behavior of the hybrid component. Therefore, it is important to identify process parameters that produce a low surface roughness in both materials. Furthermore, it is important to reduce the shape deviation in the transition zone in order to maintain the dimension tolerances of the component.

In Figure 4, the influence of different feeds and microgeometries for turning and overlaps for deep rolling on surface roughness is presented. To investigate the effects of the varied parameters, the surfaces after friction welding and after pre-turning were measured and evaluated. It was ensured that the shafts were pre-turned in the same machining direction and with the same process parameters, in order to achieve a constant initial state before the finishing step of the shafts. The surface analysis after pre-turning showed throughout the investigations an unchanged step in the material transition zone in the range of 12–15 µm, whereby the material range SAE5140 was always higher than SAE1020. The roughness after pre-turning was in the amount of Ra = 0.5–0.6 µm in both material ranges. With decreasing feed, f, it is clearly visible that the roughness decreases, too. The reason for this is that, with the reduction of feed, consecutive roughness peaks are reduced. Consequently, the surface roughness values decrease. However, it can still be observed that the shape deviation in the material transition zone increases with decreasing feed rate. At a feed rate of f = 0.05 mm, the minimum chip thickness is not exceeded in the major part of the chip cross-section, especially in the case of large cutting-edge roundings. As a result, the material is accumulated in front of the cutting edge, and the material is diverted either under the cutting edge into the base material or over the rake face into the chip. This effect, which occurs especially when the chip thickness gets below the minimum chip thickness, is known as the ploughing effect. It is assumed that the properties of the microstructure in the joining zone area of the hybrid shafts are changed by the friction-welding process.

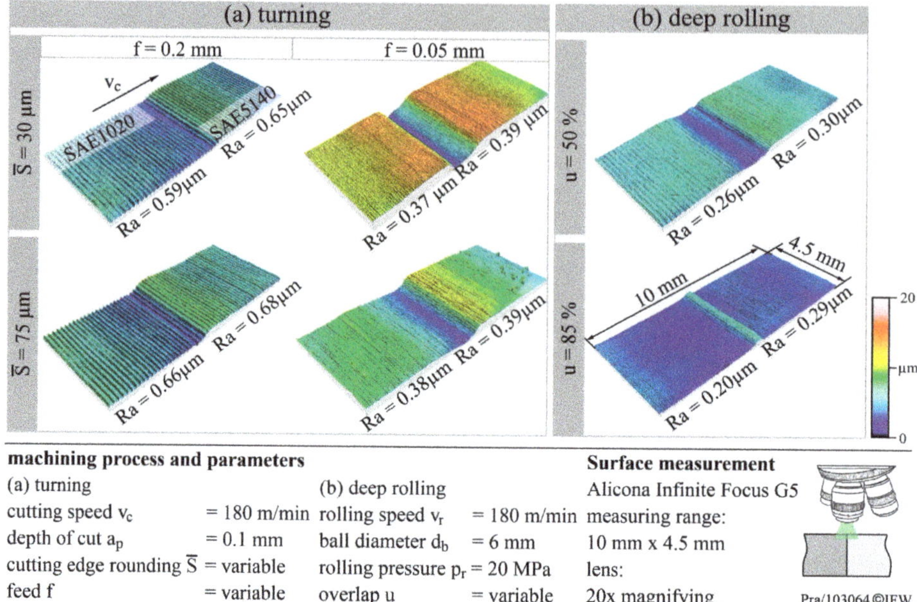

Figure 4. Surface topography and roughness, depending on the machining process (**a**) turning and (**b**) deep rolling.

At low feed rates, this leads to an amplification of the ploughing effect, resulting in a higher indentation depth of the transition zone during turning. In the harder material area of SAE5140, the passive forces increase and leads to a tool displacement. This again increases the workpiece height accordingly. This explains the strongly formed groove for $\overline{S}$ = 30 μm at a feed rate of f = 0.05 mm. With increasing feed, the "weaker" transition area during turning with the higher feed is mainly skipped, so that a step is created, but this is not strongly trough-shaped as with the smaller one. Concerning the influence of the cutting-edge microgeometry on surface roughness, a significant difference between both is not visible. Nearly the same surface roughness values are apparent for both. The shape deviation is smaller by using a cutting-edge rounding of $\overline{S}$ = 75 μm, compared to the cutting-edge rounding of $\overline{S}$ = 30 μm, at a feed of f = 0.05 mm. Here, the mechanical effect exceeds the thermal effect and leads to a smaller shape deviation. The results of the surface roughness measurements for deep rolling show that, with increasing overlap, the surface roughness remains nearly the same, but the shape deviation is significantly reduced. A small shape deviation is only visible in the transition zone. It is apparent that, at an overlap of u = 50%, a groove is formed in the material transition zone, and at an increase of the overlap to u = 85%, in contrast, a small hump is formed. It should be considered that the initial topography before the rolling tests always corresponds to a turned surface with $\overline{S}$ = 30 μm, f = 0.2 mm, and v_c = 180 m/min. According to this, there is always a step in the material transition zone in the initial state before deep rolling. As in turning, deep rolling with an overlap of 50% also causes an indentation of the material due to a changed microstructure and microstructural properties in the material transition zone, as a result of the friction-welding process. If the overlap is increased, however, more material in the soft SAE1020 area is plastically deformed and pushed in front of the ball that the ball nearly fills the step and moreover starts to accumulate into a hump. This is compacted and hardened as the ball passes over it. In the harder SAE5140 area, the material is again densified. As a result, a small hump remains in the material transition zone. However, further, more detailed investigations are necessary to confirm this thesis.

Finally, comparing the surface roughness and topography of the different machining processes of turning and deep rolling, the results are evident. It can be clearly seen that deep rolling produces significantly better surface roughnesses and topographies than turning of hybrid components.

3.2. Microstructure and Hardness

Figure 5 shows the microstructure of the hybrid shafts produced with different machining processes for the fatigue-life investigations. The area influenced by the machining process is significantly larger in the deep rolled specimens than in the turned specimens. The joining zone is not directly located in the bearing raceway but in the area of the chamfer, about 3 mm from the rolling element. The characteristic microstructure of the joining zone due to the extrusion process is also clearly visible. Deep rolling induces higher mechanical stress in a greater material depth, z, which modifies the microstructure to a higher depth. Consequently, in addition to the residual stresses, structural properties are also influenced by different machining processes and must be considered in fatigue tests, too.

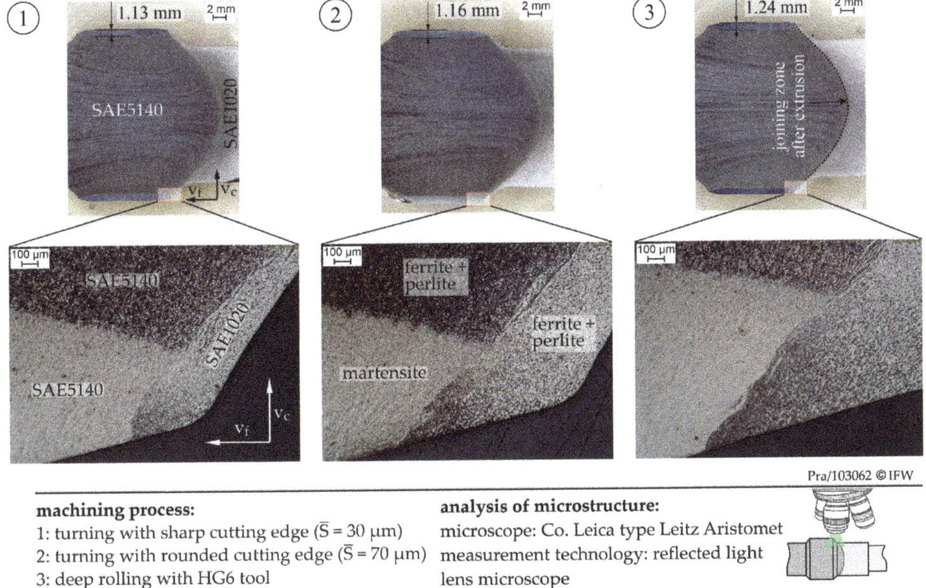

Figure 5. Influence of different manufacturing processes (1) turning with sharp cutting edge, (2) turning with rounded cutting edge and (3) deep rolling on the microstructure of hybrid components.

Hardness measurements were carried out in order to investigate the influence of different processes and their parameters on hardness. In Figure 6, the results are presented. Differences in the hardness values, depending on the machining process, are obvious. The transition to basic hardness depends on the depth of the heat-affected zone during induction heating and lies within the usual process tolerance, cf. Figure 5.

Deep rolling leads to an increase in hardness of the subsurface material. During deep rolling, the material is compacted due to the contact pressure of the rolling tool. This plastic deformation causes an increase in hardness. An influence of different cutting-edge roundings on surface hardness can also be seen. A larger cutting-edge rounding leads to a slight increase in the hardness of the component. This is also due to greater plastic deformation or compaction of the material with larger cutting-edge rounding. Consequently, the investigations show that the machining process influences surface hardness, which can also lead to a different operating behavior of the hybrid component.

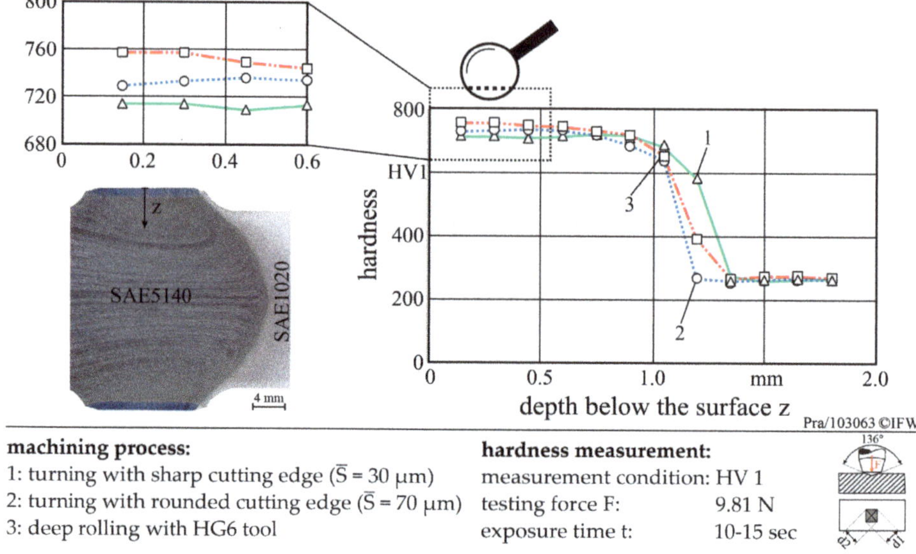

machining process:
1: turning with sharp cutting edge ($\bar{S}$ = 30 µm)
2: turning with rounded cutting edge ($\bar{S}$ = 70 µm)
3: deep rolling with HG6 tool

hardness measurement:
measurement condition: HV 1
testing force F: 9.81 N
exposure time t: 10-15 sec

Figure 6. Influence of different manufacturing processes on surface hardness.

3.3. Residual Stresses

In Figure 7, residual stress depth profiles of hybrid components for lifetime investigations, adjusted by turning with different cutting-edge microgeometries and by deep rolling, are displayed. The residual stress measurements were carried out on the bearing running surface, in order to be able to investigate the influence of the subsurface properties on the operating behavior.

The residual stresses were measured both parallel and transverse to the feed direction. For the residual-stress measurements, the energy-dispersive method, in combination with electrolytic-material removal, was used. The procedure is shown schematically on the right side in Figure 7. By first looking at the different machining processes of turning and deep rolling, it is noticeable that, parallel to the feed direction, different courses of the residual stress graphs can be seen. In turned components, on the one hand, maximum compressive residual stresses can be observed near the surface. With increasing depth, z, the residual stress profile has a positive gradient. In deep-rolled components, on the other hand, residual compressive stresses also occur near the surface, but these increase further in the direction of compression with increasing depth. The maximum compressive residual stresses are not surface near. In deep rolling, the maximum elongation is below the surface of the workpiece. Due to the mechanical coupling of the plastically expanded subsurface of the workpiece with the plastically undeformed workpiece environment, compensating strains occur which cause a surface-parallel compression of the plastically deformed subsurface and thus cause residual compressive stresses. Since the maximum strain is below the surface of the workpiece, the maximum plastic deformation also occurs here.

Therefore, the residual compressive stress maximum is also localized in this area. For deep rolling, the mechanical load is much higher than for turning processes. Therefore, higher values for compressive residual stresses are obtained by deep rolling. The residual-stress depth profiles obtained by the variation of cutting-edge roundings ($\bar{S}$ = 30 µm and $\bar{S}$ = 75 µm), show no significant differences. The influence of these different residual-stress depth profiles on the application behavior, especially between turned and deep-rolled samples, was further investigated during fatigue testing.

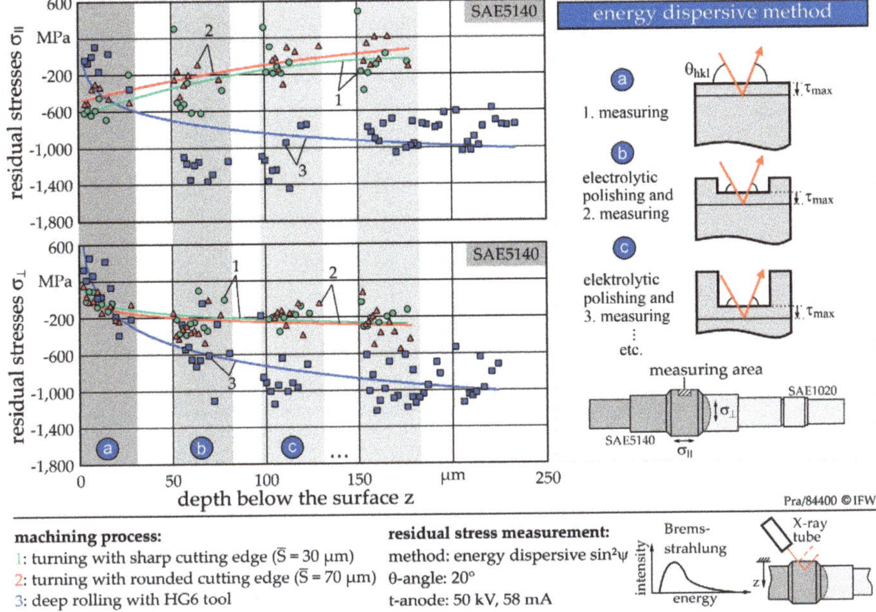

machining process:
1: turning with sharp cutting edge ($\bar{S} = 30$ μm)
2: turning with rounded cutting edge ($\bar{S} = 70$ μm)
3: deep rolling with HG6 tool

residual stress measurement:
method: energy dispersive $\sin^2\psi$
θ-angle: 20°
t-anode: 50 kV, 58 mA

Figure 7. Influence of different manufacturing processes on the residual-stress depth profile measured with the energy dispersive method combined with electrolytic polishing (**a–c**).

3.4. Fatigue Life

Figure 8 shows the two-parameter Weibull probability plot [35–38] for bearing fatigue-life testing, using maximum likelihood estimation. Here, the failure probability of four series is plotted against the number of load cycles in terms of inner-ring rotations in a double logarithmic diagram. The three production strategies are illustrated, which had a sample size of $n = 3$ each: turning with sharp and rounded cutting edges, respectively, and deep rolling. The tailored forming shafts were compared to $n = 15$ conventionally produced samples with the same geometry made only from SAE5140 from Reference [36], which were ground after heat treatment. Due to restrictions in test bench operation, a reduced speed of 1250 min^{-1} at 60 °C ($\kappa = 1.6$) was used for the reference series. The slope of the regression line is named as shape parameter β of the Weibull distribution. This parameter can be understood as a constant that depends on the damage mechanism or fatigue behavior of a system. The characteristic lifetime, L_{63}, also called location parameter, $\hat{T}$, of a series is the time period after which approximately 63.2% of the samples have failed.

For the aforementioned load parameters, a bearing life of $L_{63} = 12 \times 10^6$ revolutions could be determined for the reference series made from monolithic SAE5140. The shape parameter of this Weibull distribution is $\beta = 1.39$, which indicates that the failure rate slightly increases with time, suggesting rolling contact fatigue as the damage mechanism. As expected, the density function shows negative skew. For the roller bearing life of SAE52100 bearing steel under good lubrication conditions ($\kappa \geq 2$), the size of the Weibull exponent ranges from $\beta = 1$ to 1.5 [28,37]. According to Reference [17], a general shape parameter of $\beta = 1.35$ for roller bearings is known, which shows that the conventional reference tests correspond well to the expected values, leading to rolling-contact fatigue.

The hybrid samples manufactured with a sharp cutting edge achieved a bearing life of $L_{63} = 10.4 \times 10^6$ revolutions. This is within an 86% margin of the reference series. A shape parameter of $\beta = 1$ suggests a constant failure rate, which usually indicates random failures. Gleß [39] also carried out fatigue tests for bearing surfaces with different mechanical surface treatments. For rough surfaces, he achieved significantly lower lifetimes than for surfaces with subsequent finishing.

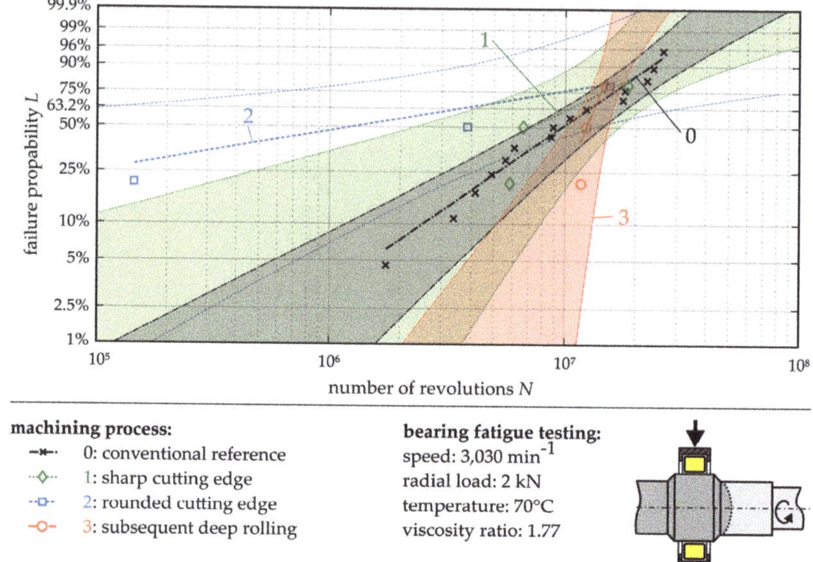

Figure 8. Weibull plots of fatigue-test failures for different machining processes.

During the test series with a rounded cutting edge, an early failure after only 0.15×10^6 revolutions occurred. This results in a bearing fatigue life of $L_{63} = 3.62 \times 10^6$ revolutions, which is only 30% of the reference lifetime. The Weibull slope of $\beta = 0.35$ also indicates early failures with a different mechanism for this series. Due to the small sample number of $n = 3$, this early failure has a strong impact, leading to a very wide confidence interval. The sample set manufactured with rounded cutting edge can therefore not be used for further considerations. Due to comparable lubrication conditions (comparable roughness values), as well as the similar residual stress states in relation to the sample set manufactured with sharp cutting edge, a very similar service life was expected for both test series.

The Weibull plot of the deep-rolled samples shows a relatively steep slope of $\beta = 5$. The samples fail within a short period, resulting in a relatively narrow confidence interval. The confidence intervals of reference dataset and the deep-rolled samples overlap for an interval width in abscissa direction of 4.21×10^6 revolutions at L_{63}. A bearing fatigue life of $L_{63} = 13.34 \times 10^6$ revolutions was achieved. This is 28.5% above the sample set manufactured with sharp cutting-edge geometry and 10% above the reference. Reasons for a lifetime-increasing effect of the rolling process are believed to be improved surface properties, such as higher hardness and especially lower roughness. Microcontacts, which lead to local stress maxima and thus influence surface degradation, can then result in early failures. Various studies have also shown that compressive residual stresses induced by the manufacturing process can extend the fatigue life of roller bearings by a factor of 2.5 [30]. Table 3 summarizes the results of bearing-fatigue testing.

Table 3. Weibull distribution parameters and relative fatigue life to reference samples.

Weibull Parameter	Unit	0: Monolithic Reference	1: Sharp Edge	2: Round Edge	3: Deep Rolling
characteristic life, L_{63}	(10^6 revs.)	12.09	10.38	3.62	13.34
shape parameter, β	(-)	1.39	1	0.35	4.99
relative fatigue life, $\hat{T}_{rel}$	(-)	100%	86%	30%	110%

3.5. Damage Analysis

For each tailored forming production strategy, the sample lying in the middle of the Weibull distribution was examined for damage analysis. Figure 9a shows each spalling damage on the surface which led to the stop of the test with use of laser scanning microscopy. Rolling direction is from bottom to top. The spalls have a size of around 1000×500 µm^2 and a maximum depth of -90 to -140 µm. Subsequently, the samples were cut in the longitudinal direction for damage analysis. Figure 9b depicts micrographs of the joining zone and the microstructure in the rolling contact area after fatigue testing. For this, the samples were etched with 2% alcoholic nitric acid.

Figure 9. Damage analysis: (**a**) surface laser scanning microscopy of each event-relevant spalling damage, and (**b**) microstructural details in axial section plane.

The depth of the damage lies within the area of maximum orthogonal shear stress induced by the roller bearing. Together with the subsurface crack propagation parallel to the raceway in Figure 9b, this indicates rolling-contact fatigue as the damage mechanism. However, micropittings are visible to the right side of each spall in Figure 9a. Due to the different material properties of the hybrid component, a slightly unevenly shaped deflection of the shaft could cause skewing in the bearing, which, in turn, would result in edge loading. The conventional rolling element features a higher hardness of > 850 HV0.5, which can lead to surface disruption on the shaft. The martensitic microstructure in the hardened area and the ferritic–pearlitic microstructure in the base material are clearly visible in Figure 9b. In the transition zone of these two microstructure areas, some sporadic ferrite grains are visible in the martensitic area. This could be a possible reason for the slightly lower hardness of the martensitic structure. However, structural failure within the joining zone or in the transition zone from the surface-hardened area to the base material did not occur in any of the samples. The test bench-induced alternating bending stress in the joining zone is therefore not critical for the operating parameters presented here.

4. Conclusions and Outlook

The machining and operating behavior of friction welded and extruded hybrid shafts of SAE1020-SAE5140 were investigated. Turning and deep rolling experiments were carried out for manufacturing of the shafts. Subsequently, the influence of different process control variables on the subsurface properties (residual stresses, microstructure, and hardness) was analyzed. For bearing raceways, surface quality is very important in order to exclude failure of the component due to surface defects. Due to this demand, the influence of process parameters on surface roughness and topography were examined first. The feed for turning and overlap for deep rolling have a significant influence on surface roughness. In particular, certain effects occur in the material transition zone, depending on these variables. Depending on the feed rate during turning and overlap during deep rolling, a step, groove, or hump can be observed in the transition area. These are mainly caused by changes in the material properties in the transition area and enhanced by effects such as tool deflection and plastic material deformation. To exclude component failure caused by insufficient surface quality, for further investigations, process parameters which produce the lowest possible surface roughness were selected. In turning experiments, a feed of 0.05 mm, and in deep-rolling experiments an overlap of 85%, were chosen.

The cutting-edge rounding of the tool has a significant influence on the subsurface properties. With increasing cutting-edge rounding, compressive residual stresses are induced. However, deep rolling significantly increases compressive residual stresses regarding induction depth and in terms of residual stress values. In order to investigate the influence of residual stresses on the operating behavior, three different process parameters were used. The hybrid shafts were produced by variation of the cutting-edge rounding in case of turning and by deep rolling. Different cutting-edge roundings led to different sizes of the influenced microstructure area and hardness values in the subsurface. The reason for this is that, with increasing cutting-edge rounding, more material is deformed under the cutting-edge rounding. Deep rolling leads to a further increase of the influenced microstructure area and of surface hardness. The stronger compression of the material results in work hardening and increased hardness. In deep rolling, the observed effects increase even further, since there is hardly any thermal load in this process and the component is thus almost exclusively subjected to mechanical load.

In this first preliminary study to evaluate finishing strategies for tailored forming components, three samples per production strategy were manufactured. Due to the complex production process, a larger sample number, of more than three samples per finishing strategy, was currently not achievable. With the produced samples, fatigue tests were carried out on a bearing fatigue test bench. The aim of the tests was to investigate the fatigue behavior of the hybrid shaft compared to mono-material shafts and, in particular, to highlight the influence of manufacturing processes on the fatigue life regarding rolling-contact fatigue. Based on the sample series manufactured with sharp cutting-edge geometry, it could be shown that tailored forming machine elements can have a comparable fatigue life to conventional components. An early failure of a sample manufactured with a rounded cutting edge did not allow an evaluation of this series. It seems that an increase in service life can be achieved by additional deep rolling. The reason for this can be improved surface and subsurface properties, such as roughness and hardness. Furthermore, the compressive residual stresses induced seem to have a positive effect on the material microstructure with regard to its fatigue strength behavior. However, it is absolutely necessary to repeat the examinations to confirm this thesis.

In order to further refine the results and, in particular, to increase the empirical significance, the fatigue tests will be continued in the further course of the project. This includes further tests with the same load parameters, as well as other load levels. Thus, different mechanisms of fatigue damage can be provoked, and the manufacturing strategy can be adapted to the later load case. In addition, the production parameters are to be transferred to other tailored forming components with different geometries. This will ensure optimum operating and performance characteristics for future applications of these new high-performance components.

Author Contributions: Designed the machining process, performed the turning and deep rolling experiments, and carried out the analysis of surface and subsurface properties, V.P.; supervised the work, B.D., B.B. and A.K.; designed the samples and carried out the fatigue tests and accompanying analyses, T.C.; supervised the work, G.P. and F.P. All authors have read and agreed to the published version of the manuscript.

Funding: This research was funded by the Deutsche Forschungsgemeinschaft (DFG, German Research Foundation)—CRC 1153—grant number 252662854, in the subprojects B4 and C3.

Acknowledgments: The results presented in this paper were obtained within the Collaborative Research Centre 1153 "Process chain to produce hybrid high-performance components by tailored forming" in subproject B4 and C3. The authors would like to thank the German Research Foundation (DFG) for the financial support of this project. The authors would also like to thank the subprojects B2, B3, A2, and A4 for their support in producing samples through the use of friction welding, impact extrusion, and hardening, as well as in metallographic analysis.

Conflicts of Interest: The authors declare no conflict of interest.

References

1. Giampieri, A.; Ling-Chin, J.; Ma, Z.; Smallbone, A.; Roskilly, A. A review of the current automotive manufacturing practice from an energy perspective. *Appl. Energy* **2020**, *261*, 114074. [CrossRef]
2. Huang, J.; Chang, Q.; Arinez, J.; Xiao, G. A Maintenance and Energy Saving Joint Control Scheme for Sustainable Manufacturing Systems. *Procedia CIRP* **2019**, *80*, 263–268. [CrossRef]
3. Schmidt, M.; Bauer, J.; Haubach, C. Ressourceneffiziente Herstellung mechanischer Verbindungselemente. In *100 Betriebe für Ressourceneffizienz-Band 1*; Springer Spektrum: Berlin/Heidelberg, Germany, 2017; pp. 138–141. [CrossRef]
4. Immarigeon, J.-P.; Holt, R.; Koul, A.; Zhao, L.; Wallace, W.; Beddoes, J. Lightweight materials for aircraft applications. *Mater. Charact.* **1995**, *35*, 41–67. [CrossRef]
5. Sköck-Hartmann, B.; Gries, T. Automotive applications of non-crimp fabric composites. *Non-Crimp Fabr. Compos.* **2011**, 461–480. [CrossRef]
6. Jurczak, P.; Witkowska, J.; Rodziewicz-Motowidlo, S.; Lach, S. Proteins, peptides and peptidomimetics as active agents in implant surface functionalization. *Adv. Colloid Interface Sci.* **2020**, *276*, 102083. [CrossRef]
7. Cole, G.; Sherman, A. Light weight materials for automotive applications. *Mater. Charact.* **1995**, *35*, 3–9. [CrossRef]
8. Padmanabhan, R.; Oliveira, M.C.; Menezes, L.F. *Tailor Welded Blanks for Advanced Manufacturing*; Woodhead Publishing Series in Welding and Other Joining Technologies; Woodhead Publishing Limited: Cambridge, UK, 2011; pp. 97–117. [CrossRef]
9. Fiebig, S.; Sellschopp, J.; Manz, H.; Vietor, T.; Axmann, K.; Schumacher, A. Future challenges for topology optimization for the usage in automotive lightweight design technologies. In Proceedings of the 11th World Congress on Structural and Multidisciplinary Optimization, Sydney, Australia, 7–12 June 2015.

10. Behrens, B.-A.; Bouguecha, A.; Frischkorn, C.; Huskic, A.; Stakhieva, A.; Duran, D. Tailored forming technology for three dimensional components: Approaches to heating and forming. In Proceedings of the 5th Conference on Thermomechanical Processing, Milan, Italy, 26–28 October 2016.

11. Jawahir, I.S.; Brinksmeier, E.; M'Saoubi, R.; Aspinwall, D.; Outeiro, J.; Meyer, D.; Umbrello, D.; Jayal, A.D. Surface integrity in material removal processes: Recent advances. *CIRP Ann.* **2011**, *60*, 603–626. [CrossRef]

12. Nélias, D.; Dumont, M.L.; Champiot, F.; Vincent, A.; Girodin, D.; Fougéres, R.; Flamand, L. Role of Inclusions, Surface Roughness and Operating Conditions on Rolling Contact Fatigue. *J. Tribol.* **1999**, *121*, 240–251. [CrossRef]

13. Olver, A.V. The Mechanism of Rolling Contact Fatigue: An Update. *Proc. Inst. Mech. Eng. Part J J. Eng. Tribol.* **2005**, *219*, 313–330. [CrossRef]

14. Sadeghi, F.; Jalalahmadi, B.; Slack, T.; Raje, N.; Arakere, N.K. A Review of Rolling Contact Fatigue. *J. Tribol.* **2009**, *131*, 041403. [CrossRef]

15. Harris, T.; Kotzalas, M. *Rolling Bearing Analysis*, 5th ed.; CRC Press: Boca Raton, FL, USA, 2006. [CrossRef]

16. Novovic, D.; Dewes, R.; Aspinwall, D.; Voice, W.; Bowen, P. The effect of machined topography and integrity on fatigue life. *Int. J. Mach. Tools Manuf.* **2004**, *44*, 125–134. [CrossRef]

17. Brinksmeier, E.; Cammett, J.; König, W.; Leskovar, P.; Peters, J.; Tönshoff, H. Residual Stresses—Measurement and Causes in Machining Processes. *CIRP Ann.* **1982**, *31*, 491–510. [CrossRef]

18. Sollich, A. *Verbesserung des Dauerschwingverhaltens hochfester Stähle durch gezielte Eigenspannungserzeugung, Fortschrittsberichte VDI*; VDI Verlag: Düsseldorf, Germany, 1994; Volume 5, p. 376.

19. Crețu, S.; Popinceanu, N. The influence of residual stresses induced by plastic deformation on rolling contact fatigue. *Wear* **1985**, *105*, 153–170. [CrossRef]

20. Matsumoto, Y.; Hashimoto, F.; Lahoti, G. Surface Integrity Generated by Precision Hard Turning. *CIRP Ann.* **1999**, *48*, 59–62. [CrossRef]

21. Schwach, D.; Guo, Y.B. A fundamental study on the impact of surface integrity by hard turning on rolling contact fatigue. *Int. J. Fatigue* **2006**, *28*, 1838–1844. [CrossRef]

22. Guo, Y.B.; Warren, A.; Hashimoto, F. The basic relationships between residual stress, white layer, and fatigue life of hard turned and ground surfaces in rolling contact. *CIRP J. Manuf. Sci. Technol.* **2010**, *2*, 129–134. [CrossRef]

23. Choi, Y. A study on the effects of machining-induced residual stress on rolling contact fatigue. *Int. J. Fatigue* **2009**, *31*, 1517–1523. [CrossRef]

24. Denkena, B.; Poll, G.; Maiß, O.; Neubauer, T. Affecting the Life Time of Roller Bearings by an Optimal Surface Integrity Design after Hard Turning and Deep Rolling. *Adv. Mater. Res.* **2014**, *966*, 425–434. [CrossRef]

25. Neubauer, T. Betriebs-und Lebensdauerverhalten Hartgedrehter und Festgewalzter Zylinderrollenlager. Ph.D. Thesis, Leibniz University Hannover, Hanover, Germany, 2016. (In German)

26. Pape, F.; Neubauer, T.; Maiß, O.; Denkena, B.; Poll, G. Influence of Residual Stresses Introduced by Manufacturing Processes on Bearing Endurance Time. *Tribol. Lett.* **2017**, *65*, 87. [CrossRef]

27. Maiß, O. Lebensdauererhöhung von Wälzlagern durch mechanische Bearbeitung. Ph.D. Thesis, Leibniz University Hannover, Hanover, Germany, 2019. (In German)

28. Pape, F.; Coors, T.; Poll, G. Studies on the Influence of Residual Stresses on the Fatigue Life of Rolling Bearings in Dependence on the Production Processes. *Front. Mech. Eng.* **2020**, *6*. [CrossRef]

29. Breidenstein, B.; Denkena, B.; Meyer, K.; Prasanthan, V. Influence of subsurface properties on the application behavior of hybrid components. *Procedia CIRP* **2020**, *87*, 302–308. [CrossRef]

30. Hsu, T.-K.; Zeren, E. Effects of cutting edge geometry, workpiece hardness, feed rate and cutting speed on surface roughness and forces in finish turning of hardened AISI H13 steel. *Int. J. Adv. Manuf. Technol.* **2004**, *25*, 262–269. [CrossRef]

31. Burhanuddin, Y.; Haron, C.H.C.; Ghani, J.A. The Effect of Tool Edge Geometry on Tool Performance and Surface Integrity in Turning Ti-6Al-4V Alloys. *Adv. Mater. Res.* **2011**, 1211–1221. [CrossRef]

32. Abrão, A.M.; Denkena, B.; Breidenstein, B.; Mörke, T. Surface and subsurface alterations induced by deep rolling of hardened AISI 1060 steel. *Prod. Eng.* **2014**, *8*, 551–558. [CrossRef]

33. Macherauch, E.; Müller, P. Das $\sin^2\psi$–Verfahren der röntgenographischen Spannungsmessung. *Zeitschrift für Angewandte Physik* **1961**, *13*, 305–312.

34. Breidenstein, B.; Denkena, B.; Mörke, T.; Prasanthan, V. Non-Destructive Determination of Residual Stress Depth Profiles of Hybrid Components by Energy Dispersive Residual Stress Measurement. *Key Eng. Mater.* **2017**, *742*, 613–620. [CrossRef]
35. Weibull, W. A Statistical Distribution Function of Wide Applicability. *J. Appl. Mech.* **1951**, *103*, 293–297.
36. Coors, T.; Pape, F.; Poll, G. Bearing Fatigue Life of a Multi-Material Shaft with an Integrated Raceway. *Bear. World J.* **2019**, *3*, 23–30.
37. Zaretsky, E.V.; Branzai, E.V. Rolling Bearing Service Life Based on Probable Cause for Removal—A Tutorial. *Tribol. Trans.* **2016**, *60*, 300–312. [CrossRef]
38. Denkena, B.; De Leon, L.; Bassett, E.; Rehe, M. Cutting Edge Preparation by Means of Abrasive Brushing. *Key Eng. Mater.* **2010**, *438*, 1–7. [CrossRef]
39. Gleß, M. Wälzkontaktermüdung bei Mischreibung. Ph.D. Thesis, Otto-von-Guericke-Universität Magdeburg, Magdeburg, Germany, 2009. (In German)

Article

Investigations on Tailored forming of AISI 52100 as Rolling Bearing Raceway

Timm Coors [1,*], Maximilian Mildebrath [2], Christoph Büdenbender [3], Felix Saure [1], Mohamad Yusuf Faqiri [2], Christoph Kahra [2], Vannila Prasanthan [4], Anna Chugreeva [3], Tim Matthias [3], Laura Budde [5], Florian Pape [1], Florian Nürnberger [2], Thomas Hassel [2], Jörg Hermsdorf [5], Ludger Overmeyer [5], Bernd Breidenstein [4], Berend Denkena [4], Bernd-Arno Behrens [3], Hans Jürgen Maier [2] and Gerhard Poll [1]

[1] Institut für Maschinenkonstruktion und Tribologie (Machine Design and Tribology), Leibniz University Hannover, An der Universität 1, 30823 Garbsen, Germany; saure@imkt.uni-hannover.de (F.S.); pape@imkt.uni-hannover.de (F.P.); poll@imkt.uni-hannover.de (G.P.)

[2] Institut für Werkstoffkunde (Materials Science), Leibniz University Hannover, An der Universität 2, 30823 Garbsen, Germany; mildebrath@iw.uni-hannover.de (M.M.); faqiri@iw.uni-hannover.de (M.Y.F.); kahra@iw.uni-hannover.de (C.K.); nuernberger@iw.uni-hannover.de (F.N.); hassel@iw.uni-hannover.de (T.H.); maier@iw.uni-hannover.de (H.J.M.)

[3] Institut für Umformtechnik und Umformmaschinen (Forming Technology and Machines), Leibniz University Hannover, An der Universität 2, 30823 Garbsen, Germany; buedenbender@ifum.uni-hannover.de (C.B.); chugreeva@ifum.uni-hannover.de (A.C.); tmatthias@ifum.uni-hannover.de (T.M.); behrens@ifum.uni-hannover.de (B.-A.B.)

[4] Institut für Fertigungstechnik und Werkzeugmaschinen (Production Engineering and Machine Tools), Leibniz University Hannover, An der Universität 2, 30823 Garbsen, Germany; prasanthan@ifw.uni-hannover.de (V.P.); breidenstein@ifw.uni-hannover.de (B.B.); denkena@ifw.uni-hannover.de (B.D.)

[5] Laser Zentrum Hannover e.V., Hollerithallee 8, 30419 Hannover, Germany; l.budde@lzh.de (L.B.); j.hermsdorf@lzh.de (J.H.); ludger.overmeyer@ita.uni-hannover.de (L.O.)

* Correspondence: coors@imkt.uni-hannover.de; Tel.: +49-511-762-5552

Received: 31 August 2020; Accepted: 10 October 2020; Published: 13 October 2020

Abstract: Hybrid cylindrical roller thrust bearing washers of type 81212 were manufactured by tailored forming. An AISI 1022M base material, featuring a sufficient strength for structural loads, was cladded with the bearing steel AISI 52100 by plasma transferred arc welding (PTA). Though AISI 52100 is generally regarded as non-weldable, it could be applied as a cladding material by adjusting PTA parameters. The cladded parts were investigated after each individual process step and subsequently tested under rolling contact load. Welding defects that could not be completely eliminated by the subsequent hot forming were characterized by means of scanning acoustic microscopy and micrographs. Below the surface, pores with a typical size of ten μm were found to a depth of about 0.45 mm. In the material transition zone and between individual weld seams, larger voids were observed. Grinding of the surface after heat treatment caused compressive residual stresses near the surface with a relatively small depth. Fatigue tests were carried out on an FE8 test rig. Eighty-two percent of the calculated rating life for conventional bearings was achieved. A high failure slope of the Weibull regression was determined. A relationship between the weld defects and the fatigue behavior is likely.

Keywords: tailored forming; hybrid bearing; AISI 52100; plasma transferred arc welding; residual stress; scanning acoustic microscopy; bearing fatigue life

1. Introduction

In many technical applications, functional part regions of components experience a significantly higher load than basic structural regions. This occurs, for example, due to rolling contacts, like rolling bearings or gears. The material at and just below the surface, such as the raceway of the bearing, can experience very high stresses of over 3000 MPa and exceed 10^6 load cycles. In order to achieve a long lifetime, the material needs to be of a high quality, strength, and hardness. Mounting components such as shafts or less stressed regions in gear wheels like the connecting structure between the hub and teeth are subjected to a much lower load. Depending on the application, a base material with reduced strength or a lighter material may be sufficient or even advantageous.

Especially in large size bearings, the material costs for high purity steels that are mostly free of tramp elements are very high. Thus, a compromise between costs and benefit has to be made. The localized application of a layer of a high-performance material in the highly loaded area improves the mechanical properties where these are required. A base material, which is located in the less highly loaded region, can save costs and can fulfill other functions regarding, e.g., weight reduction or a modified ductility. These multi-material components can be adapted to thermal, environmental, and mechanical stresses depending on the application.

2. State-of-the-Art

The Collaborative Research Center (SFB) 1153 tailored forming aims to develop and optimize a novel process chain for the manufacturing of hybrid solid components. This is done by thermomechanically joining semi-finished products and subsequent forming to produce components with locally adapted properties. By combining different materials within one component, it is possible to reduce the component weight by the local use of lightweight materials or to reduce cost by combining low-cost base materials with high-quality alloys at certain functional surfaces. Among others, the mechanical properties of a hybrid cylindrical roller thrust bearing are of interest. Thus, the low-alloy steel AISI 1022M (material number 1.0460) was used as base material, cladded by the steel AISI 52100 (1.3505) and produced using the tailored forming technology to manufacture hybrid bearing washers. This process chain is depicted schematically in Figure 1. In the first step, the high strength steel was cladded onto the base materials using plasma transferred arc powder deposition (PTA) welding (Figure 1a). Subsequently, the hybrid workpiece was formed into the semi-finished product with a near net-shaped contour by means of forging (Figure 1b). This thermomechanical treatment improved the quality of the interface zones, as shown for deposition welded workpieces in [1,2]. In the following, a heat treatment by means of quenching and tempering, as well as a subsequent machining (Figure 1c), were carried out to produce the axial bearings as depicted in Figure 1d. In the load direction, the two hybrid discs on which the rolling element was running featured a material transition, with the high strength material serving as the raceway.

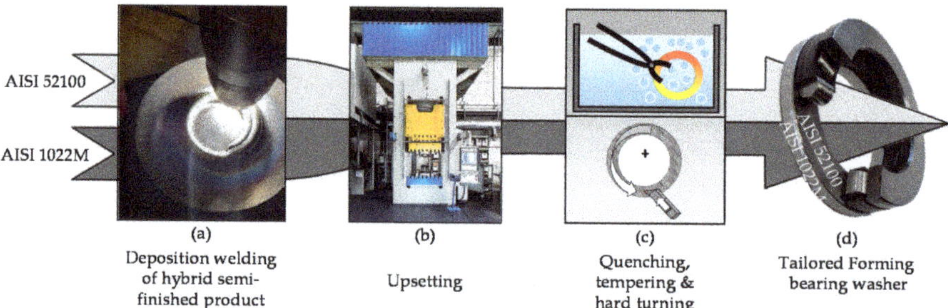

Figure 1. Process chain for manufacturing a hybrid cylindrical roller thrust bearing washer by tailored forming: (**a**) plasma-transferred arc deposition welding; (**b**) upsetting; (**c**) heat treatment by quenching, tempering, and subsequent mechanical processing; (**d**) bearing assembly for analysis in a test bench.

In the following, the term "hybrid bearing" is used for multi-material bearing components manufactured by means of tailored forming. Here, the application of two different materials with a material transition in the load direction is characteristic. In contrast, the term "hybrid" in the context of rolling bearings is usually employed for bearings in which different materials are combined for the bearing rings and rolling elements [3]. Nitrided rolling bearing steel is often used for the bearing rings and a high-strength ceramic, such as silicon nitride, for the rolling elements [4]. These bearings are suited for high performance applications and are complex regarding their proper evaluation [5].

Regarding iron-based cladded components and their fatigue behavior under complex loads due to rolling contacts, few previous works exist. Koehler et al. welded ASTM F75 powder (American Society for Testing and Materials standard for cobalt-28 chromium-6 molybdenum alloy; material number 2.4979) via deposition welding on AISI 4140 (American Iron and Steel Institute standard; 1.7225) and detected a reduced fatigue strength in 4-point bending tests compared to the mono-material, which could be explained by tensile residual stresses caused by the cladding process [6]. In further research by Koehler et al., fatigue tests were carried out on crankshafts of the same material combination repaired by deposition welding [7]. The specimen showed a reduced tensile strength and yield stress compared to newly manufactured crankshafts, resulting in a reduced fatigue life. Alam et al. cladded ASTM F75 on ASTM A572 (1.0045) and observed pores on the surfaces of cylindrical and square bars [8]. It was found that surface cracks were initiated due to these pores, with a negative effect on service life. Pores or defects below the surface were not investigated. In Chew et al., a cladding of 1 mm of ASTM F75 was applied to AISI 4340 (1.6511). Cross-section examination of the specimen showed that no macro defects such as pores, lack of fusion, or cracking appeared [9]. By further surface grinding processes of the specimen, a maximum fatigue life of 95% compared to the monolithic substrate could be achieved. At the location of direction change in the course of the welding track, early fatigue cracks were initiated due to the higher heat input and the local microstructure evolution. The positive influence of compressive residual stresses on the fatigue life of radial bearings made of AISI 52100 was investigated in [10]. In contrast to the aforementioned, tests were carried out on real components under a high Hertzian contact pressure of 2.5 GPa. By hard-turning and subsequent deep rolling, an improvement in service life was achieved in comparison to ground hybrid bearings.

Blohm et al. [1] used a subsequent forming process for shafts from AISI 1022M cladded with AISI 5140 (1.7035) by PTA welding. Due to this forming process, a defect-free interface zone could be obtained. Using the example of a hot-formed specimen manufactured by welding, Mildebrath et al. [2,11] demonstrated that the cladding completely recrystallized during the subsequent forming process and that disadvantageous microstructures present after welding were transformed. Behrens et al. [12] researched hybrid forging billets that were welded out of alloy steel AISI 5140 and carbon steel AISI 1022M. They employed a test matrix that utilized the variations of process parameters to influence the geometry and microstructure of the materials' joining zone. Pape et al. [13] investigated the manufacturability of multimaterial components by combining the high-alloy steel AISI HNV3 (1.4718) with the base substrate of ASTM A283 (1.0038) in a single component. The materials were joined by laser metal deposition by wire according to the stressed zones of the component and examined. The aim was to ensure an efficient use of resources. Behrens et al. [14] welded bearing washers of AISI 1022M with the alloyed steel AISI 5140. They researched the grain refinement depending on the degree of forming and investigated the wear resistance of the specimen. Stanford and Jain [15] found that pores in the loaded material areas had the greatest influence on the service life of the components if they had not been closed or reduced in size by a subsequent forming process. In particular, they emphasized the positive influence of the forming process on fatigue life.

3. Aim and Research Objectives

As has been shown in the State-of-the-Art Processes section, the deposition welding of a low-alloyed steel can result in a reduction in its fatigue strength. The build-up welded surfaces of low-alloyed steels do not yet have sufficient fatigue resistance to rolling bearing loads. No references are known for the welding and subsequent forming of rolling bearing steel. The aim of this study was therefore to develop a process route for the application of rolling bearing steel by means of tailored forming and to investigate the mechanical properties of the components produced in this way. For this purpose, the following questions have to be answered:

1. Which welding parameters are necessary to functionalize bearing steel that is normally declared as not weldable?
2. How do the downstream processes need to be adjusted in order to achieve a high component quality for use as rolling bearings raceways?
3. What is the microstructure of the material compound after the different manufacturing steps?
4. Were the relevant component properties for use as a rolling bearing component achieved?
5. What are the relevant mechanisms that govern the fatigue behavior?
6. What damage patterns occur when used as a rolling bearing raceway?

4. Materials and Methods

In this study, classical bearing steel was welded on a base material by means of plasma-powder-transferred arc welding (PTA) and later used as bearing raceway of a cylindrical roller thrust bearing type 81212, see Figure 2. A washer with an outer diameter of 95 mm was produced, see Figure 2a. The washers' thickness was 7.5 mm. It had a material transition in axial direction (Figure 2b), which was the direction of loading. The base material of the washer consisted of the unalloyed and heat-resistant weldable steel AISI 1022M, which is mainly used in valve construction. The cladding material for the PTA consisted of the rolling bearing steel AISI 52100, which had a CEV > 1 and was considered to be non-weldable [16]. In order to weld the material in spite of this, the material was atomized to utilize it for PTA. The material AISI 52100 has a high wear resistance and toughness, which is why components such as rolling bearings are made of it. The chemical compositions of the materials are shown in Table 1. The chemical compositions were measured with a spark analyzer LMX07 (Spectro Analytical Instruments GmbH, Kleve, Germany).

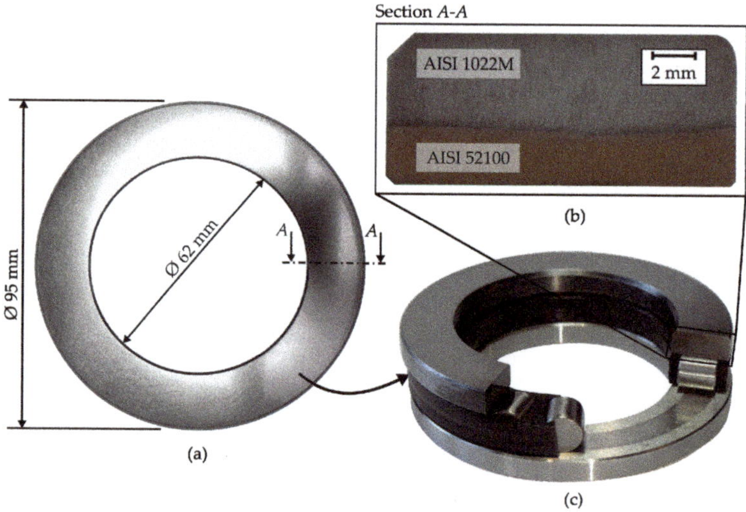

Figure 2. Tailored forming bearing: (**a**) finished specimen as roller bearing raceway; (**b**) sectional view; (**c**) cylindrical roller thrust bearing type 81212.

Table 1. Chemical composition of AISI 1022M and AISI 52100 determined by optical emission spectrometry in wt.%.

AISI	C	Si	Mn	P	S	Cr
1022 M	0.22	0.29	0.84	0.01	0.001	0.04
52100	0.97	0.25	0.31	0.01	0.004	1.38

After PTA welding, the component was formed to improve the material properties. By means of detailed analytical methods, the processes necessary for production were investigated step by step. The components that were produced, as described in Figure 1, later served as rolling bearings and will be characterized in the following with regard to their material properties. Subsequently, fatigue tests were carried out on a rolling bearing test rig, which served as a baseline for further investigations. The roller and cage assembly was taken from conventional bearings, in order to assemble a cylindrical roller thrust bearing consisting of two discs, see Figure 2c.

4.1. Plasma Transferred Arc Deposition Welding

PTA is a thermal process for applying wear and corrosion resistant layers on surfaces of metallic materials. During the PTA welding process, a tungsten electrode created a plasma arc with high energy density, which melted the surface of the base material. At the same time, the cladding material was inserted into the arc by a powder stream and was molten. During solidification, a substance-to-substance bond between the cladding and base material was created. The welding process was shielded by argon gas. The advantages of this process were a low dilution rate, a small heat affected zone, and deposition rates up to 10 kg/h. Further benefits were a high degree of automation and reproducibility [17].

The welding process was carried out on a six-axis REIS RV industrial robot (KUKA AG, Augsburg, Germany), where two additional axes were realized by a turn and tilt table. The welding torch was a Kennametal Stellite HPM 302 (Kennametal Stellite, Pittsburgh, PA, USA), which was water-cooled. The equipment is shown in Figure 3a. To obtain the optimum grain size with a diameter of a minimum 63 µm to maximum 200 µm, the powder was filtered with sieving units (Retsch, Haan, Germany). This corresponds to the current industrial standard for additional materials in powder form that are used for PTA welding [18]. Powders with the standard grain size were used because of a regular melting behavior in the arc as well as a good transportability in the gas flow of the transport gas. Prior to the welding process, the surface of the discs was cleaned with acetone (E-Coll, Wuppertal, Germany). The substrate was at room temperature at the start of welding and was not preheated. Welding without preheating normally makes a welding process more difficult, especially when high carbon equivalent steels are processed. However, by choosing a slow welding speed of 0.12 m/min, the resulting high heat input enabled the discs to be heated up to suitable joining temperatures in a short time by the welding process itself. This improved the weldability of the steel alloys used. The disc was placed on the additional axis, which was aligned parallel to the ground. Due to the spiral-shaped seam geometry (Figure 3a), the beginning and the end of the seam were not located in the area where the rolling contact will later take place. Therefore, any defects that occur in the beginning or the end of the weld, e.g., because of too low or too high temperatures, were not relevant for the later function of the components. During the welding process, the torch oscillated with a frequency of 2.0 Hz and amplitude of 4 mm at an angle of 90° to the welding direction. The oscillation increased the dynamics of the weld pool and allowed the degassing of the melt, preventing pores. The welding process took about 10 min and 30 s, whereby the disc heated up to 650 °C. To keep the dilution between the base material and cladding metal nearly constant, the welding current was dynamically adapted during the welding process (see Figure 3b). Starting with a current of 180 A, the current was gradually reduced to 130 A. A suitable current curve was empirically determined in preliminary tests by temperature measurements accompanying the welding process. A short working distance of 10 mm was selected. This allows a very precise control of the seam geometry. This was set with a seam-overlap of 1.5 mm.

Due to the short working distance, the plasma gas was set as low as possible at 1.5 min^{-1} to avoid spattering of the melt pool. The remaining gas values for transport and protective gas correspond to standard values. An overview of the general welding parameters is given in Table 2.

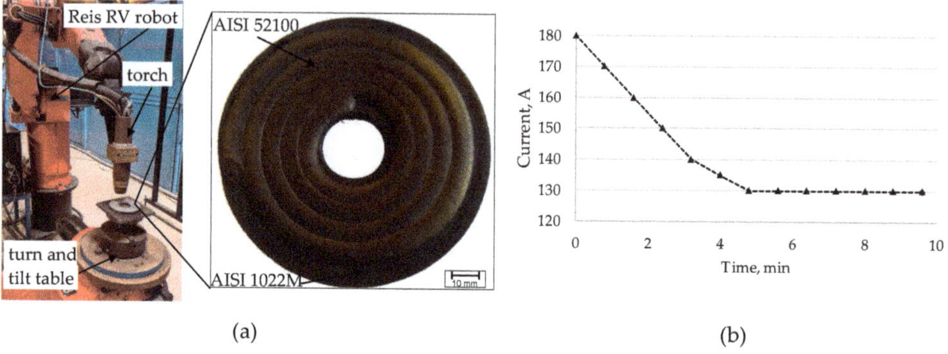

(a) (b)

Figure 3. (**a**) Experimental setup of the welding process; (**b**) adjustment of current intensity.

Table 2. Welding parameters.

Parameter	Value	Parameter	Value
Shielding gas flow (argon)	10 L/min	Current	180–130 A
Plasma gas flow (argon)	1.5 L/min	Voltage	25–27 V
Transport gas flow (argon)	6 L/min	Powder material	AISI 52100
Welding velocity	0.12 m/min	Particle size	0.06–0.2 mm
Working distance	10 mm	Deposition rate	0.9 kg/h

4.2. Upsetting

After deposition welding, the hybrid bearing washers were formed in a single-stage upsetting process. The application of forming allowed to improve the welded microstructure by thermomechanical treatment and to close the pores in the cladding layer [19]. The upsetting was performed on a hydraulic press (SPS Schirmer-Plate Siempelkamp GmbH, Krefeld, Germany) with a maximum capacity of 12,500 kN, as depicted in Figure 4a. The required forging temperature of 1050 °C was achieved by heating in a chamber furnace (Nabertherm, Lilienthal, Germany) in an inert gas atmosphere (Figure 4b) in order to prevent the surface decarburization and oxidation of the weld material. After the individual workpieces were put in the shielding gas container, the air was displaced by argon within a timeframe of 15 min. Then, the container was placed in the furnace until the workpieces were heated up within a timeframe of 20 min. Subsequently, the hot preforms for bearing washers were manually transferred to the forming tool and formed by upsetting from 15 mm to 9 mm. After forging, the bearing washers were cooled in air. A bearing washer after forming is shown in Figure 4c. The main parameters of the forging are summarized in Table 3.

Table 3. Parameters of upsetting process.

Parameter	Value
Forming temperature	1050 °C
Furnace shielding gas	argon
Forging force	1700 kN
Height reduction	35%
Final height	9 mm

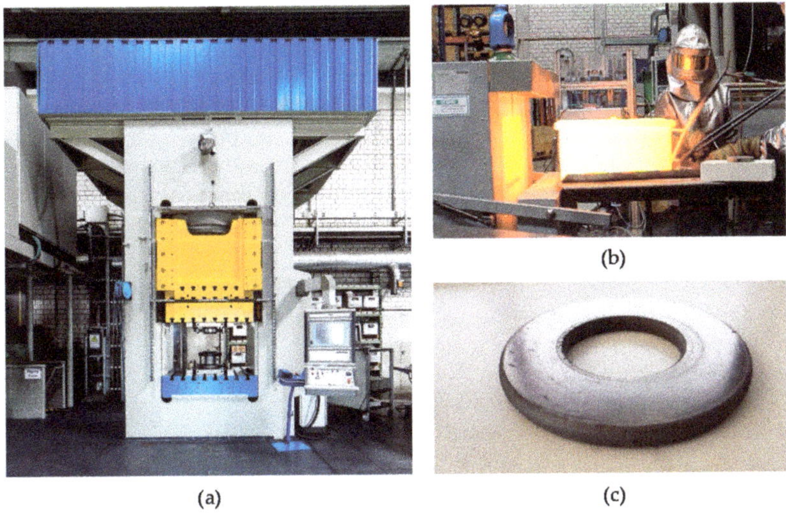

Figure 4. Forming setup: (**a**) hydraulic press; (**b**) removal of the shielding gas container from the furnace after heating of the workpieces; (**c**) forged bearing washer.

The macrographs depicted in Figure 5 show the effect of the die forging on the material distribution for the example of a bearing washer preform that is welded with a single seam. Both samples were prepared from the same workpiece, which was separated in two parts. One of them was subsequently formed; the second one remained without any further treatment. The mentioned parts were metallographically investigated at several positions. The representative results of this examination are illustratively demonstrated in Figure 5. After upsetting, the weld material was pressed into the substrate. A flattening of the cladding surface took place as a result of the forming process. After the deposition welding, some pores marked with red circles in Figure 5a were macroscopically observed close to the joining zone area. After forging, no pores at the investigated positions could be seen at the macroscopic scale (Figure 5b). A quantitative comparison of the porosity before and after forging was, however, not possible because the cross-sections of cladded and forged parts were not extracted from the same position. Forming is not only able to close pores; non-metallic inclusions are also pressed further into the depth of the surface, and thus out of the intended loaded functional area. This is also a desirable effect, as pores or defects directly below the surface are the most critical ones [6].

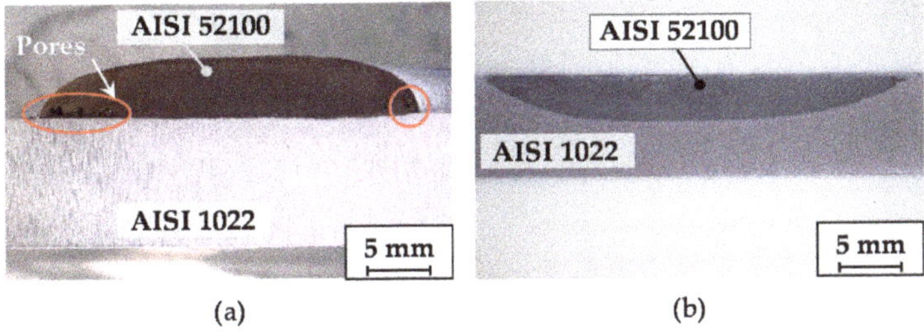

Figure 5. (**a**) Section view of the cladded workpiece welded with a single seam and (**b**) of a near-net-shaped bearing washers after forging (etched with nitric acid solution).

4.3. Heat Treatment

To adjust the required strength and hardness of the bearing washers, a heat treatment in terms of quenching and tempering was carried out after the machining. For this purpose, the bearing washers were placed in a self-designed hardening box with neutral annealing coal. The box was equipped with a thermocouple and preheated to 850 °C in an electrically heated chamber furnace. The bearing washers were austenitized at this temperature for 45 min (heating + holding time). In addition to the neutral annealing coal, argon was used to further prevent decarburization of the parts. To avoid stress cracks, quenching was carried out in an oil bath at room temperature. Subsequently, the bearing washers were tempered at 150 °C for 1 h to reduce brittleness and internal stresses, and achieve a target hardness of 60 according to Rockwell C (HRC) (measured on industrial bearing raceway).

4.4. Machining

After the upsetting process in the process chain of manufacturing hybrid bearing washers, water jet cutting was employed to produce the annular geometry and the component was pre-machined to tolerance with a 1-mm off-cut. After the subsequent heat treatment, the final geometry was obtained by a grinding process (Figure 6). This was conducted on a grinding machine with computerized numerical control (CNC) of type Blohm Orbit 48 (Blohm Jung GmbH, Hamburg, Germany). A surface grinding wheel type 517A 54/11 G7H 2020 V341A (Butzbacher Schleifmittel-Werke GmbH, Butzbach, Germany) was used here for the finishing operation step. In order to fulfill the surface quality requirements, a cutting speed of v_c = 25 m/s, a feed of f = 200 mm, and a depth of cut of a_e = 100 μm were applied.

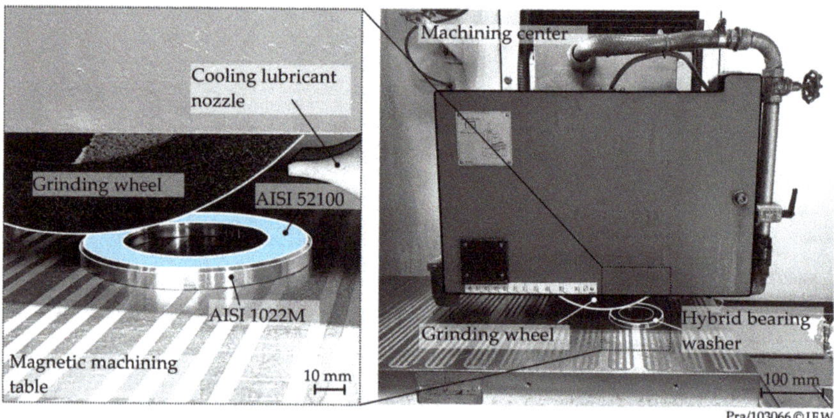

Figure 6. Machining setup in the CNC grinding machine.

4.5. Analytical Methods

4.5.1. Metallographic Investigations and Hardness Measurements

For microstructural characterization of the interface zone, cross-sections were extracted from the bearing washers in a radial direction. Metallographic investigations were carried out after deposition welding, hot forming, and heat treatment in order to examine the microstructural evolution. The samples were prepared metallographically by grinding, polishing, and subsequently etched with the reagent Beraha I or 2% nitric acid solution (CRIDA Chemie, Wenden, Germany). To investigate the hardness distribution across the joining zone, hardness measurements were carried out after each process step. Hardness testing on a ATM Q10A+ (ATM Qness GmbH, Mammelzen, Germany) according to Vickers HV 0.5 standard was used for this purpose [20]. Hardness profiles were measured from the AISI 52100 surface of the bearing disc in the AISI 1022M base material. To compare the surface hardness of the tailored forming bearing washer with an industrial bearing washer, hardness measurements were carried out after the bearing washer was manufactured. The Rockwell HRC hardness test according to DIN EN ISO 6508 was used for this purpose [21].

4.5.2. Residual Stress Measurements

Due to different thermal expansion coefficients of the materials, delayed cooling of the subsurface and core, as well as local structural transformations limited to the surface layer; residual stresses were caused after forging and the heat treatment process, which were modified by subsequent machining of the hybrid component [22,23]. Residual stresses have a significant influence on the operating behavior and therefore the lifetime of a component. The fatigue life of a cyclically loaded component is reduced by high internal tensile residual stresses because they lead to failure of the component by crack formation and propagation. Compressive residual stresses, however, can increase the component's service life by reducing crack initiation and propagation [24–26]. For this reason, it was necessary to analyze the residual stress condition of a component before the tribological investigations.

The residual stresses were determined radiographically using the $\sin^2\psi$-method [27,28]. A Seifert XRD 3003TT two circle X-ray diffractometer (Röntgenwerk Rich. Seifert & Co., Ahrensburg, Germany), equipped with a Cr tube and a spatially resolved detector, was used for this purpose. The point focus measuring spot was delimited with a 2-mm point collimator. The ω-axis of the diffractometer served as tilting axis ψ. For measurement on the martensitic phase, the tilt range of ψ varied from $-45°$ to $+45°$ with a total of nine tilt positions. To determine the net plane spacing d_{211} of the α-iron, the intensity was recorded over 2θ in the range between $144.0°$ and $164.8°$ with a step size of $0.2°$. The measurement time per step was 36 s. The maximum information depth of the X-radiation was $\tau_{max} = 5.5$ μm. For measurements at a greater distance from the surface, material was successively removed by electrolytic polishing. For the X-ray elasticity constants s_1, $\frac{1}{2}s_2$, and the reference values for the unstressed material, the values of the pure α-iron grid were assumed [29].The evaluation of the residual stress measurements was carried out with the RayfleX software (Version 2.501, Röntgenwerk Rich. Seifert & Co., Ahrensburg, Germany) from General Electric. After data reduction, the position of the diffraction reflections was determined using the parabolic fit method. The data reduction was carried out according to the following procedure: smoothing according to the Savitzky and Golay algorithm, left-sided background correction, intensity corrections, and a parabola fit with a threshold value of 70% of the maximum intensity. The measuring accuracy of the X-ray diffractometer specified by the manufacturer for flat sample geometries was $\sigma = \pm10$ MPa.

4.5.3. Non-Destructive Examination

The nanomechanical properties of the bearing washers were investigated before forming and after finishing. The aim of the tests was to determine the frictional and mechanical surface properties of the hybrid bearing washer. Nano-scratch tests were carried out with a Hysitron TriboIndenter TI950 (Bruker Corporation, Billerica, MA, USA). A cono-spherical diamond tip with a radius of 300 nm was used. The measured values were used to determine the elastic and plastic deformation behavior and the penetration-dependent friction coefficient μ during ploughing. The measuring tip traveled along the specimen surface with linearly increasing normal load and constant speed. It was moved over the sample with electrostatic force resulting in a tangential force. As the normal force increased during ploughing (Scratch test), the tangential force also increased. By the change in capacitance, the system detected the tangential force in dependence on the normal force and the depth displacement. According to these data, the coefficient of friction μ based on Coulomb friction was calculated. In the tests, a maximum load of the tip of 1 mN was used. The surface profile was analyzed before and after the scratch test using Scanning Probe Microscopy (SPM). The travel length of the tip was about 8 μm. The test was divided in three steps. First, the initial surface profile was scanned by the tip under a low load on the scratch path (prescan). Next, the scratch was performed on the same route with a load magnitude of 1 mN (scratch). To determine the resulting plastic deformation, the scratch route was scanned again with a low force magnitude (postscan). By analyzing the prescan, scratch, and postscan profile, it was possible to record the elastic and plastic deformation by calculating the different penetration depths. The plastic behavior was calculated based on five measurements and the standard deviation was determined.

Non-destructive testing using scanning acoustic microscopy (SAM) with a modified PVA TePla SAM 301 system (PVA TePla, Wettenberg, Germany) was used to examine the components for sub-surface damages after production and after fatigue testing. Here, the scanner-mounted transducer generates a pulse into the sample in the ultrasonic frequency range. The sound pulse is partially or completely scattered and reflected by inhomogeneities in the material, which are then measured (pulse-echo method). Distilled water served as a coupling medium between transducer and samples. Penetrating oil was used as a corrosion inhibitor. The surface roughness of the raceway was measured tactilely with a Mahr Perthometer PCV (Mahr-Gruppe, Göttingen, Germany). The measurement was carried out in radial and in circumferential direction of the disc at five points each.

4.5.4. Bearing Fatigue Testing

In order to investigate the performance of the hybrid bearing washers, bearing fatigue tests were carried out on a self-designed FE8 test rig according to DIN 51819 [30]. The test rig was originally used for the dynamic-mechanical testing of automotive and industrial lubricants. In the FE8 test head, two cylindrical roller thrust bearings of type 81212 were mounted on a single shaft, see Figure 7. Each bearing consisted of a bearing washer that was fixed in the housing and a washer that rotated with the shaft, as well as the rolling elements and cage. Only one tailored forming housing washer was used for bearing fatigue testing, while the other washers were taken from conventional 81212 bearings. For these investigations, 19 rolling elements mounted in a fiberglass-reinforced polyamide cage were used. The conventional rolling elements and washers were made from martensitic through-hardened AISI 52100. The force was applied by a disc spring assembly in axial direction and measured by a HBM C2 load cell (Hottinger Baldwin Messtechnik GmbH, Darmstadt, Germany) during mounting. For an initial functional test, the test parameters from previous tests with other cladding materials were applied [13,14]. Here, the axial load was 40 kN. Due to very high run times, the load was increased to 60 kN for the subsequent fatigue tests. The test parameters are specified in chapter 0. A circulating lubrication system with a filtration quotient of $\beta_{10} = 200$ according to ISO 16889 supplied each bearing with lubricant at a rate of 0.1 L/min via the housing side [31]. The lubricant used was a commercial gear oil based on synthetic polyalphaolefins with a viscosity grade of ISO VG 68 (Fuchs Petrolub SE, Mannheim, Germany). The oil contained, among other additives, extreme-pressure and anti-wear additives. Failure by spalling on the washer's raceway was manifested as a sudden increase in vibration, which was used as a termination criterion. A threshold of 150% of the steady state signal had to be exceeded for the test to shut down. The tests were carried out using the sudden death method.

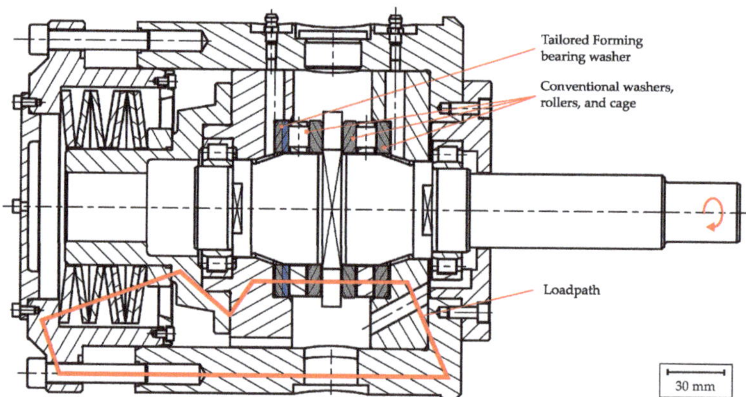

Figure 7. FE8 test head with type 81212 cylindrical roller thrust bearing.

5. Results and Discussion

5.1. Microstructure Evolution during Processing

The micrographs, showing the transition of the cladding layer into the base material, are illustrated in Figure 8. Figure 8a shows the microstructure after deposition welding. The base material AISI 1022M has a ferritic-pearlitic microstructure with large grains. Close to the joining zone, needle-shaped Widmanstätten ferrite could be observed. The microstructure of the cladding material was mainly pearlitic. Figure 8b shows the microstructure of the hot-formed and air-cooled bearing washers. Here, the base material also consisted of a ferritic-pearlitic microstructure and the cladded material of a mainly pearlitic microstructure. However, the microstructure was more homogeneous and fine-grained compared to the welded state due to hot forming-induced recrystallization. The micrographs taken after hardening, i.e., quenching and tempering, are shown in Figure 8c. The microstructure of the cladding material was completely transformed into martensite during hardening by quenching, featuring a needle-shaped morphology. Martensite was also formed close to the interface zone in the base material, though the remainder of the base material was ferritic-pearlitic. The results from the hardness measurements (Figure 9) were in accordance with the metallographic investigations. The hardness profiles of the bearing washers after deposition welding and after forming were very similar, since the components were cooled in air after the respective process step and thus have not been subjected to any specific heat treatment to increase the hardness. After the targeted heat treatment by quenching and tempering of the bearing washers, a hardness of 880 HV0.5 was determined in the cladding layer and of 250 HV0.5 in the base material. The base material remained ductile after hardening, while the cladding featured high hardness values as desired to withstand the rolling contact loads. The lower hardness in the base material resulted from the low carbon content of the steel AISI 1022M. The hardness was examined again on the cutting surface after the process steps of deposition welding, forming, and hardening (see Figure 9). The results show that the mean hardness after deposition welding and the forming process did not differ in the cladding material. It can be observed that there was a gradient from the cladding material to the base material. The hardness in the cladding material increased again after targeted heat treatment of the bearing washer. While an almost defect-free material transition between the base material and the cladding material was achieved by upsetting, the subsequent heat treatment allowed to achieve hardness values for the tailored forming bearings that are similar to those of the industrial bearings, which were also measured at 60 HRC.

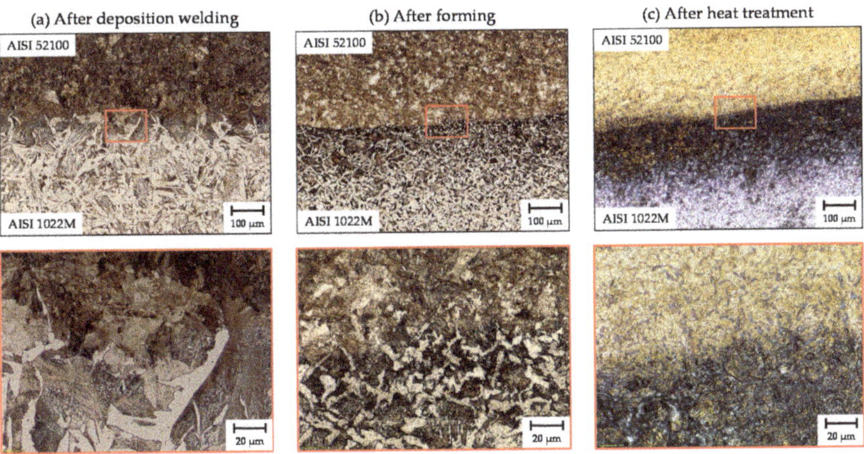

Figure 8. Micrographs of the joining zone: (**a**) After deposition welding; (**b**) after forming; (**c**) after a quenching and tempering heat treatment; (**a**,**c**) etched with 2% nitric acid solution; (**b**) etched with Beraha I reagent.

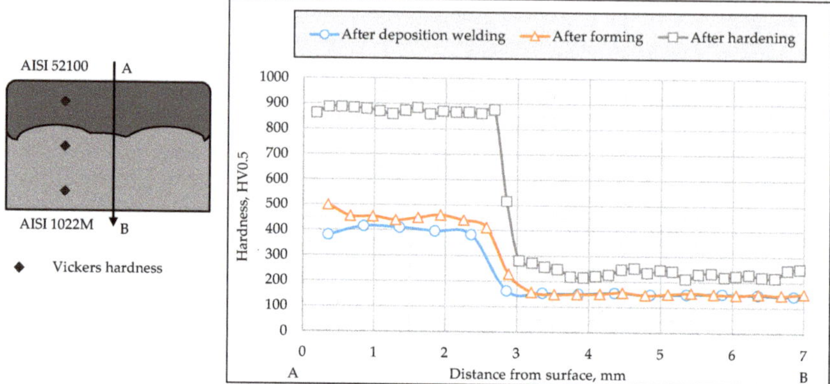

Figure 9. Hardness profiles of the hybrid bearing washers.

The chemical composition measured by spark spectroscopy showed that the carbon content was slightly below the values specified for AISI 52100 (0.87% instead of at least 0.93%, see EN 10132-4), c.f. Table 1. The reduced carbon content is attributed to the welding process, as the elemental content was slightly reduced due to the material dilution between the base material and the cladding material.

5.2. Residual Stress State

In Figure 10, the residual stress depth profiles and the corresponding peak halfwidths (full width at half maximum, abbreviated as FWHM) depth profiles of a machined hybrid axial-bearing washer are presented. The residual stress depth curves show that in both directions (circumference $\varphi = 0°$ of the bearing washers as well as radially $\varphi = 90°$) maximum compressive residual stresses occured near the surface of the hybrid component. With increasing surface distance, a very abrupt shift in the direction of tensile residual stresses was noticeable. A significant difference in the residual stress profile depending on the measuring direction was not visible in the surface-near area. Only from a surface depth of approx. 50 µm did deviations occur in the residual stress profiles between radial and circumferential direction.

Taking a closer look at the associated FWHM in connection with the residual stress measurement, it is clear that a relatively similar course of the profiles can be seen here as well. The differences between the two measuring directions are not significant. It is apparent that the FWHM at the surface was high, then decreased with further increasing surface distance and reached a minimum at approx. 4 µm depth. With further increasing surface distance, the FWHM value increased again and reached the value of the basic structure. The area that was influenced by the grinding process was only minimal and lies approximately in the range between 0 and 50 µm. The FWHM was an indication of the plastic deformation and thus of the hardening state of the surface as a result of increased or decreased dislocation density. There was a proportional relationship between plastic deformation and FWHM, which means that, with increasing plastic deformation, the FWHM would also increase. The grinding process caused a softening of the material in the surface-near area. The reason for this could be the heat transfer during grinding. In recent research projects, references were made to the high gradient in the residual stress depth curve caused by the high temperature development during grinding of hard materials [32–34]. This would also explain the abrupt drop of the high compressive residual stresses.

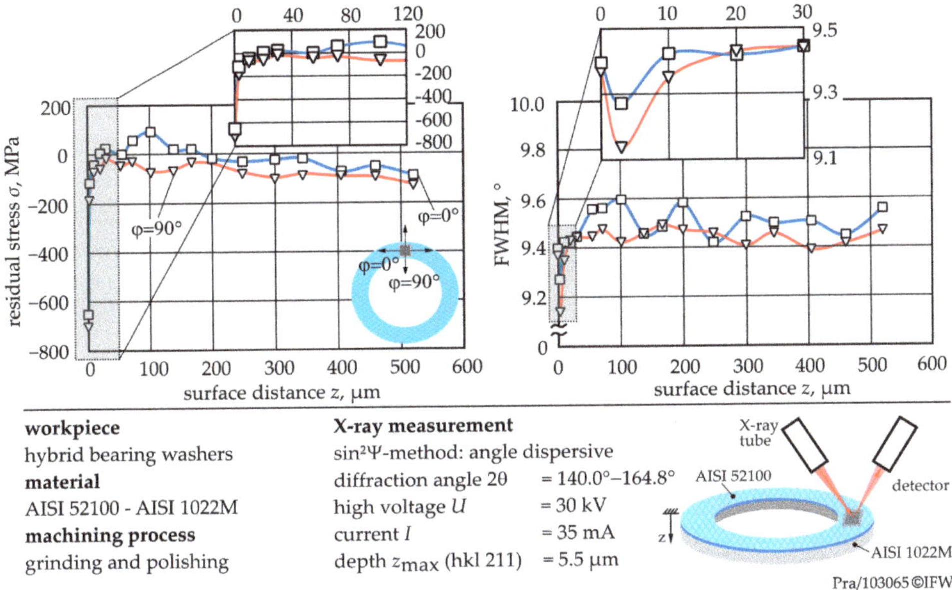

Figure 10. Residual stress and full width at half maximum (FWHM) depth profile after machining.

5.3. Microtribological Investigations

The coefficient of friction was calculated based on the performed scratch test for the cladding material. The microtribological investigations showed a coefficient of friction (CoF) of 0.4 for the considered hybrid bearing washer. This was in a good agreement to the CoF of industrial axial bearing made of AISI 52100, which was investigated in [14].

Figure 11 show the plastic behavior calculated by the nano-scratch tests. Without an additional forming and hardening of the axial bearing washer, a mean difference of 10.6 nm (standard deviation $SD = 2.46$) was achieved. The mean difference of the indentation depth with additional forming and hardening was about 9.5 nm ($SD = 2.55$), which corresponds to a decrease of about 11% between bearing washer with an additional forming step and without. However, the results of the industrial bearing washer made of AISI 52100 showed an even lower tendency to plastic deformation (8.6 nm with $SD = 3.13$) than the bearing washer made by means of the tailored forming process chain after the forming stage and a target heat treatment.

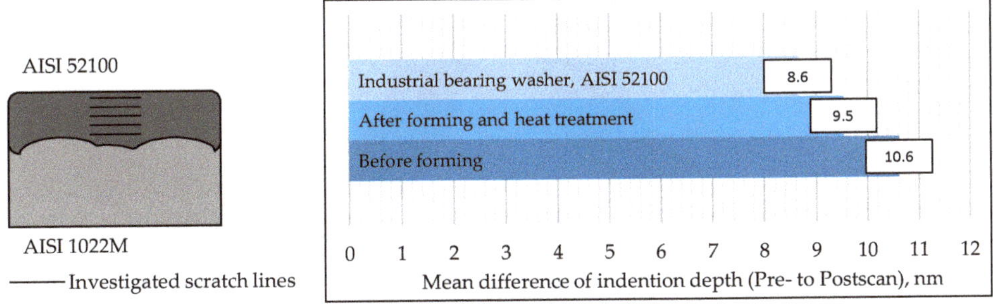

Figure 11. Evaluation of the scratch test of the cladding at different stages of the process chain.

These results indicate that the additional forming and heat treatment steps reduced the tendency to plastic deformation of the cladding material of the bearing washer made by the tailored forming process chain. However, industrially produced bearing washers have a lower tendency to plastic deformation than the hybrid bearing washer, so that individual process steps in the tailored forming process chain should be further optimized.

5.4. Optical and Acoustic Microscopy

Before fatigue testing, the washers were optically inspected. For the hybrid washers, the arithmetic mean value of the surface roughness Ra was measured as 95 ± 7 nm. The industrial standard is 80 nm. In applications with high speeds and high loads, bearing manufacturers usually recommend a mean roughness value of Ra < 0.2 µm for running surfaces in direct bearing arrangements, which was exceeded.

Via acoustic sectional images in axial direction (C-scan), various welding defects could be detected, as shown in Figure 12. At a depth between 150 and 300 µm from the surface, smaller pores are visible as white dots in the C-scan (circled in red in Figure 12a left). They have a typical size of 10–20 µm in diameter. The pores appeared randomly distributed. For the parameters investigated here with an axial load of 60 kN, the maximum of the Tresca equivalent stress is at a depth of approx. 125 µm. Since these pores occur in the zone of highest stress, they are rated as potentially critical. SAM images with higher magnification at this depth are shown in Figure 13 (bottom). The material transition zone between AISI 52100 and AISI 1022M lies in a depth of around 2.1 mm, which is shown in Figure 12a right. Here, larger cavities beyond 1 mm can be seen. These are arranged along the helical welding tracks (red arrows), which are about 10 mm apart. Concentric interference patterns are partially visible at the cavities, as there are repetitive echoes of the ultrasound signal. In order to support these findings, the sample was ground down layer by layer in the axial direction and compared to SAM imaging, see Figure 12b. The region of interest, marked in Figure 12a) on the right, is a 45° cutout. The cavities (circled in red in Figure 12b) open and close again in the course of 1.2 to 1.5 mm below the surface. Accordingly, the axial elongation is > 0.3 mm. By means of SAM (Figure 12a left), a greater degree of detail of the cavities is visible, as information was partially lost during the grinding process.

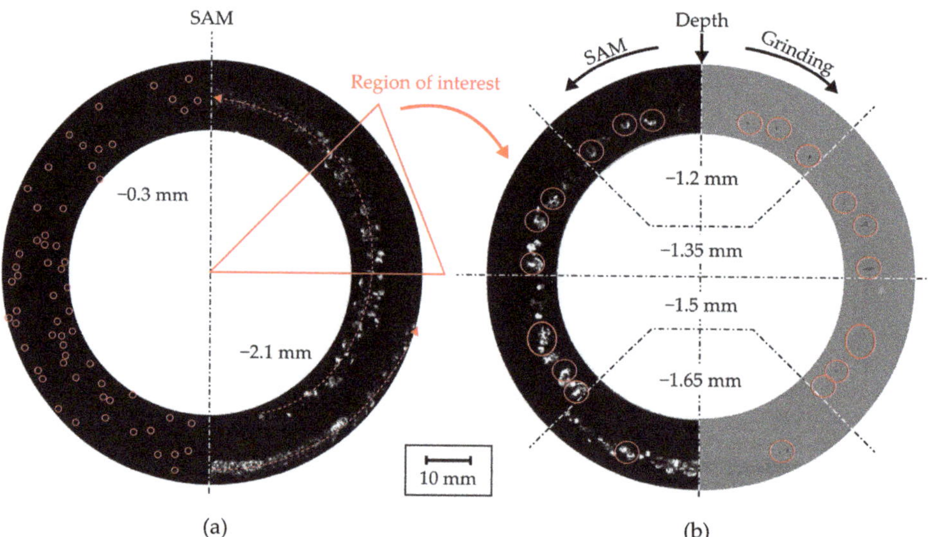

Figure 12. Scanning acoustic microscope images before testing: (**a**) Smaller pores near the surface (left) and larger cavities in the material transition zone (right); (**b**) comparison of scanning acoustic microscopy (SAM) (left) and optical microscopy after grinding (right) for different depths.

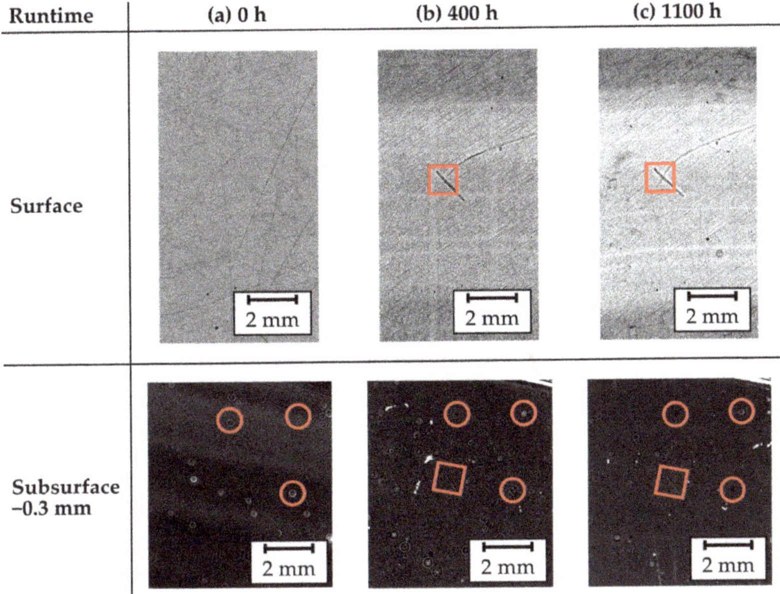

Figure 13. Optical (top) and acoustic (bottom) microscope images after: (**a**) machining; (**b**) 400 h runtime at 250 min^{-1} with surface damage; (**c**) test termination with 1100 h runtime at 250 min^{-1}.

5.5. Rolling Contact Fatigue Performance

5.5.1. Screening Test

First, a hybrid bearing washer with AISI 52100 cladding was tested on the FE8 test rig with 40 kN at 250 min^{-1} (p_H = 1.85 GPa, κ = 0.55), in accordance with previous investigations [13,14]. In these investigations, maximum running times until failure of 260 h for a cladding with a martensitic chromium-silicon steel (AISI HNV3, 1.4718), and 332 h for a cladding with AISI 5140, were achieved.

The tailored forming washer with AISI 52100 cladding, on the other hand, ran 1100 h without a critical damage event. After this, the test was aborted and considered passed. The test was interrupted every 100 to 200 h to document the possible progression of the damage. The disc was extracted and examined visually, microscopically, and by means of SAM, see Figure 13. After 400 h, a surface defect with a maximum depth of 2.6 µm was detected.

The form of damage is unusual for damage that could have occurred during operation of the rolling bearing. It is more likely to be associated with a scrape mark during reassembly of the test. However, it is noticeable that this damage did not lead to the failure of the bearing. Even by means of ultrasonic microscopic images, no damage could be detected below the surface during continuous testing.

5.5.2. Pitting Tests

In order to shorten the test time, the test conditions were changed to the standardized test according to VW PV 1483 [35]. In the so-called pitting test, the axial load was 60 kN and was thus 50% higher than the previously mentioned conditions. This rolling bearing fatigue test usually aims to differentiate gear oils with regard to their pitting resistance. The test parameters are summarized in Table 4. The main criterion is a target running time of 200 h, divided into two phases:

1. Run-in procedure for 24 h at a reduced speed of 500 min^{-1},
2. Fatigue test for the remaining 176 h at 750 min^{-1}.

Table 4. Test conditions for fatigue testing of 81212 type tailored forming bearings.

Parameter		Value	Parameter		Value
Speed	n	500–750 min^{-1}	Axial load	F_{ax}	60 kN
Viscosity at 40 °C	v_{40}	68 mm^2/s	Load equivalent	C/P	2.87
Oil temperature	T_{Oil}	95 °C	Contact load	F_1	3.16 kN
Viscosity ratio	k	0.44	Contact pressure	p_{max}	1.8 GPa

The fatigue life evaluation was carried out using Weibull analysis. Accordingly, the failure probability was calculated for $n = 6$ samples using maximum likelihood estimation and plotted in a double logarithmic diagram over the lifetime in Figure 14. A two-parameter Weibull distribution was assumed. The probability of error was 10%. The shape parameter, which indicates the slope of the regression line, was $\beta = 3.47$. The characteristic lifetime parameter indicated the lifetime at a failure probability of 63.2% and had a value of $B_{63} = 6.32 \times 10^6$ revolutions. The experimentally determined nominal life, which corresponds to a 10% probability of failure, is $B_{10} = 3.88 \times 10^6$ revolutions. This result serves as a baseline for future research with tailored forming bearings.

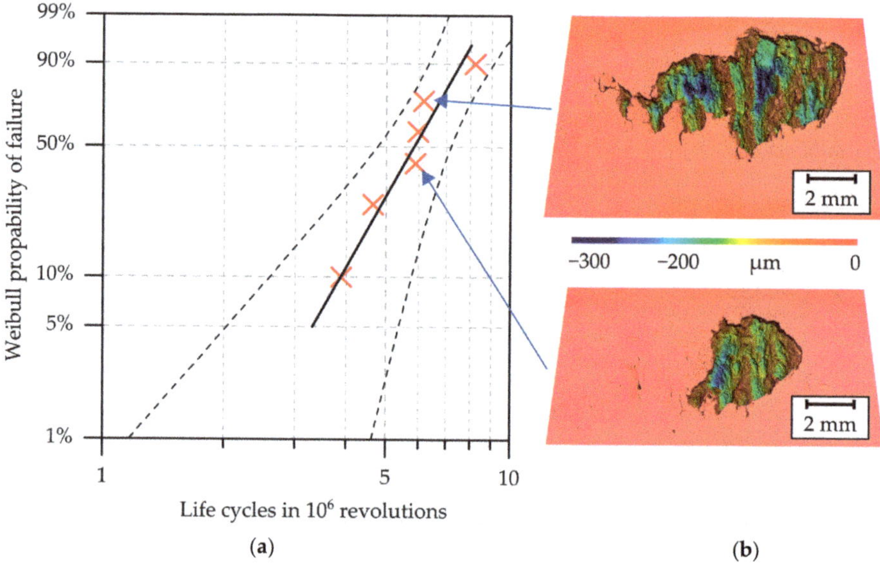

Figure 14. Bearing fatigue testing results: (**a**) Weibull plot with 95% confidence band for AISI 52100 tailored forming washers; (**b**) fatigue damage analysis by laser-scanning microscopy (rolling direction: right to left).

The theoretical rolling bearing life for conventional bearings can be calculated according to DIN ISO 281 [36], based on the work of Ioannides and Harris [37]. Since the operating speed of 500–750 min^{-1} exceeds the thermally permissible speed, an adjusted reference speed of 300 min^{-1} was assumed for the lifetime calculation of conventional bearings as a reference. The additional factors are summarized in Table 5. The extended modified lifetime at a 90% survival probability was calculated to $L_{10m} = 4.75 \times 10^6$ revolutions. The deviation of the experimentally and mathematically determined lifetime is $B_{10}/L_{10m} = 81.68\%$. Thus, the bearing failures of AISI 52100 tailored forming bearings can be considered as premature failures due to small pores below the surface.

Table 5. Results of the fatigue life of tailored forming bearings and comparison with the calculated life of conventional bearings.

Variable		Value
Weibull shape parameter	β	3.47
Experimental life	B_{10}	3.88×10^6 revs.
Adjusted speed	n_{ref}	300 min^{-1}
Life exponent	p	10/3
Life factor	a_{ISO}	0.14
Modified rating life	L_{10m}	4.75×10^6 revs.
Deviation factor	B_{10}/L_{10m}	0.82

6. Conclusions

In this study, the tailored forming technology was successfully implemented for manufacturing of bearing components. Using plasma-transferred arc welding, classical bearing steel AISI 52100 was cladded on AISI 1022M basic steel and subsequently formed by upsetting. It served as cladding material for a rolling bearing raceway, which was investigated in bearing fatigue tests. The following conclusions can be drawn:

- AISI 52100 is considered as non-weldable, but it was possible to weld by means of PTA. The challenge of reducing pores during the welding process in order to increase the bearing fatigue life still remains.
- The forming process demonstrated a positive impact on the microstructure in the interface zone. Besides a grain refinement in the clad material, the Widmanstätten structure was fully recrystallized. Moreover, the porosity occurring after PTA was significantly reduced and a defect-free material transition between the base material and the cladding material was observed.
- Hardness values similar to those of industrial bearings were achieved by the applied heat treatment of quenching and tempering.
- Residual stress measurements demonstrated that no critical tensile residual stresses occurred at the surface of the hybrid bearing washers like in industrial bearing washers. High compressive residual stresses on the subsurface avoided premature failure of the bearing washers.
- By using AISI 52100 as cladding material, the load-bearing capacity of the tailored forming bearing washers was increased compared to AISI 5140 and AISI HNV3 claddings. Eighty-two percent of the calculated modified rating life for conventional bearings L_{10m} was achieved. The slope of the determined Weibull curve indicates premature damage. By means of SAM, pores and cavities could be detected, which were probably responsible for this.

If the defects could be reduced or even completely prevented, a very high potential for improving the performance of tailored forming components opens up. In this context, follow-up investigations should consider in particular an optimization of the welding process by means of preheating and adapted welding materials as well as an increase in the degree of forming.

Author Contributions: Conceptualization, C.B., M.M., T.C.; investigation, C.B., F.S., M.Y.F., C.K., V.P., M.M., A.C., T.M., T.C.; writing—original draft preparation, C.B., F.S., M.Y.F., C.K., V.P., M.M., A.C., T.C.; writing—review and editing, F.P., F.N., T.H., B.B., L.B., J.H., B.-A.B., H.J.M., B.D., L.O., G.P.; visualization, C.B., F.S., M.Y.F., C.K., V.P., M.M., A.C., T.C.; supervision, F.P., F.N., T.H., B.B., B.-A.B., H.J.M., B.D., L.O., G.P. All authors have read and agreed to the published version of the manuscript.

Funding: This research was funded by the Deutsche Forschungsgemeinschaft (DFG, German Research Foundation) under grant number 252662854.

Acknowledgments: The results presented in this paper were obtained within the Collaborative Research Centre 1153 "Process chain to produce hybrid high performance components by Tailored forming"—252662854—in the subprojects A2, A4, B2, B4, C1, and C3. The authors thank the German Research Foundation (DFG) for their financial support of this project.

Conflicts of Interest: The authors declare no conflict of interest.

References

1. Blohm, T.; Mildebrath, M.; Stonis, M.; Langner, J.; Hassel, T.; Behrens, B.-A. Investigation of the cladding thickness of plasma-transferred arc deposition welded and cross wedge rolled hybrid parts. *Prod. Eng.* **2017**, *11*, 255–263. [CrossRef]
2. Mildebrath, M.; Blohm, T.; Hassel, T.; Stonis, M.; Langner, J.; Maier, H.J.; Behrens, B.-A. Influence of Cross Wedge Rolling on the Cladding Quality of Plasma-Transferred Arc Deposition Welded Hybrid Steel Parts. *Int. J. Emerg. Technol. Adv. Eng.* **2017**, *7*, 1–7.
3. Cento, P.; Dareing, D.W. Ceramic Materials in Hybrid Ball Bearings. *Tribol. Trans.* **1999**, *42*, 707–714. [CrossRef]
4. Wan, G.T.Y.; Gabelli, A.; Ioannides, E. Increased Performance of Hybrid Bearings with Silicon Nitride Balls. *Tribol. Trans.* **1997**, *40*, 701–707. [CrossRef]
5. Dill, J.F. Hybrid bearing technology for advanced turbomachinery. *J. Eng. Gas Turbines Power* **1996**, *118*, 173–178. [CrossRef]
6. Koehler, H.; Schumacher, J.; Schuischel, K.; Partes, K.; Bomas, H.; Jablonski, F.; Vollertsen, F.; Kienzler, R. An approach to calculate fatigue properties of laser cladded components. *Prod. Eng.* **2012**, *6*, 137–148. [CrossRef]
7. Koehler, H.; Partes, K.; Seefeld, T.; Vollertsen, F. Influence of laser reconditioning on fatigue properties of crankshafts. *Phys. Procedia* **2011**, *12*, 512–518. [CrossRef]
8. Alam, M.M.; Kaplan, A.; Tuominen, J.; Vuoristo, P.; Miettinen, J.; Poutala, J.; Näkki, J.; Junkala, J.; Peltola, T.; Barsoum, Z. Analysis of the stress raising action of flaws in laser clad deposits. *Mater. Des.* **2013**, *46*, 328–337. [CrossRef]
9. Chew, Y.; Hock Lye Pang, J.; Bi, G.; Song, B. Effects of laser cladding on fatigue performance of AISI 4340 steel in the as-clad and machine treated conditions. *J. Mater.* **2017**, *243*, 246–257. [CrossRef]
10. Pape, F.; Coors, T.; Poll, G. Studies on the Influence of Residual Stresses on the Fatigue Life of Rolling Bearings in Dependence on the Production Processes. *Front. Mech. Eng.* **2020**, *6*. [CrossRef]
11. Mildebrath, M.; Coors, T.; Barroi, A.; Pape, F.; Lammers, M.; Hermsdorf, J.; Overmeyer, L.; Poll, G.; Hassel, T. Herstellungsprozess und Wälzfestigkeit von hybriden Hochleistungsbauteilen. *Konstruktion* **2018**, *9*, 84–89.
12. Behrens, B.-A.; Amiri, A.; Duran, D.; Nothdurft, S.; Hermsdorf, J.; Kaierle, S.; Ohrdes, H.; Wallaschek, J.; Hassel, T. Improving the mechanical properties of laser beam welded hybrid workpieces by deformation processing. *AIP Conf. Proc.* **2019**, *2113*. [CrossRef]
13. Pape, F.; Coors, T.; Barroi, A.; Hermsdorf, J.; Mildebrath, M.; Hassel, T.; Kaierle, S.; Matthias, T.; Chugreev, A.; Chugreeva, A.; et al. Tribological Study on Tailored-Formed Axial Bearing Washers. *Tribol. Online* **2018**, *12*. [CrossRef]
14. Behrens, B.-A.; Chugreev, A.; Matthias, T.; Poll, G.; Pape, F.; Coors, T.; Hassel, T.; Maier, H.J.; Mildebrath, M. Manufacturing and Evaluation of Multi-Material Axial-Bearing Washers by Tailored forming. *Metals* **2019**, *9*, 232. [CrossRef]
15. Stanford, M.K.; Jain, V.K. Friction and wear characteristics of hard cladding. *Wear* **2001**, *251*, 990–996. [CrossRef]
16. Matthes, K.; Schneider, W. *Schweißtechnik–Schweißen von metallischen Konstruktionswerkstoffen*, 6th ed.; Hanser Verlag: München, Germany, 2016; pp. 33–34.
17. Schneider, W.; Twrdek, J. *Praxiswissen Schweißtechnik*, 6th ed.; Springer: Wiesbaden, Germany, 2019; p. 268.
18. Dilthey, U. *Schweißtechnische Fertigungsverfahren 2—Verhalten der Werkstoffe beim Schweißen*, 3rd ed.; Springer: Berlin, Germany, 2005; p. 130.
19. Kittner, K.; Wiesner, J.; Kawalla, R. A new approach for void closure in bulk metal forming. *Key Eng. Mater.* **2016**, *716*, 595–604. [CrossRef]
20. *DIN EN ISO 6507, Metallic Materials—Vickers Hardness Test*; International Organization for Standardization: Geneva, Switzerland, 2005.
21. *DIN EN ISO 6508, Metallic Materials—Rockwell Hardness Test*; International Organization for Standardization: Geneva, Switzerland, 2005.
22. Krektuleva, R.V.; Cherepanov, O.I.; Cherepanov, R.O. Numerical investigation of residual thermal stresses in welded joints of heterogeneous steels with account of technological features of multi-pass welding. *Appl. Math. Model.* **2016**, *42*, 244–256. [CrossRef]

23. Kanakaraju, V.; Gunjal, S.; Rathinam, K.; Sharma, P. Optimization of cutting parameters of residual stresses in simultaneous turning technique. *Mater. Today Proc.* **2020**, *28*, 2108–2115. [CrossRef]

24. Jawahir, I.S.; Brinksmeier, E.; M'Saoubi, R.; Aspinwall, D.K.; Outeiro, J.C.; Meyer, D.; Umbrello, D.; Jayal, A.D. Surface integrity in material removal processes: Recent advances. *CIRP Ann.* **2011**, *60*, 603–626. [CrossRef]

25. Breidenstein, B. Oberflächen und Randzonen hoch Belasteter Bauteile. Habilitation Thesis, Leibniz University Hannover, Hannover, Germany, 2011.

26. Novovic, D.; Dewes, R.C.; Aspinwall, D.K.; Voice, W.; Bowen, P. The effect of machined topography and integrity on fatigue life. *Int. J. Mach. Tools Manuf.* **2004**, *44*, 125–134. [CrossRef]

27. Guo, J.; Fu, H.; Pan, B.; Kang, R. Recent progress of residual stress measurement methods: A review. *Chin. J. Aeronaut.* **2020**. [CrossRef]

28. Macherauch, E.; Müller, P. Das sin2psi-Verfahren der röntgenographischen Spannungsmessung. *Z. Angew. Phys.* **1961**, *13*, 305–312.

29. Eigenmann, B.; Macherauch, E. Röntgenographische Untersuchung von Spannungszuständen in Werkstoffen, Teil III. *Mater. Werkst.* **1996**, *27*, 426–437. [CrossRef]

30. *DIN 51819, Testing of Lubricants—Mechanical-Dynamic Testing in the Roller Bearing Test Apparatus FE8—Part 1: General working principles*; Deutsches Institut für Normung e.V.: Berlin, Germany, 2016.

31. *ISO 16889, Hydraulic Fluid Power—Filters—Multi-Pass Method for Evaluating Filtration Performance of a Filter Element*; International Organization for Standardization: Geneva, Switzerland, 2008.

32. Jamshidi, H.; Budak, E. Grinding Temperature Modeling Based on a Time Dependent Heat Source. *Procedia CIRP* **2018**, *77*, 299–302. [CrossRef]

33. Zhao, Z.; Qian, N.; Ding, W.; Wang, Y.; Fu, Y. Profile grinding of DZ125 nickel-based superalloy: Grinding heat, temperature field, and surface quality. *J. Manuf. Process.* **2020**, *57*, 10–22. [CrossRef]

34. Rabiey, M.; Maerchy, P. Investigation on Surface Integrity of Steel DIN 100Cr6 by Grinding Using CBN Tool. *Procedia CIRP CSI* **2020**, *87*, 192–197. [CrossRef]

35. *VW PV 1483, Factory Standard, Gear Oils—Testing the Pitting Load Capacity in Rolling Bearings (in German)*; Volkswagen AG: Wolfsburg, Germany, 2007.

36. *DIN ISO 281, Rolling Bearings—Dynamic Load Ratings and Rating Life*; International Organization for Standardization: Geneva, Switzerland, 2007.

37. Ioannides, E.; Bergling, G.; Gabelli, A. An Analytical Formulation for the Life of Rolling Bearings. In *Acta Polytechnica Scandinavica*; Mechanical Engineering Series 137; Finnish Academies of Technology: Espoo, Finland, 1999.

Article

Structural Characteristics of Multilayered Ni-Ti Nanocomposite Fabricated by High Speed High Pressure Torsion (HSHPT)

Gheorghe Gurau [1], Carmela Gurau [1,*], Francisco Manuel Braz Fernandes [2], Petrica Alexandru [1], Vedamanickam Sampath [3], Mihaela Marin [1] and Bogdan Mihai Galbinasu [4]

[1]   Faculty of Engineering, "Dunărea de Jos" University of Galati, Domnească Street, 47, 800008 Galati, Romania; gheorghe.gurau@ugal.ro (G.G.); Petrica.Alexandru@ugal.ro (P.A.); Mihaela.Marin@ugal.ro (M.M.)

[2]   CENIMAT/I3N, Departamento de Ciências dos Materiais, Faculdade de Ciências e Tecnologia, Universidade NOVA de Lisboa, 2829-516 Caparica, Portugal; fbf@fct.unl.pt

[3]   Department of Metallurgical and Materials Engineering, Indian Institute of Technology Madras, Chennai 600036, India; vsampath@iitm.ac.in

[4]   Faculty of Dentistry, U.M.F. University of Medicine and Pharmacy "Carol Davila", Bucharest, Dionisie Lupu, Street, 37, 020021 Bucharest, Romania; bogdan.galbinasu@umfcd.ro

*   Correspondence: carmela.gurau@ugal.ro; Tel.: +40-720-044084

Received: 30 October 2020; Accepted: 29 November 2020; Published: 4 December 2020

Abstract: It is generally accepted that severe plastic deformation (SPD) has the ability to produce ultrafinegrained (UFG) and nanocrystalline materials in bulk. Recent developments in high pressure torsion (HPT) processes have led to the production of bimetallic composites using copper, aluminum or magnesium alloys. This article outlines a new approach to fabricate multilayered Ni-Ti nanocomposites by a patented SPD technique, namely, high speed high pressure torsion (HSHPT). The multilayered composite discs consist of Ni-Ti alloys of different composition: a shape memory alloy (SMA) Ti-rich, whose Mf > RT, and an SMA Ni-rich, whose Af < RT. The composites were designed to have 2 to 32 layers of both alloys. The layers were arranged in different sequences to improve the shape recovery on both heating and cooling of nickel-titanium alloys. The manufacturing process of Ni-Ti multilayers is explained in this work. The evolution of the microstructure was traced using optical, scanning electron and transmission electron microscopes. The effectiveness of the bonding of the multilayered composites was investigated. The shape memory characteristics and the martensitic transition of the nickel-titanium nanocomposites were studied by differential scanning calorimetry (DSC). This method opens up new possibilities for designing various layered metal-matrix composites achieving the best combination of shape memory, deformability and tensile strength.

Keywords: composites; HSHPT; nano multilayers; Ni-Ti; SPD

---

1. Introduction

Multilayered composites have attracted much attention in engineering design as a promising technique to develop a novel combination of physical and mechanical properties acquired from the individual characteristics of the incorporated materials [1–5]. Bimetallic shape memory composites are among the most widely investigated class of composites, offering better shape memory properties for the design of new decomplex applications [6].

Shape memory alloys are stimulus-responsive materials with two universal properties: superelasticity and shape memory effect [7,8]. The occurrence of martensite-to-austenite and austenite-to-martensite transitions gives rise to shape memory and superelastic responses [9,10]. Among these alloys, Ni-Ti SMAs are among the most interesting thermo-responsive SMAs that are capable of exhibiting reliable shape memory characteristics, in addition to presenting high ductility and strength [11]. Research on Ni-Ti shape memory alloys has been revived by controlling the "size-effect". Grain size reduction to the nano range greatly increases recovery stress [12]. The nanocrystalline (NC) or ultrafinegrained (UFG) microstructure significantly enhances the mechanical and shape memory characteristics in comparison with the coarse grained alloy of the same composition [13]. One way to refine the microstructure of the fabricated bulk UFG and nanostructured SMAs is severe plastic deformation (SPD) processing [14]. Most metal-matrix composites are obtained by accumulative roll bonding, a variant of the SPD technique, which also makes it possible to obtain multiple layers [15]. In addition, bimetallic "Ni-Ti/Ni-Ti" shape memory composites obtained by welding present large recoverable strain on heating and cooling [16], and are potential candidates for use as thermomechanical actuators [7,9,17]. In earlier work, the major problem of the formation of a thick brittle intermetallic layer was encountered, in particular in high temperature bonding processes [18]. Recent developments have led to the use of high pressure torsion (HPT) as an appropriate severe plastic deformation technique in the manufacturing of bimetallic composites [18].

The aim of our research is to study the structure and phase transformations of smart Ni-Ti multilayered composites obtained using the HSHPT technique. This process combines HPT and friction stir processing, and is capable of fabricating UFG discs of about 40 mm in diameter from different types of alloys [19–21]. In addition, a more dependable method to fabricate bulk UFG metallic composites is HSHPT. We fabricated an "Ni-Ti/Ni-Ti" composite with 2 to 32 multilayered discs. The HSHPT process helps achieve very good bonding, high-quality interfaces with no intermetallic layers and ultrafine-grained microstructure from individual Ni-Ti layers.

2. Experimental Procedure

For this study, multilayered shape memory $Ni_{50.3}Ti/Ni_{49.6}Ti$ composites were produced using HSHPT severe deformation at ambient temperature. The materials used in this investigation were cut from commercial shape memory $Ni_{49.6}Ti_{50.4}$ (at.%) sheets and superelastic $Ni_{50.3}Ti_{49.7}$ (at.%) rods. The martensitic transformation temperature Ms for the SMA that is rich in titanium is 51 °C, while it is −16.5 °C for Ni-rich alloy.

The details of the HSHPT procedure and the ad hoc machine used in the present work are given in our earlier papers [19,21]. To enable microstructural refinement concurrent with bonding of layers, the SPD process variables were chosen utilizing an EATON SVX024A1-4A1B1 frequency converter via PLC XC 200. The speed of rotation of the upper punch was maintained at 900 rpm. Initially, a pressure of 20 bars was applied using the bottom punch. The pressure levels monitored making use of the Hottinger Spider 8 equipment were between 0.01 GPa and 0.68 GPa, depending on the number of layers. The maximum torque reached was 42 Nm. The processing time lasted between 11 and 28 s. The maximum pressure was applied for less than 5 s.

HSHPT was first applied on each Ni-rich sample (about 9.5 mm × 7.4 mm and 2.35 mm in thickness) and Ti-rich sample (9.5 mm in diameter and ~2.35 mm in thickness) with austenitic and martensitic structures, respectively, at room temperature. The second step involved was fabricating the composites. To obtain two- and three-layer composites, discs with different chemical compositions were made to overlap alternatively in different successions. In the third step, these modules were cut in half and assembled as sandwich stacks. Four-layered composites were obtained by halving the two-layered composite and overlapping the parts in the HSHPT machine. The same procedure was used to obtain 8, 16 and 32 layers. Three-layered composites were obtained by overlapping the obtained Ti-rich, Ni-rich and T-rich disks. Five-layered composites were obtained by overlapping half of the obtained three- and two-layered composites. Six-layered composites were obtained by overlapping the obtained three-layered composite (Ti-rich, Ni-rich and T-rich half disks) with another three-layered composite (Ni-rich, Ti-rich and Ni-rich half disks). Nine-layered composites were obtained by overlapping the half of obtained four- and five-layered composites. Twelve-layered composites were obtained by overlapping the obtained 6-layered composite, and 24-layered composites were obtained by overlapping the obtained 12-layered composite (Figure 1). The cumulative degrees of deformation of multilayered bimetallic composites, calculated using the formula $\varepsilon = \frac{h_0}{h_1}$, (where h_0 is initial thickness of the sample and h_1 is the final thickness of the sample), ranged from 0.95 to 4.65. The SPD discs produced were with d $\leq$ 40 mm and t = 1.5–0.15 mm.

Microstructural examinations highlighted the ability to manufacture multilayered composites and revealed the reliable bond of layers, as well the reduction in grain diameter accomplished using the HSHPT technique. Investigation of the multilayered $Ni_{50.3}Ti/Ni_{49.6}Ti$ microstructure was done using an OLYMPUS BX51 (manufactured by Olympus microscopes, Tokyo, Japan) optical microscope, with the QCapture (QuickPHOTO MICRO 2.3, Prague, Czech Republic) software package, under bright and dark field modes. The microstructure was studied using a Zeiss (ZEISS EVO MA15, manufactured by Carl Zeiss Microscopy GmbH, Jena, Germany SEM/EDX (Scanning Electron Microscope coupled with Energy Dispersive X-ray analyzer) to study the grain structure and the quality of the joints.

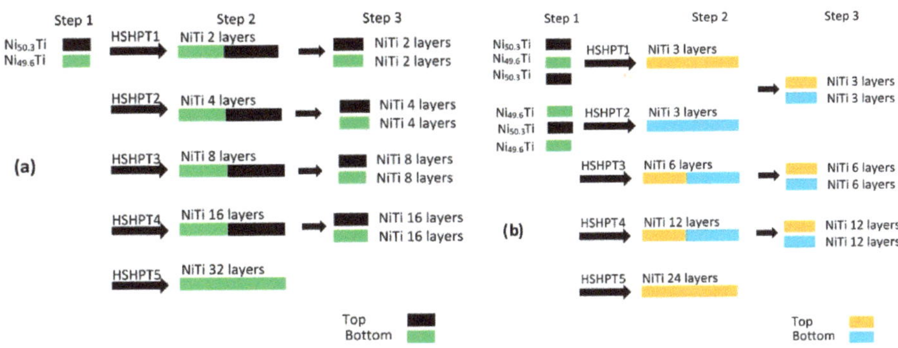

Figure 1. Processing route of $Ni_{50.3}Ti/Ni_{49.6}Ti$ composite discs cut in half and assembled as sandwich stacks with: (**a**) number of layers multiple of 2 and (**b**) number of layers multiple of 3.

An in-depth microstructural analysis was also carried out using a TEM (Transmission Electron Microscope, Model Tecnai 20G2, FEI, Hillsboro, OR, USA), operating at a voltage of 200 kV. The martensitic transformation temperatures were measured using a differential scanning calorimeter. The DSC tests were run using a DSC 204 F1 Phoenix model from Netzsch (Selb, Germany). The tests were performed between −150 °C and 150 °C using a cooling and heating rate of 10 °C/min, under a protective gaseous nitrogen atmosphere. Specimens (15–20 mg) were obtained from the HSHPT discs for 4, 16, 32 layered composites:

- one set of samples from mid-radius of the discs,
- another set from the edge and the center of each disc.

Prior to DSC, an etching solution of HF:HNO3: H20 (1:5:10 in volume) was used to remove the oxidation of the surface layer and the regions affected by the cutting process.

3. Results and Discussion

3.1. Optical Microscopy

Figure 2 illustrates bright and dark field optical micrographs (OM) of the multilayers observed in the $Ni_{50.3}Ti/Ni_{49.6}Ti$ alloy composite, whose sample was cut as a cross-section around the middle of the discs. The bright field micrograph of the three-layered $Ni_{50.3}Ti/Ni_{49.6}Ti/Ni_{50.3}Ti$ alloy composite is shown in Figure 2a.

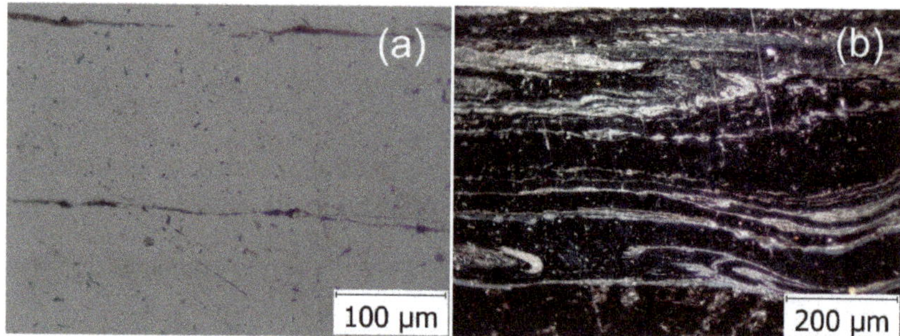

Figure 2. Optical micrograph of $Ni_{50.3}Ti/Ni_{49.6}Ti$ composite discs: (**a**) bright field image of three layers and (**b**) dark field image of 24 layers.

This feature is consistent with the refined microstructure resulting from severe plastic deformation by HSHPT. While some precipitate particles could be observed, the grain boundaries were not resolved, as the size of the microstructural features was beyond the resolution range of the optical microscope. The bonding between the sheets was discontinuous, narrow and could hardly be detected, since the chemical compositions were almost the same. The dark field optical micrograph of the 24-layered composite (Figure 2b) showed flowlines and waviness, as highlighted by the color contrast. This is in good agreement with HPT findings for bi-layers in the Al/Mg composite [18]. The thickness of the layers in this composite was about 22 μm. The interfacial layers were less obvious as the number of layers of the composite increased. The quality of the bonding was evident.

3.2. SEM/EDX Analysis

The SEM microstructure of the 12-layered specimen demonstrates a typical SPD structure (Figure 3a). The microstructural modification by HSHPT results from three opposing effects. In the first stage, low pressure and high speed of the punch act together, leading to an increase in temperature of the material to almost 800 °C (estimated using a temperature sensor-CT laser radiation pyrometer T2 MHCF OPTC).

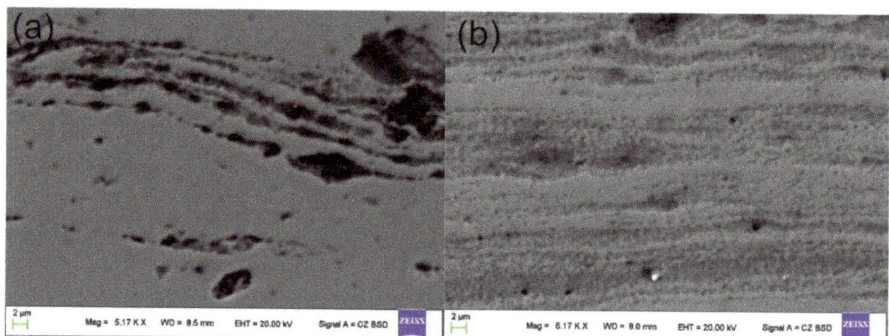

Figure 3. SEM image of HSHPT-processed metallic composite: (**a**) 12 layers, (**b**) 24 layers.

At this point, high pressure was exerted on the discs, leading to severe deformation. Besides shearing, recovery and recrystallization of grains took place due to the samples attaining high temperature. Grain refining during HSHPT should result from dynamic recrystallization that takes place at high temperature; it could be said that rapid cooling to room temperature "freezes" the UFG structure produced at high temperature. On the surface of the sample, only curved lines could be observed, and not grain boundaries. The materials were welded without a detectable intermediate area. The adhesion of layers was noticeable. The presence of smooth interfaces between layers was attributed to the specific condition created during HSHPT. Other methods of manufacturing the metallic composite led to the occurrence of an intermetallic layer at the interface.

To investigate the distribution of elements in the Ni rich/Ti-rich areas, a line scan by EDX was run on the nine layers of the disc (Supplementary material: Figure S1). The nine layers were emphasized by the variation of the Ti (a) and Ni (b) content, respectively. Across the layers (quasi 20 μm), alternating areas Ni-richer or Ti-richer could be identified. EDX characterization was performed in an area comprising the nine-layers of the composite. Figure S2 of the Supplementary Materials presents the opposing variations of Ni and Ti contents in the successive nine layers of the composite. The severity of plastic deformation introduced by HSHPT produced rotation and plastic flow of large volumes of material caused by upper punch rotation at high speeds [19]. The 3-D images of the surfaces (seen in Figure S2 of Supplementary Material) suggest the arrangement of the distinct layers.

3.3. Transmission Electron Microscopy

Figure 4 illustrates a TEM micrograph (bright field) of the four-layered $Ni_{50.3}Ti/Ni_{49.6}Ti$ composite. The UFG structure with an equiaxed morphology prevailed after HSHPT. The average size of the grains was about 200–300 nm.

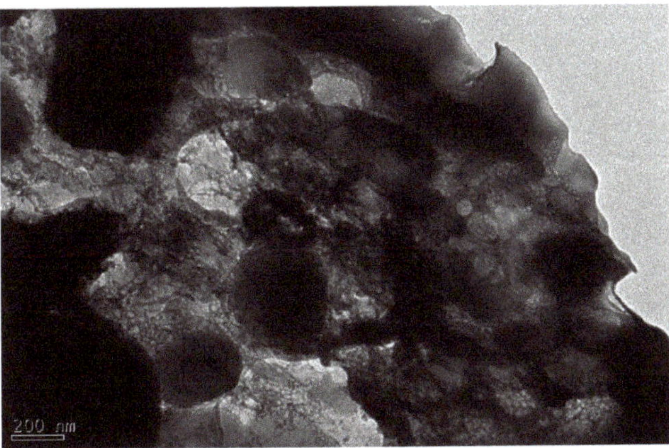

Figure 4. TEM image of the 4 layered $Ni_{50.3}Ti/Ni_{49.6}Ti$ composite.

The image highlights equiaxed subgrains isolated by dislocation cells. Between them, several nanograins were also interspersed. These grains, with a size of under 50 nm, formed clear boundaries. Various dislocation cell configurations, including high-density dislocations, zones of dislocation tangles or condensed dislocation boundaries, were characteristic of the rather inhomogeneous microstructures of the sample.

3.4. DSC

The DSC curves for samples extracted from the midradius of the discs of the $Ni_{50.3}Ti/Ni_{49.6}Ti$ composite with 4, 16 and 32 layers after HSHPT are shown in Figure 5. The equipment is capable of measuring temperatures in the range of −150 °C to 150 °C, covering the martensite transformation temperature range for both alloys. The thermograms of severe plastic-deformed composites revealed two peaks upon heating and two peaks after cooling. It was possible to identify the transformation temperatures corresponding to the Ni-rich (lower temperatures) and Ti-rich (higher temperatures) SMAs. The peaks between 80 °C and 120 °C represented the martensite (B19′)-to-austenite (B2) reversible transformation for the shape memory $Ni_{49.6}Ti_{50.4}$ (Ti-rich) alloy.

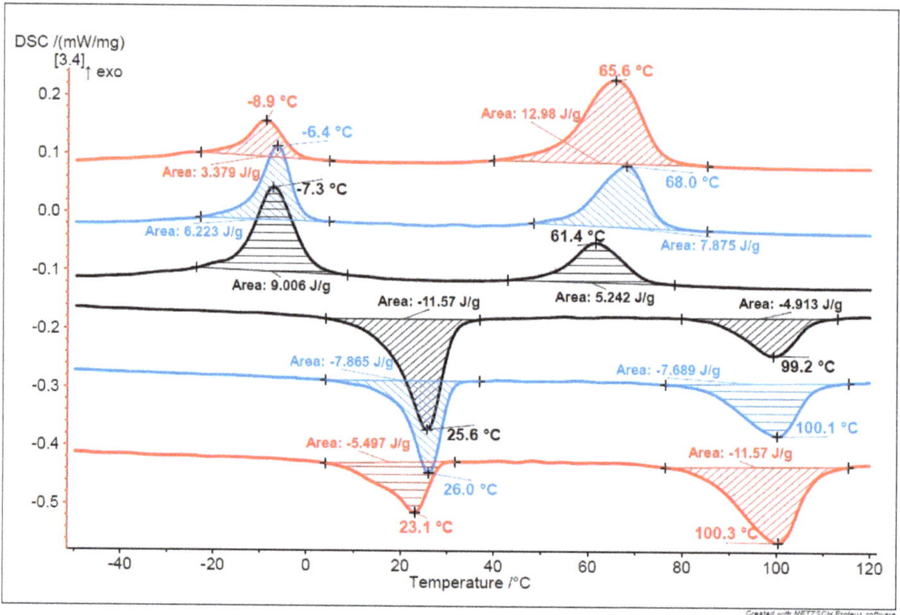

Figure 5. DSC curves of the 4 (black), 16 (blue) and 32 (red) layered Ni$_{50.3}$Ti/Ni$_{49.6}$Ti composites after HSHPT.

The peaks between 10 °C and –30 °C represented the austenite (B2)-to-martensite (B19′) reversible transformation for Ni-rich Ni$_{50.3}$Ti$_{49.7}$ alloy. Making successive deformation to produce 4, 16, 32 layers of composite did not significantly change the transformation temperatures. This result was obtained despite the fact that the degrees of deformation of the multilayered bimetallic composites with 4, 16 and 32 layers ranged from 1.17, 2.41 to 4.65. Likewise, the grain size was significantly reduced, while the cumulative degree of deformation increased. The reasonably stable transformation temperatures may be explained by the fact that the HSHPT process imposed a complex strain due to the high pressure concomitant with high rotational speed of the upper punch. The friction between punches and sample heated the severely plastic-deformed disc, thereby contributing to the rearrangement and decrement of the lattice defects. However, the intensity of the peaks was seen to decrease slightly with the increasing number of layers.

On the other hand, the Ms temperature slightly increased in the initial state in both alloys, as compared to the Ni-Ti SMAs that were rich in titanium and nickel. These results can be attributed to the grain refinement brought about by severe plastic deformation.

The Ni$_{50.3}$Ti/Ni$_{49.6}$Ti composite showed reversible martensitic transformation subsequent to SPD. Postdeformation annealing was not required in contrast to deformation using other severe plastic deformation methods. The HSHPT technique combines SPD imposed on the sample at RT by HPT with PDA caused due to the heat generated because of friction occurring between the anvils and the sample.

Another important element that could be observed was the absence of the second step of the martensitic transformation, even in the Ni-Ti SMA that was rich in nickel. In the initial state, this alloy showed a two-stage phase transition of B19′ ↔ R-phase ↔ B2 (Figure S3 of Supplementary Materials). But the DSC curves corresponding to severely plastic-deformed discs exhibited just one strong transformation stage, namely B19′ ↔ B2. After HSHPT, the intermediate R-phase transition was suppressed, as observed in our earlier results from research on a Ni-rich Ni-Ti alloy [22].

The DSC curves at the center and edge of the disc, which had 32 layers after processing, are illustrated in Figure 6.

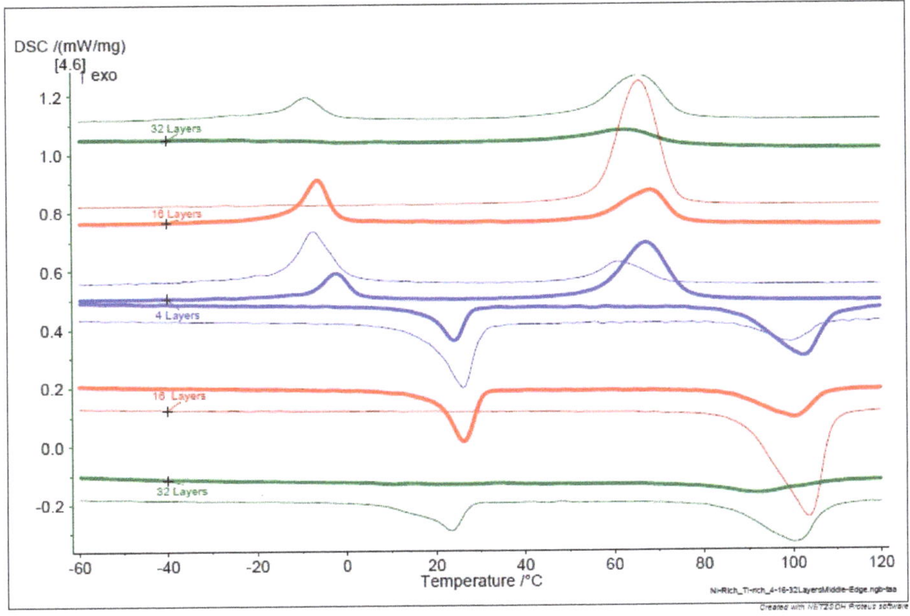

Figure 6. DSC curves for the 32 layers of $Ni_{50.3}Ti/Ni_{49.6}Ti$ composite after HSHPT. Thicker lines: center sample; thinner lines: edge sample.

In the HSHPT condition, the DSC curves revealed two peaks upon cooling and two upon heating, both for the edge sample and the middle sample only in the four-layered composite; for the 16- and 32-layered samples, the DSC peaks associated with the Ni-rich layers were observed only for the center and edge samples, respectively. The exothermic and endothermic peaks of the Ti-rich alloy broadened, perhaps because of the increased density of the dislocations. The severe plastic deformation of Ni-rich alloy led to a considerable broadening of the exothermic peaks which were, however, still visible. Processing by HSHPT is complex because the imposed strain varies, both with the rotational speed of superior punch applied to the composite and with the position within the disc. Consequently, the thickness of the discs was slightly higher at the center where the cooling speed would be expected to decrease. The microstructure developed more rapidly at the edge than at the center of the disc in all plastic deformation processes by torsion at high pressure. The absence of transformation peaks associated with the Ni-rich alloy could be related to heterogeneous deformation of the HSHPT material [23].

4. Conclusions

In summary, a new SPD method, HSHPT, can successfully be used to manufacture bimetallic composites with an ultrafine grained structures. The achieved shape memory multilayer composites were composed of 2 to 32 layers of $Ni_{50.3}Ti$ and $Ni_{49.6}Ti$ alloys. The bonding of layers was achieved mainly by the high pressure imposed and short time high temperature reached during the high speed rotation of upper anvil. A very good joint was obtained, regardless of the number of layers or alternating them. The microstructure of composites was uniform where the bonding of adjacent layers could not be detected by optical microscopy and SEM. The SPD process was effective at refining both the alloys that made up the composite. As-HSHPT processed composite discs revealed martensitic transformation after deformation. The module diameter varied between 20 and 40mm, and was dependent on the number of layers and the degree of deformation applied. The thickness of the layers varied from 300 μm to 20 μm. HSHPT yields multilayered, fine structures and may, in future study, be adapted for use for two-way SME. Future work will be also focus on the fabrication the new metallic UFG composites by HSHPT. This technology is capable of inducing thermocompression bonding starting from dissimilar materials. Multilayer composites with desirable mechanical, electrical, magnetic and biocompatibility properties provide the opportunity to create effective functional applications.

Supplementary Materials: The following are available online at http://www.mdpi.com/2075-4701/10/12/1629/s1.

Author Contributions: Conceptualization, C.G. and G.G.; methodology, C.G., G.G., F.M.B.F., P.A., V.S., M.M. and B.M.G. investigation, C.G., P.A., M.M. and B.M.G. (OM, SEM), F.M.B.F. (DSC) and V.S. (TEM); writing—original draft preparation, C.G. and G.G.; writing—review and editing, V.S., G.G., C.G. and P.A. supervision, F.M.B.F. and V.S.; project administration, C.G. and G.G.; All authors have read and agreed to the published version of the manuscript.

Funding: This research was funded by 47PCCDI/2018 project and MANUNET 3 PN3-P3-302, grant number 99-2019. F.M. Braz Fernandes acknowledges the funding of CENIMAT/I3N by National Funds through the FCT - Fundação para a Ciência e a Tecnologia, I.P., within the scope of the project refª UIDB/50025/2020-2023. Edgar Camacho (CENIMAT, FCT/UNL) is acknowledged for running the DSC tests.

Conflicts of Interest: The authors declare no conflict of interest. The funders had no role in the design of the study; in the collection, analyses, or interpretation of data; in the writing of the manuscript, or in the decision to publish the results.

References

1. Ji, X.; Wang, Q.; Yin, F.; Cui, C.; Ji, P.; Hao, G. Fabrication and properties of novel porous CuAlMn shape memory alloys and polymer/CuAlMn composites. *Compos. Part A Appl. Sci. Manuf.* **2018**, *107*, 21–30. [CrossRef]

2. Yi, X.; Sun, K.; Gao, W.; Meng, X.; Cai, W.; Zhao, L. Microstructure design of the excellent shape recovery properties in (Ti,Hf) 2Ni/Ti-Ni-Hf high temperature shape memory alloy composite. *J. Alloys Compd.* **2017**, *729*, 758–763. [CrossRef]

3. Huang, G.Q.; Yan, Y.F.; Wu, J.; Shen, Y.F.; Gerlich, A.P. Microstructure and mechanical properties of fine-grained aluminum matrix composite reinforced with nitinol shape memory alloy particulates produced by underwater friction stir processing. *J. Alloys Compd.* **2019**, *786*, 257–271. [CrossRef]

4. Oliveira, J.P.; Duarte, J.F.; Inácio, P.; Schell, N.; Miranda, R.M.; Santos, T.G. Production of Al/NiTi composites by friction stir welding assisted by electrical current. *Mater. Des.* **2017**, *113*, 311–318. [CrossRef]

5. Wang, E.; Tian, Y.; Wang, Z.; Jiao, F.; Guo, C.; Jiang, F. A study of shape memory alloy NiTi fiber/plate reinforced (SMAFR/SMAPR) Ti-Al laminated composites. *J. Alloys Compd.* **2017**, *696*, 1059–1066. [CrossRef]

6. Belyaev, S.; Rubanik, V.; Resnina, N.; Rubanik, V.; Demidova, E.; Lomakin, I. Bimetallic shape memory alloy composites produced by explosion welding: Structure and martensitic transformation. *J. Mater. Process. Technol.* **2016**, *234*, 323–331. [CrossRef]

7. Lohan, N.M.; Pricop, B.; Popa, M.; Matcovschi, E.; Cimpoeşu, N.; Cimpoeşu, R.; Istrate, B.; Bujoreanu, L.G. Hot Rolling Effects on the Microstructure and Chemical Properties of NiTiTa Alloys. *J. Mater. Eng. Perform.* **2019**, *28*, 7273–7280. [CrossRef]

8. Cimpoesu, N.; Mihalache, E.; Lohan, N.M.; Suru, M.G.; Comăneci, R.I.; Özkal, B.; Bujoreanu, L.G.; Pricop, B. Structural-Morphological Fluctuations Induced by Thermomechanical Treatment in a Fe–Mn–Si Shape Memory Alloy. *Met. Sci. Heat Treat.* **2018**, *60*, 471–477. [CrossRef]

9. Oliveira, J.P.; Miranda, R.M.; Fernandes, F.M.B. Welding and Joining of NiTi Shape Memory Alloys: A Review. *Prog. Mater. Sci.* **2017**, *88*, 412–466. [CrossRef]

10. Sun, L.; Huang, W.M.; Ding, Z.; Zhao, Y.; Wang, C.C.; Purnawali, H.; Tang, C. Stimulus-responsive shape memory materials: A review. *Mater. Des.* **2012**, *33*, 577–640. [CrossRef]

11. Soejima, Y.; Motomura, S.; Mitsuhara, M.; Inamura, T.; Nishida, M. In situ scanning electron microscopy study of the thermoelastic martensitic transformation in Ti–Ni shape memory alloy. *Acta Mater.* **2016**, *103*, 352–360. [CrossRef]

12. Tsuchiya, K.; Hada, Y.; Koyano, T.; Nakajima, K.; Ohnuma, M.; Koike, T.; Todaka, Y.; Umemoto, M. Production of TiNi amorphous/nanocrystalline wires with high strength and elastic modulus by severe cold drawing. *Scr. Mater.* **2009**, *60*, 749–752. [CrossRef]

13. Jiang, D.; Jiang, J.; Shi, X.; Jiang, X.; Jiao, S.; Cui, L. Constrained martensitic transformation in nanocrystalline TiNi/NbTi shape memory composites. *J. Alloys Compd.* **2013**, *577*, S749–S751. [CrossRef]

14. Tohidi, A.A.; Ketabchi, M.; Hasannia, A. Nanograined Ti–Nb microalloy steel achieved by Accumulative Roll Bonding (ARB) process. *Mater. Sci. Eng. A* **2013**, *577*, 43–47. [CrossRef]

15. Ana, S.V.A.; Reihanian, M.; Lotfi, B. Accumulative roll bonding (ARB) of the composite coated strips to fabricate multi-component Al-based metal matrix composites. *Mater. Sci. Eng. A* **2015**, *647*, 303–312. [CrossRef]

16. Belyaev, S.; Rubanik, V.; Resnina, N.; Lomakin, I.; Demidova, E. Reversible strain in bimetallic TiNi-based shape memory composites produced by explosion welding. *Mater. Today Proc.* **2017**, *4*, 4696–4701. [CrossRef]

17. Oliveira, J.P.; Zeng, Z.; Andrei, C.; Braz Fernandes, F.M.; Miranda, R.M.; Ramirez, A.J.; Omori, T.; Zhou, N. Dissimilar laser welding of superelastic NiTi and CuAlMn shape memory alloys. *Mater. Des.* **2017**, *128*, 166–175. [CrossRef]

18. Qiao, X.; Li, X.; Zhang, X.; Chen, Y.; Zheng, M.; Golovin, I.S.; Gao, N.; Starink, M.J. Intermetallics formed at interface of ultrafine grained Al/Mg bi-layered disks processed by high pressure torsion at room temperature. *Mater. Lett.* **2016**, *181*, 187–190. [CrossRef]

19. Gurau, G.; Gurau, C.; Bujoreanu, L.G.; Sampath, V. A Versatile Method for Nanostructuring Metals, Alloys and Metal Based Composites. In Proceedings of the IOP Conference Series: Materials Science and Engineering, Iasi, Romania, 25–26 May 2017; pp. 24–27.

20. Gurau, G.; Gurau, C.; Sampath, V.; Bujoreanu, L.G. Investigations of a nanostructured FeMnSi shape memory alloy produced via severe plastic deformation. *Int. J. Miner. Metall. Mater.* **2016**, *23*, 1315–1322. [CrossRef]

21. Gurău, G.; Gurău, C.; Potecaşu, O.; Alexandru, P.; Bujoreanu, L.-G. Novel high-speed high pressure torsion technology for obtaining Fe-Mn-Si-Cr shape memory alloy active elements. *J. Mater. Eng. Perform.* **2014**, *23*, 2396–2402. [CrossRef]

22. Fernandes, F.M.B.; Mahesh, K.K.; Silva, R.J.C.; Gurau, C.; Gurau, G. XRD study of the transformation characteristics of severely plastic deformed Ni-Ti SMAs. *Phys. Status Solidi C* **2010**, *7*, 1348–1350. [CrossRef]

23. Ghat, M.; Mehtedi, M.E.; Ciccarelli, D.; Paoletti, C.; Spigarelli, S.; Mehtedi, M.E.; Ciccarelli, D.; Paoletti, C.; High, S.S. Materials at High Temperatures High temperature deformation of IN718 superalloy: Use of basic creep modelling in the study of Nickel and single-phase Ni-based superalloys. *Mater. High Temp.* **2019**, *3409*, 1–10.

Publisher's Note: MDPI stays neutral with regard to jurisdictional claims in published maps and institutional affiliations.

MDPI
St. Alban-Anlage 66
4052 Basel
Switzerland
Tel. +41 61 683 77 34
Fax +41 61 302 89 18
www.mdpi.com

Metals Editorial Office
E-mail: metals@mdpi.com
www.mdpi.com/journal/metals